Free PDF eBook
for this BNi 2023 COSTBOOK!

You can now have access to the data in this book anywhere you go.

With the companion PDF eBook, you can easily search for and find all the cost items in this book right from your laptop, tablet or smartphone.

If you ordered this book from our bnibooks.com website, you should have already received your PDF eBook. If you have not received it, please call us (M-F, 9 am - 6 pm Eastern) at **1.888.264.2665** and we'll email you a link to download it.

Please note that the PDF eBook you received is a personally licensed copy for your individual use. Creating or distributing unauthorized copies is a violation of Federal copyright law.

BNi® Building News

A **DESIGN COST DATA** COMPANY

DATA YOU CAN TRUST

30TH
EDITION

BNi. *Building News*
A DESIGN COST DATA COMPANY
DATA YOU CAN TRUST

ELECTRICAL
2023 COSTBOOK

CSI. MasterFormat®

EDITOR-IN-CHIEF

William D. Mahoney, P.E.

TECHNICAL SERVICES

Joan Hamilton
Eric Mahoney, AIA
Ana Varela

GRAPHIC DESIGN

Robert O. Wright Jr.

BNi Publications, Inc.

VISTA
990 PARK CENTER DRIVE, SUITE E
VISTA, CA 92081

1-888-BNI-BOOK (1-888-264-2665)
www.bnibooks.com

ISBN 978-1-58855-239-6

Table of Contents

Preface

For over 75 years, BNi Building News has been dedicated to providing construction professionals with timely and reliable information. Based on this experience, our staff has researched and compiled thousands of up-to-the-minute costs for the **BNi Costbooks**. This book is an essential reference for contractors, engineers, architects, facility managers — any construction professional who must provide an estimate for any type of building project.

Whether working up a preliminary estimate or submitting a formal bid, the costs listed here can be quickly and easily tailored to your needs. All costs are based on prevailing labor rates. Overhead and profit should be included in all costs. Man-hours are also provided.

All data is categorized according to the CSI division format. This industry standard provides an all-inclusive checklist to ensure that no element of a project is overlooked. In addition, to make specific items even easier to locate, there is a complete alphabetical index.

The "Features of this Book" section presents a clear overview of the many features of this book. Included is an explanation of the data, sample page layout and discussion of how to best use the information in the book.

Of course, all buildings and construction projects are unique. The information provided in this book is based on averages from well-managed projects with good labor productivity under normal working conditions (eight hours a day). Other circumstances affecting costs, such as overtime, unusual working conditions, savings from buying bulk quantities for large projects, and unusual or hidden costs, must be factored in as they arise.

The data provided in this book is for estimating purposes only. Check all applicable federal, state and local codes and regulations for local requirements.

Format

All data is categorized according to the *CSI MASTERFORMAT.* This industry standard provides an all-inclusive checklist to ensure that no element of a project is overlooked.

DIVISION 00 ...PROCUREMENT & CONTRACTING REQUIREMENTS
00 10 00 SOLICITATION
00 20 00 INSTRUCTIONS FOR PROCUREMENT
00 30 00 AVAILABLE INFORMATION
00 40 00 PROCUREMENT FORMS AND SUPPLEMENTS
00 50 00 CONTRACTING FORMS AND SUPPLEMENTS
00 60 00 PROJECT FORMS
00 70 00 CONDITIONS OF THE CONTRACT
00 80 00 Reserved
00 90 00 REVISIONS, CLARIFICATIONS, AND MODIFICATIONS

DIVISION 01 .... GENERAL REQUIREMENTS
01 10 00 SUMMARY
01 20 00 PRICE AND PAYMENT PROCEDURES
01 30 00 ADMINISTRATIVE REQUIREMENTS
01 40 00 QUALITY REQUIREMENTS
01 50 00 TEMPORARY FACILITIES AND CONTROLS
01 60 00 PRODUCT REQUIREMENTS
01 70 00 EXECUTION AND CLOSEOUT REQUIREMENTS
01 80 00 PERFORMANCE REQUIREMENTS
01 90 00 LIFE CYCLE ACTIVITIES

DIVISION 02 .... EXISTING CONDITIONS
02 30 00 SUBSURFACE INVESTIGATION
02 40 00 DEMOLITION AND STRUCTURE MOVING
02 50 00 SITE REMEDIATION
02 60 00 CONTAMINATED SITE MATERIAL REMOVAL
02 70 00 WATER REMEDIATION
02 80 00 FACILITY REMEDIATION

DIVISION 03 .... CONCRETE
03 10 00 CONCRETE FORMING AND ACCESSORIES
03 20 00 CONCRETE REINFORCING
03 30 00 CAST-IN-PLACE CONCRETE
03 40 00 PRECAST CONCRETE
03 50 00 CAST DECKS AND UNDERLAYMENT
03 60 00 GROUTING
03 70 00 MASS CONCRETE
03 80 00 CONCRETE CUTTING AND BORING

DIVISION 04 .... MASONRY
04 20 00 UNIT MASONRY
04 30 00 Reserved
04 40 00 STONE ASSEMBLIES
04 50 00 REFRACTORY MASONRY
04 60 00 CORROSION-RESISTANT MASONRY
04 70 00 MANUFACTURED MASONRY
04 80 00 Reserved
04 90 00 Reserved

DIVISION 05 .... METALS
05 10 00 STRUCTURAL METAL FRAMING
05 12 00 STRUCTURAL STEEL FRAMING
05 20 00 METAL JOISTS
05 30 00 METAL DECKING
05 40 00 COLD-FORMED METAL FRAMING
05 50 00 METAL FABRICATIONS
05 60 00 Reserved
05 70 00 DECORATIVE METAL
05 80 00 Reserved
05 90 00 Reserved

DIVISION 06 .... WOOD, PLASTICS, AND COMPOSITES
06 10 00 ROUGH CARPENTRY
06 30 00 Reserved
06 40 00 ARCHITECTURAL WOODWORK
06 50 00 STRUCTURAL PLASTICS
06 60 00 PLASTIC FABRICATIONS
06 70 00 STRUCTURAL COMPOSITES
06 80 00 COMPOSITE FABRICATIONS
06 90 00 Reserved

DIVISION 07 .... THERMAL AND MOISTURE PROTECTION
07 10 00 DAMPPROOFING AND WATERPROOFING
07 30 00 STEEP SLOPE ROOFING
07 40 00 ROOFING AND SIDING PANELS
07 50 00 MEMBRANE ROOFING
07 60 00 FLASHING AND SHEET METAL
07 70 00 ROOF AND WALL SPECIALTIES AND ACCESSORIES
07 80 00 FIRE AND SMOKE PROTECTION
07 90 00 JOINT PROTECTION

DIVISION 08 .... OPENINGS
08 10 00 DOORS AND FRAMES
08 20 00 Reserved
08 30 00 SPECIALTY DOORS AND FRAMES
08 40 00 ENTRANCES, STOREFRONTS, AND CURTAIN WALLS
08 50 00 WINDOWS
08 60 00 ROOF WINDOWS AND SKYLIGHTS
08 70 00 HARDWARE
08 80 00 GLAZING
08 90 00 LOUVERS AND VENTS

DIVISION 09 .... FINISHES
09 20 00 PLASTER AND GYPSUM BOARD
09 30 00 TILING
09 40 00 Reserved
09 50 00 CEILINGS
09 60 00 FLOORING
09 70 00 WALL FINISHES
09 80 00 ACOUSTIC TREATMENT
09 90 00 PAINTING AND COATING

Format (Continued)

Features of this Book

Sample pages with graphic explanations are included before the Costbook pages. These explanations, along with the discussions below, will provide a good understanding of what is included in this book and how it can best be used in construction estimating.

Material Costs

The material costs used in this book represent national averages for prices that a contractor would expect to pay plus an allowance for freight (if applicable) and handling and storage. These costs reflect neither the lowest or highest prices, but rather a typical average cost over time. Periodic fluctuations in availability and in certain commodities (e.g. copper, conduit) can significantly affect local material pricing. In the final estimating and bidding stages of a project when the highest degree of accuracy is required, it is best to check local, current prices.

Labor Costs

Labor costs include the basic wage, plus commonly applicable taxes, insurance and markups for overhead and profit. The labor rates used here to develop the costs are typical average prevailing wage rates. Rates for different trades are used where appropriate for each type of work.

Fixed government rates and average allowances for taxes and insurance are included in the labor costs. These include employer-paid Social Security/Medicare taxes (FICA), Worker's Compensation insurance, state and federal unemployment taxes, and business insurance.

Please note, however, most of these items vary significantly from state to state and within states. For more specific data, local agencies and sources should be consulted.

Man-Hours

These productivities represent typical installation labor for thousands of construction items. The data takes into account all activities involved in normal construction under commonly experienced working conditions such as site movement, material handling, start-up, etc.

Equipment Costs

Costs for various types and pieces of equipment are included in Division 1 - General Requirements and can be included in an estimate when required either as a total "Equipment" category or with specific appropriate trades. Costs for equipment are included when appropriate in the installation costs in the Costbook pages.

Overhead and Profit

Included in the labor costs are allowances for overhead and profit for the contractor/employer whose workers are performing the specific tasks. No cost allowances or fees are included for management of subcontractors by the general contractor or construction manager. These costs, where appropriate, must be added to the costs as listed in the book.

The allowance for overhead is included to account for office overhead, the contractors' typical costs of doing business. These costs normally include in-house office staff salaries and benefits, office rent and operating expenses, professional fees, vehicle costs and other operating costs which are not directly applicable to specific jobs. It should be noted for this book that office overhead as included should be distinguished from project overhead, the General Requirements (Division 1) which are specific to particular projects. Project overhead should be included on an item by item basis for each job.

Depending on the trade, an allowance of 10-15 percent is incorporated into the labor/installation costs to account for typical profit of the installing contractor. See Division 1, General Requirements, for a more detailed review of typical profit allowances.

Features of this Book *(Continued)*

Adjustments to Costs

The costs as presented in this book attempt to represent national averages. Costs, however, vary among regions, states and even between adjacent localities.

In order to more closely approximate the probable costs for specific locations throughout the U.S., a table of Geographic Multipliers is provided. These adjustment factors are used to modify costs obtained from this book to help account for regional variations of construction costs. Whenever local current costs are known, whether material or equipment prices or labor rates, they should be used if more accuracy is required.

Editor's Note: This **Costbook** is intended to provide accurate, reliable, average costs and typical productivities for thousands of common construction components. The data is developed and compiled from various industry sources, including government, manufacturers, suppliers and working professionals. The intent of the information is to provide assistance and guidelines to construction professionals in estimating. The user should be aware that local conditions, material and labor availability and cost variations, economic considerations, weather, local codes and regulations, etc., all affect the actual cost of construction. These and other such factors must be considered and incorporated into any and all construction estimates.

Sample Costbook Page

In order to best use the information in this book, please review this sample page and read the "Features in this Book" section.

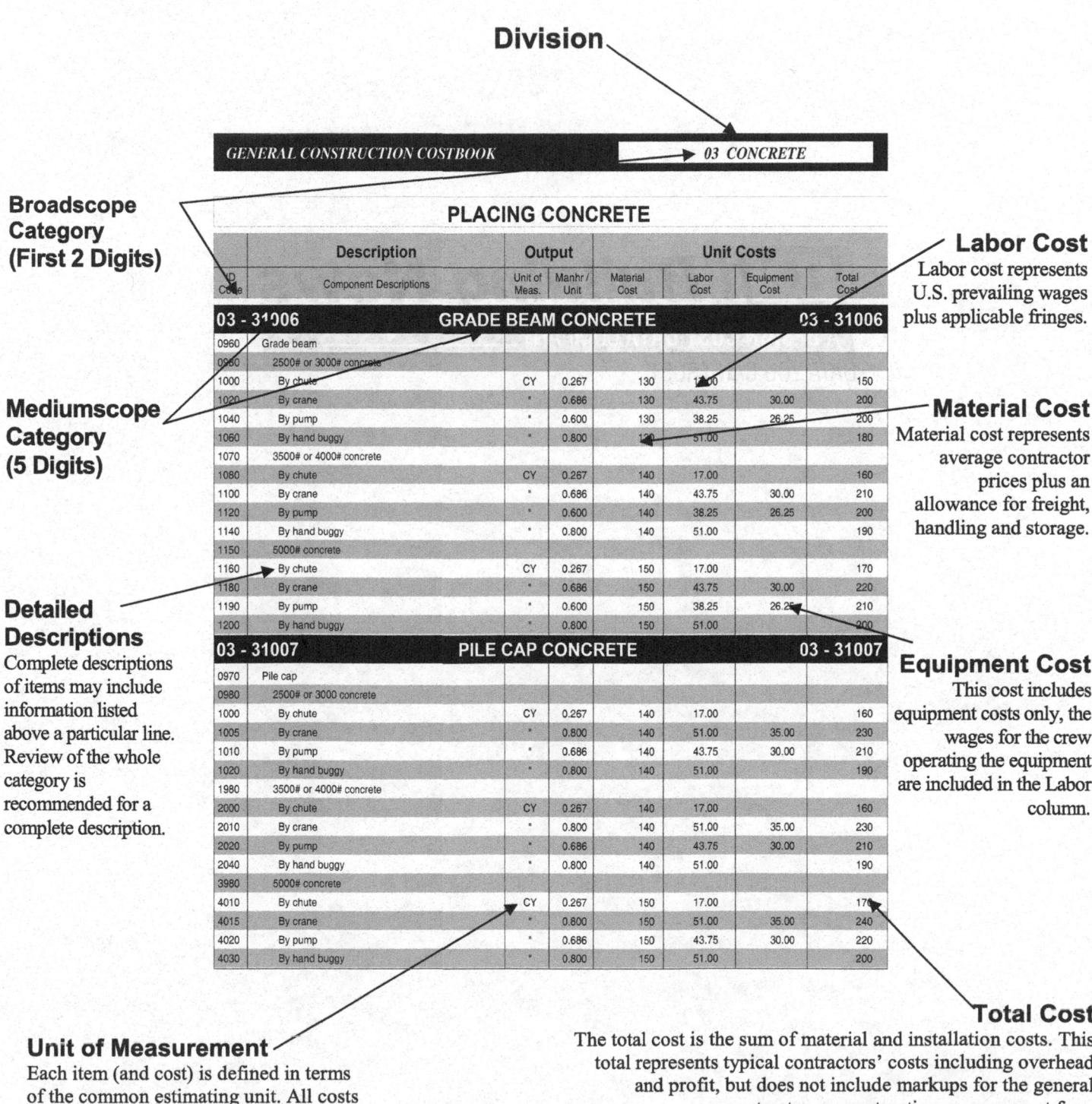

Division

Broadscope Category (First 2 Digits)

Mediumscope Category (5 Digits)

Detailed Descriptions
Complete descriptions of items may include information listed above a particular line. Review of the whole category is recommended for a complete description.

Labor Cost
Labor cost represents U.S. prevailing wages plus applicable fringes.

Material Cost
Material cost represents average contractor prices plus an allowance for freight, handling and storage.

Equipment Cost
This cost includes equipment costs only, the wages for the crew operating the equipment are included in the Labor column.

GENERAL CONSTRUCTION COSTBOOK — 03 CONCRETE

PLACING CONCRETE

ID Code	Component Descriptions	Unit of Meas.	Manhr / Unit	Material Cost	Labor Cost	Equipment Cost	Total Cost
03 - 31006	**GRADE BEAM CONCRETE**						**03 - 31006**
0960	Grade beam						
0960	2500# or 3000# concrete						
1000	By chute	CY	0.267	130	17.00		150
1020	By crane	"	0.686	130	43.75	30.00	200
1040	By pump	"	0.600	130	38.25	26.25	200
1060	By hand buggy	"	0.800	130	51.00		180
1070	3500# or 4000# concrete						
1080	By chute	CY	0.267	140	17.00		160
1100	By crane	"	0.686	140	43.75	30.00	210
1120	By pump	"	0.600	140	38.25	26.25	200
1140	By hand buggy	"	0.800	140	51.00		190
1150	5000# concrete						
1160	By chute	CY	0.267	150	17.00		170
1180	By crane	"	0.686	150	43.75	30.00	220
1190	By pump	"	0.600	150	38.25	26.25	210
1200	By hand buggy	"	0.800	150	51.00		200
03 - 31007	**PILE CAP CONCRETE**						**03 - 31007**
0970	Pile cap						
0980	2500# or 3000 concrete						
1000	By chute	CY	0.267	140	17.00		160
1005	By crane	"	0.800	140	51.00	35.00	230
1010	By pump	"	0.686	140	43.75	30.00	210
1020	By hand buggy	"	0.800	140	51.00		190
1980	3500# or 4000# concrete						
2000	By chute	CY	0.267	140	17.00		160
2010	By crane	"	0.800	140	51.00	35.00	230
2020	By pump	"	0.686	140	43.75	30.00	210
2040	By hand buggy	"	0.800	140	51.00		190
3980	5000# concrete						
4010	By chute	CY	0.267	150	17.00		170
4015	By crane	"	0.800	150	51.00	35.00	240
4020	By pump	"	0.686	150	43.75	30.00	220
4030	By hand buggy	"	0.800	150	51.00		200

Unit of Measurement
Each item (and cost) is defined in terms of the common estimating unit. All costs are listed in dollars per unit.

Total Cost
The total cost is the sum of material and installation costs. This total represents typical contractors' costs including overhead and profit, but does not include markups for the general contractor or construction management fees.

DIVISION 01
GENERAL

REQUIREMENTS

ID Code	Description — Component Descriptions	Output — Unit of Meas.	Manhr / Unit	Unit Costs — Material Cost	Labor Cost	Equipment Cost	Total Cost
01 - 21001	**ALLOWANCES**						**01 - 21001**
0090	Overhead						
1000	$20,000 project						
1020	Minimum	PCT					15.00
1040	Average	"					20.00
1060	Maximum	"					40.00
1080	$100,000 project						
1100	Minimum	PCT					12.00
1120	Average	"					15.00
1140	Maximum	"					25.00
1160	$500,000 project						
1170	Minimum	PCT					10.00
1180	Average	"					12.00
1200	Maximum	"					20.00
1480	Profit						
1500	$20,000 project						
1520	Minimum	PCT					10.00
1540	Average	"					15.00
1560	Maximum	"					25.00
1580	$100,000 project						
1600	Minimum	PCT					10.00
1620	Average	"					12.00
1640	Maximum	"					20.00
1660	$500,000 project						
1680	Minimum	PCT					5.00
1700	Average	"					10.00
1720	Maximum	"					15.00
4080	Taxes						
5000	Sales tax						
5020	Minimum	PCT					4.00
5040	Average	"					5.00
5060	Maximum	"					10.00
5080	Unemployment						
5100	Minimum	PCT					3.00
5120	Average	"					6.50
5140	Maximum	"					8.00
5200	Social security (FICA)	"					7.85

REQUIREMENTS

ID Code	Description — Component Descriptions	Output — Unit of Meas.	Output — Manhr / Unit	Unit Costs — Material Cost	Unit Costs — Labor Cost	Unit Costs — Equipment Cost	Unit Costs — Total Cost
01 - 31130	**FIELD STAFF**						**01 - 31130**
1000	Superintendent						
1020	Minimum	YEAR					105,179
1040	Average	"					131,491
1060	Maximum	"					157,948
1160	Foreman						
1180	Minimum	YEAR					69,908
1200	Average	"					111,801
1220	Maximum	"					130,897
1240	Bookkeeper/timekeeper						
1260	Minimum	YEAR					40,439
1280	Average	"					52,808
1300	Maximum	"					68,323
1320	Watchman						
1340	Minimum	YEAR					30,131
1360	Average	"					40,306
1380	Maximum	"					50,879
01 - 32130	**SCHEDULING**						**01 - 32130**
0090	Scheduling for						
1000	$100,000 project						
1020	Minimum	PCT					1.00
1040	Average	"					2.00
1060	Maximum	"					6.00
1080	$500,000 project						
1100	Minimum	PCT					0.50
1120	Average	"					1.00
1140	Maximum	"					2.00
4000	Scheduling software						
4020	Minimum	EA					720
4040	Average	"					4,090
4060	Maximum	"					81,790
01 - 32330	**JOB REQUIREMENTS**						**01 - 32330**
1000	Job photographs, small jobs						
1020	Minimum	EA					150
1040	Average	"					230
1060	Maximum	"					540
1080	Large projects						
1100	Minimum	EA					780

REQUIREMENTS

ID Code	Description — Component Descriptions	Unit of Meas.	Manhr / Unit	Material Cost	Labor Cost	Equipment Cost	Total Cost
01 - 32330	**JOB REQUIREMENTS, Cont'd...**						**01 - 32330**
1120	Average	EA					1,160
1140	Maximum	"					3,880
1240	Welding, per test						
1260	Minimum	EA					25.25
1280	Average	"					42.00
1300	Maximum	"					170
01 - 54001	**CONSTRUCTION AIDS**						**01 - 54001**
1000	Scaffolding/staging, rent per month						
1020	Measured by lineal feet of base						
1040	10' high	LF					15.00
1060	20' high	"					27.00
1080	30' high	"					37.75
1100	40' high	"					43.50
1120	50' high	"					52.00
1140	Measured by square foot of surface						
1160	Minimum	SF					0.65
1180	Average	"					1.13
1200	Maximum	"					2.03
1220	Safety nets, heavy duty, per job						
1240	Minimum	SF					0.44
1260	Average	"					0.53
1280	Maximum	"					1.16
01 - 54008	**MOBILIZATION**						**01 - 54008**
1000	Equipment mobilization						
1020	Bulldozer						
1040	Minimum	EA					240
1060	Average	"					510
1080	Maximum	"					850
1100	Backhoe/front-end loader						
1120	Minimum	EA					150
1140	Average	"					260
1160	Maximum	"					560
1180	Crane, crawler type						
1200	Minimum	EA					2,690
1220	Average	"					6,600
1240	Maximum	"					14,170
1260	Truck crane						

REQUIREMENTS

ID Code	Description — Component Descriptions	Output — Unit of Meas.	Output — Manhr / Unit	Unit Costs — Material Cost	Unit Costs — Labor Cost	Unit Costs — Equipment Cost	Total Cost
01 - 54008	**MOBILIZATION, Cont'd...**						**01 - 54008**
1280	Minimum	EA					610
1300	Average	"					950
1320	Maximum	"					1,640
01 - 54009	**EQUIPMENT**						**01 - 54009**
0080	Air compressor						
1000	60 cfm						
1020	By day	EA					110
1030	By week	"					330
1040	By month	"					1,000
1300	Air tools, per compressor, per day						
1310	Minimum	EA					45.50
1320	Average	"					57.00
1330	Maximum	"					80.00
1400	Generators, 5 kw						
1410	By day	EA					110
1420	By week	"					340
1430	By month	"					1,040
1500	Heaters, salamander type, per week						
1510	Minimum	EA					140
1520	Average	"					190
1530	Maximum	"					410
1600	Pumps, submersible						
1605	50 gpm						
1610	By day	EA					91.00
1620	By week	"					270
1630	By month	"					820
1900	Diaphragm pump, by week						
1920	Minimum	EA					140
1930	Average	"					240
1940	Maximum	"					510
2000	Pickup truck						
2020	By day	EA					150
2030	By week	"					450
2040	By month	"					1,390
2080	Dump truck						
2100	6 c.y. truck						
2120	By day	EA					410

REQUIREMENTS

ID Code	Component Descriptions	Unit of Meas.	Manhr / Unit	Material Cost	Labor Cost	Equipment Cost	Total Cost
01 - 54009	**EQUIPMENT, Cont'd...**						**01 - 54009**
2130	By week	EA					1,230
2140	By month	"					3,680
3000	Backhoe/loader, rubber tired						
3005	1/2 c.y. capacity						
3010	By day	EA					570
3020	By week	"					1,700
3030	By month	"					5,110
3200	Bulldozer						
3205	75 hp						
3210	By day	EA					790
3220	By week	"					2,380
3230	By month	"					7,160
4000	Cranes, crawler type						
4005	15 ton capacity						
4010	By day	EA					1,020
4020	By week	"					3,070
4030	By month	"					9,200
4145	Truck mounted, hydraulic						
4150	15 ton capacity						
4160	By day	EA					960
4170	By week	"					2,890
4180	By month	"					8,350
5380	Loader, rubber tired						
5385	1 c.y. capacity						
5390	By day	EA					680
5400	By week	"					2,040
5410	By month	"					6,140
01 - 56230	**TEMPORARY FACILITIES**						**01 - 56230**
1000	Barricades, temporary						
1010	Highway						
1020	Concrete	LF	0.080	39.50	5.90		45.50
1040	Wood	"	0.032	9.07	2.36		11.50
1060	Steel	"	0.027	8.25	1.96		10.25
1090	Pedestrian barricades						
1100	Plywood	SF	0.027	8.25	1.96		10.25
1120	Chain link fence	"	0.027	6.60	1.96		8.56
1130	Trailers, general office type, per month						

REQUIREMENTS

ID Code	Description — Component Descriptions	Output — Unit of Meas.	Output — Manhr / Unit	Unit Costs — Material Cost	Unit Costs — Labor Cost	Unit Costs — Equipment Cost	Unit Costs — Total Cost
01 - 56230	**TEMPORARY FACILITIES, Cont'd...**						**01 - 56230**
2020	Minimum	EA					250
2040	Average	"					420
2060	Maximum	"					840
2070	Crew change trailers, per month						
2100	Minimum	EA					150
2120	Average	"					170
2140	Maximum	"					250
01 - 58130	**SIGNS**						**01 - 58130**
0080	Construction signs, temporary						
1000	Signs, 2' x 4'						
1020	Minimum	EA					42.00
1040	Average	"					100
1060	Maximum	"					360
1080	Signs, 4' x 8'						
1100	Minimum	EA					88.00
1120	Average	"					230
1140	Maximum	"					990
1160	Signs, 8' x 8'						
1180	Minimum	EA					110
1200	Average	"					360
1220	Maximum	"					3,580
01 - 78330	**BONDS**						**01 - 78330**
1000	Performance bonds						
1020	Minimum	PCT					0.62
1040	Average	"					1.93
1060	Maximum	"					3.07

DIVISION 02
SITE CONSTRUCTION

DEMOLITION

ID Code	Description / Component Descriptions	Output		Unit Costs			
		Unit of Meas.	Manhr / Unit	Material Cost	Labor Cost	Equipment Cost	Total Cost
02 - 51190	**SELECTIVE BUILDING DEMOLITION**						**02 - 51190**
9200	Cut-outs						
9230	Concrete, elevated slabs, mesh reinforcing						
9240	Under 5 cf	CF	0.800		59.00		59.00
9260	Over 5 cf	"	0.667		49.25		49.25
9270	Bar reinforcing						
9280	Under 5 cf	CF	1.333		98.00		98.00
9290	Over 5 cf	"	1.000		74.00		74.00
9500	Rubbish handling						
9519	Load in dumpster or truck						
9520	Minimum	CF	0.018		1.31		1.31
9540	Maximum	"	0.027		1.96		1.96
9550	For use of elevators, add						
9560	Minimum	CF	0.004		0.29		0.29
9570	Maximum	"	0.008		0.59		0.59
9600	Rubbish hauling						
9640	Hand loaded on trucks, 2 mile trip	CY	0.320		23.50	27.25	51.00
9660	Machine loaded on trucks, 2 mile trip	"	0.240		17.50	15.00	32.50

HAZARDOUS WASTE

ID Code	Description / Component Descriptions	Output		Unit Costs			
		Unit of Meas.	Manhr / Unit	Material Cost	Labor Cost	Equipment Cost	Total Cost
02 - 82001	**ASBESTOS REMOVAL**						**02 - 82001**
1000	Enclosure using wood studs & poly, install & remove	SF	0.020	590	1.47		590
1020	Trailer (change room)	DAY					130
1100	Disposal suits (4 suits per man day)	"					53.00
1120	Type C respirator mask, includes hose & filters, per	"					26.25
1130	Respirator mask & filter, light contamination	"					10.50
1980	Air monitoring test, 12 tests per day						
2000	Off job testing	DAY					1,470
2020	On the job testing	"					1,960
6000	Asbestos vacuum with attachments	EA					850
6500	Hydraspray piston pump	"					1,140
6600	Negative air pressure system	"					1,140
6800	Grade D breathing air equipment	"					2,560
6900	Glove bag, 44" x 60" x 6 mil plastic	"					8.23
7980	40 CY asbestos dumpster						
8000	Weekly rental	EA					950
8100	Pick up/delivery	"					430
8400	Asbestos dump fee	"					270

HAZARDOUS WASTE

ID Code	Description — Component Descriptions	Unit of Meas.	Manhr / Unit	Material Cost	Labor Cost	Equipment Cost	Total Cost
02 - 82002	**DUCT INSULATION REMOVAL**						**02 - 82002**
0080	Remove duct insulation, duct size						
1000	6" x 12"	LF	0.044	290	3.28		290
1020	x 18"	"	0.062	210	4.54		210
1040	x 24"	"	0.089	150	6.56		160
1060	8" x 12"	"	0.067	200	4.92		200
1080	x 18"	"	0.073	180	5.36		190
1100	x 24"	"	0.100	130	7.38		140
1120	12" x 12"	"	0.067	200	4.92		200
1140	x 18"	"	0.089	150	6.56		160
1160	x 24"	"	0.114	110	8.43		120
02 - 82003	**PIPE INSULATION REMOVAL**						**02 - 82003**
0060	Removal, asbestos insulation						
0080	2" thick, pipe						
1000	1" to 3" dia.	LF	0.067		4.92		4.92
1020	4" to 6" dia.	"	0.076		5.62		5.62
1030	3" thick						
1040	7" to 8" dia.	LF	0.080		5.90		5.90
1060	9" to 10" dia.	"	0.084		6.21		6.21
1070	11" to 12" dia.	"	0.089		6.56		6.56
1080	13" to 14" dia.	"	0.094		6.94		6.94
1090	15" to 18" dia.	"	0.100		7.38		7.38

DIVISION 03
CONCRETE

FORMWORK

ID Code	Component Descriptions	Unit of Meas.	Manhr / Unit	Material Cost	Labor Cost	Equipment Cost	Total Cost
	Description	**Output**		**Unit Costs**			
03 - 11134	**ELEVATED SLAB FORMWORK**						**03 - 11134**
0100	Elevated slab formwork						
1000	Slab, with drop panels						
1020	1 use	SF	0.064	5.49	6.03		11.50
1100	5 uses	"	0.055	1.96	5.20		7.16
2000	Floor slab, hung from steel beams						
2020	1 use	SF	0.062	4.41	5.80		10.25
2100	5 uses	"	0.053	1.62	5.02		6.64
3000	Floor slab, with pans or domes						
3020	1 use	SF	0.073	7.88	6.85		14.75
3100	5 uses	"	0.062	3.94	5.80		9.74
9030	Equipment curbs, 12" high						
9035	1 use	LF	0.080	4.06	7.54		11.50
9100	5 uses	"	0.067	1.77	6.28		8.05
03 - 11135	**EQUIPMENT PAD FORMWORK**						**03 - 11135**
1000	Equipment pad, job built						
1020	1 use	SF	0.100	4.84	9.42		14.25
1040	2 uses	"	0.094	2.90	8.87		11.75
1060	3 uses	"	0.089	2.33	8.37		10.75
1080	4 uses	"	0.084	1.81	7.93		9.74
1100	5 uses	"	0.080	1.44	7.54		8.98
03 - 11136	**FOOTING FORMWORK**						**03 - 11136**
2000	Wall footings, job built, continuous						
2040	1 use	SF	0.080	2.59	7.54		10.25
2090	5 uses	"	0.067	1.15	6.28		7.43
3000	Column footings, spread						
3020	1 use	SF	0.100	2.74	9.42		12.25
3100	5 uses	"	0.080	1.12	7.54		8.66
03 - 11139	**SLAB / MAT FORMWORK**						**03 - 11139**
3980	Edge forms						
3990	6" high						
4000	1 use	LF	0.073	4.08	6.85		11.00
4004	5 uses	"	0.062	1.17	5.80		6.97
4006	12" high						
4010	1 use	LF	0.080	3.85	7.54		11.50
4014	5 uses	"	0.067	1.07	6.28		7.35
5000	Formwork for openings						

FORMWORK

	Description	Output		Unit Costs			
ID Code	Component Descriptions	Unit of Meas.	Manhr / Unit	Material Cost	Labor Cost	Equipment Cost	Total Cost
03 - 11139	**SLAB / MAT FORMWORK, Cont'd...**						**03 - 11139**
5020	1 use	SF	0.160	5.51	15.00		20.50
5100	5 uses	"	0.114	1.71	10.75		12.50
03 - 11141	**WALL FORMWORK**						**03 - 11141**
2980	Wall forms, exterior, job built						
3000	Up to 8' high wall						
3120	1 use	SF	0.080	4.36	7.54		12.00
3190	5 uses	"	0.067	1.60	6.28		7.88
3200	Over 8' high wall						
3220	1 use	SF	0.100	4.79	9.42		14.25
3290	5 uses	"	0.080	1.96	7.54		9.50
5000	Column pier and pilaster						
5020	1 use	SF	0.160	4.58	15.00		19.50
5090	5 uses	"	0.114	2.06	10.75		12.75
6980	Interior wall forms						
7000	Up to 8' high						
7020	1 use	SF	0.073	4.36	6.85		11.25
7100	5 uses	"	0.062	1.55	5.80		7.35
7200	Over 8' high						
7220	1 use	SF	0.089	4.79	8.37		13.25
7290	5 uses	"	0.073	1.98	6.85		8.83

REINFORCEMENT

	Description	Output		Unit Costs			
03 - 21004	**ELEVATED SLAB REINFORCING**						**03 - 21004**
0980	Elevated slab						
1000	#3 - #4	TON	10.000	1,880	960		2,840
1020	#5 - #6	"	8.889	1,650	860		2,510
1980	Galvanized						
2000	#3 - #4	TON	10.000	3,060	960		4,020
2020	#5 - #6	"	8.889	3,020	860		3,880
03 - 21005	**EQUIP. PAD REINFORCING**						**03 - 21005**
0980	Equipment pad						
1000	#3 - #4	TON	16.000	1,880	1,540		3,420
1020	#5 - #6	"	14.545	1,650	1,400		3,050
1040	#7 - #8	"	13.333	1,610	1,280		2,890
1060	#9 - #10	"	12.308	1,610	1,180		2,790
1080	#11	"	11.429	1,610	1,100		2,710

REINFORCEMENT

ID Code	Component Descriptions	Unit of Meas.	Manhr / Unit	Material Cost	Labor Cost	Equipment Cost	Total Cost
03 - 21006	**FOOTING REINFORCING**						**03 - 21006**
1000	Footings						
1020	#3 - #4	TON	13.333	1,880	1,280		3,160
1030	#5 - #6	"	11.429	1,650	1,100		2,750
4980	Straight dowels, 24" long						
5000	1" dia. (#8)	EA	0.020	6.41	1.92		8.33
5040	3/4" dia. (#6)	"	0.019	5.76	1.79		7.55
5050	5/8" dia. (#5)	"	0.017	4.98	1.67		6.65
5060	1/2" dia. (#4)	"	0.016	3.75	1.54		5.29
03 - 21009	**SLAB / MAT REINFORCING**						**03 - 21009**
0900	Bars, slabs						
1000	#3 - #4	TON	13.333	1,880	1,280		3,160
1020	#5 - #6	"	11.429	1,650	1,100		2,750
1040	#7 - #8	"	10.000	1,610	960		2,570
1090	Galvanized						
2000	#3 - #4	TON	13.333	3,210	1,280		4,490
2020	#5 - #6	"	11.429	3,030	1,100		4,130
4990	Wire mesh, slabs						
5000	Galvanized						
5010	4x4						
5020	W1.4xW1.4	SF	0.005	0.51	0.51		1.02
5040	W2.0xW2.0	"	0.006	0.66	0.55		1.21
5060	W2.9xW2.9	"	0.006	0.93	0.59		1.52
5080	W4.0xW4.0	"	0.007	1.38	0.64		2.02
5090	6x6						
5100	W1.4xW1.4	SF	0.004	0.47	0.38		0.85
5120	W2.0xW2.0	"	0.004	0.66	0.42		1.08
5140	W2.9xW2.9	"	0.005	0.90	0.45		1.35
5150	W4.0xW4.0	"	0.005	0.97	0.51		1.48
03 - 21011	**WALL REINFORCING**						**03 - 21011**
0980	Walls						
1000	#3 - #4	TON	11.429	1,880	1,100		2,980
1020	#5 - #6	"	10.000	1,650	960		2,610
1980	Galvanized						
2000	#3 - #4	TON	11.429	3,210	1,100		4,310
2020	#5 - #6	"	10.000	3,030	960		3,990
8980	Masonry wall (horizontal)						
9000	#3 - #4	TON	32.000	1,880	3,080		4,960

REINFORCEMENT

ID Code	Description Component Descriptions	Output Unit of Meas.	Output Manhr / Unit	Unit Costs Material Cost	Unit Costs Labor Cost	Unit Costs Equipment Cost	Unit Costs Total Cost
03 - 21011	**WALL REINFORCING, Cont'd...**						**03 - 21011**
9020	#5 - #6	TON	26.667	1,650	2,570		4,220
9030	Galvanized						
9040	#3 - #4	TON	32.000	3,210	3,080		6,290
9060	#5 - #6	"	26.667	3,030	2,570		5,600
9180	Masonry wall (vertical)						
9200	#3 - #4	TON	40.000	1,880	3,850		5,730
9220	#5 - #6	"	32.000	1,650	3,080		4,730
9230	Galvanized						
9240	#3 - #4	TON	40.000	3,210	3,850		7,060
9260	#5 - #6	"	32.000	3,030	3,080		6,110

PLACING CONCRETE

ID Code	Description Component Descriptions	Output Unit of Meas.	Output Manhr / Unit	Unit Costs Material Cost	Unit Costs Labor Cost	Unit Costs Equipment Cost	Unit Costs Total Cost
03 - 31003	**ELEVATED SLAB CONCRETE**						**03 - 31003**
0980	Elevated slab						
0990	2500# or 3000# concrete						
1000	By crane	CY	0.480	160	35.50	23.50	220
1010	By pump	"	0.369	160	27.25	18.00	210
1020	By hand buggy	"	0.800	160	59.00		220
4000	5000# concrete						
4010	By crane	CY	0.480	180	35.50	23.50	240
4020	By pump	"	0.369	180	27.25	18.00	230
4040	By hand buggy	"	0.800	180	59.00		240
03 - 31004	**EQUIPMENT PAD CONCRETE**						**03 - 31004**
0960	Equipment pad						
0980	2500# or 3000# concrete						
1000	By chute	CY	0.267	160	19.75		180
1020	By pump	"	0.686	160	51.00	33.50	240
1040	By crane	"	0.800	160	59.00	39.25	260
1050	3500# or 4000# concrete						
1060	By chute	CY	0.267	170	19.75		190
1080	By pump	"	0.686	170	51.00	33.50	250
1100	By crane	"	0.800	170	59.00	39.25	270
1110	5000# concrete						
1120	By chute	CY	0.267	180	19.75		200
1140	By pump	"	0.686	180	51.00	33.50	260
1160	By crane	"	0.800	180	59.00	39.25	280

PLACING CONCRETE

ID Code	Component Descriptions	Unit of Meas.	Manhr / Unit	Material Cost	Labor Cost	Equipment Cost	Total Cost
	Description	**Output**		**Unit Costs**			
03 - 31005	**FOOTING CONCRETE**						**03 - 31005**
0980	Continuous footing						
0990	2500# or 3000# concrete						
1000	By chute	CY	0.267	160	19.75		180
1010	By pump	"	0.600	160	44.25	29.25	230
1020	By crane	"	0.686	160	51.00	33.50	240
4000	5000# concrete						
4010	By chute	CY	0.267	180	19.75		200
4020	By pump	"	0.600	180	44.25	29.25	250
4030	By crane	"	0.686	180	51.00	33.50	260
4980	Spread footing						
5000	2500# or 3000# concrete						
5010	Under 5 c.y.						
5020	By chute	CY	0.267	160	19.75		180
5040	By pump	"	0.640	160	47.25	31.25	240
5060	By crane	"	0.738	160	54.00	36.25	250
7200	5000# concrete						
7205	Under 5 c.y.						
7210	By chute	CY	0.267	190	19.75		210
7220	By pump	"	0.640	190	47.25	31.25	270
7230	By crane	"	0.738	190	54.00	36.25	280
03 - 31008	**SLAB / MAT CONCRETE**						**03 - 31008**
0960	Slab on grade						
0980	2500# or 3000# concrete						
1000	By chute	CY	0.200	160	14.75		170
1010	By crane	"	0.400	160	29.50	19.50	210
1020	By pump	"	0.343	160	25.25	16.75	200
1030	By hand buggy	"	0.533	160	39.25		200
3980	5000# concrete						
4010	By chute	CY	0.200	180	14.75		190
4020	By crane	"	0.400	180	29.50	19.50	230
4030	By pump	"	0.343	180	25.25	16.75	220
4040	By hand buggy	"	0.533	180	39.25		220

PLACING CONCRETE

ID Code	Description — Component Descriptions	Output — Unit of Meas.	Manhr / Unit	Unit Costs — Material Cost	Labor Cost	Equipment Cost	Total Cost
03 - 31009	**WALL CONCRETE**						**03 - 31009**
0940	Walls						
0960	2500# or 3000# concrete						
0980	To 4'						
1000	By chute	CY	0.229	160	16.75		180
1005	By crane	"	0.800	160	59.00	39.25	260
1010	By pump	"	0.738	160	54.00	36.25	250
1020	To 8'						
1030	By crane	CY	0.873	160	64.00	42.75	270
1040	By pump	"	0.800	160	59.00	39.25	260
8480	Filled block (CMU)						
8490	3000# concrete, by pump						
8500	4" wide	SF	0.034	1.98	2.53	1.67	6.18
8510	6" wide	"	0.040	2.95	2.95	1.95	7.86
8520	8" wide	"	0.048	3.97	3.54	2.35	9.86
8530	10" wide	"	0.056	2.93	4.16	2.76	9.86
8540	12" wide	"	0.069	3.78	5.06	3.35	12.25
8560	Pilasters, 3000# concrete	CF	0.960	8.30	71.00	47.00	130
8700	Wall cavity, 2" thick, 3000# concrete	SF	0.032	1.54	2.36	1.56	5.46

CONCRETE BORING

ID Code	Description — Component Descriptions	Output — Unit of Meas.	Manhr / Unit	Unit Costs — Material Cost	Labor Cost	Equipment Cost	Total Cost
03 - 82131	**CORE DRILLING**						**03 - 82131**
0100	Concrete						
0110	6" thick						
0120	3" dia.	EA	0.571		42.25	14.25	56.00
0140	4" dia.	"	0.667		49.25	16.75	66.00
0160	6" dia.	"	0.800		59.00	20.00	79.00
0180	8" dia.	"	1.333		98.00	33.25	130
0300	8" thick						
0320	3" dia.	EA	0.800		59.00	20.00	79.00
0360	4" dia.	"	1.000		74.00	25.00	99.00
0380	6" dia.	"	1.143		84.00	28.50	110
0400	8" dia.	"	1.600		120	40.00	160
0420	10" thick						
0440	3" dia.	EA	1.000		74.00	25.00	99.00
0460	4" dia.	"	1.143		84.00	28.50	110
0480	6" dia.	"	1.333		98.00	33.25	130
0490	8" dia.	"	2.000		150	50.00	200

CONCRETE BORING

ID Code	Description		Output		Unit Costs			
	Component Descriptions		Unit of Meas.	Manhr / Unit	Material Cost	Labor Cost	Equipment Cost	Total Cost
03 - 82131		**CORE DRILLING, Cont'd...**						**03 - 82131**
0520	12" thick							
0540	3" dia.		EA	1.333		98.00	33.25	130
0560	4" dia.		"	1.600		120	40.00	160
0580	6" dia.		"	2.000		150	50.00	200
0600	8" dia.		"	2.667		200	67.00	260

DIVISION 04
MASONRY

MASONRY RESTORATION

ID Code	Description — Component Descriptions	Output — Unit of Meas.	Output — Manhr / Unit	Unit Costs — Material Cost	Unit Costs — Labor Cost	Unit Costs — Equipment Cost	Unit Costs — Total Cost
04 - 01201	**RESTORATION AND CLEANING**						**04 - 01201**
1080	Masonry cleaning						
1090	Washing brick						
1120	Smooth surface	SF	0.013	0.26	1.19		1.45
1130	Rough surface	"	0.018	0.36	1.59		1.95
1360	Pointing masonry						
1420	Brick	SF	0.032	1.37	2.87		4.24
1430	Concrete block	"	0.023	0.60	2.05		2.65
8010	Brick removal and replacement						
8020	Minimum	EA	0.100	0.82	8.99		9.81
8040	Average	"	0.133	1.07	12.00		13.00
8060	Maximum	"	0.400	2.17	36.00		38.25

MORTAR, GROUT AND ACCESSORIES

ID Code	Component Descriptions	Unit of Meas.	Manhr / Unit	Material Cost	Labor Cost	Equipment Cost	Total Cost
04 - 05161	**MASONRY GROUT**						**04 - 05161**
0100	Grout, non-shrink, non-metallic, trowelable	CF	0.016	6.05	1.17	1.00	8.22
3070	Grout-filled individual CMU cells						
3090	4" wide	LF	0.012	0.41	0.87	0.75	2.03
3120	8" wide	"	0.012	1.38	0.87	0.75	3.00
04 - 05235	**MASONRY FLASHING**						**04 - 05235**
0080	Through-wall flashing						
1000	5 oz. coated copper	SF	0.067	4.18	5.99		10.25
1020	0.030" elastomeric	"	0.053	1.32	4.79		6.11

UNIT MASONRY

ID Code	Component Descriptions	Unit of Meas.	Manhr / Unit	Material Cost	Labor Cost	Equipment Cost	Total Cost
04 - 21131	**BRICK MASONRY**						**04 - 21131**
0100	Standard size brick, running bond						
1000	Face brick, red (6.4/sf)						
1020	Veneer	SF	0.133	6.51	12.00		18.50
1030	Cavity wall	"	0.114	6.51	10.25		16.75
1040	9" solid wall	"	0.229	13.00	20.50		33.50
3000	Chimney, standard brick, including flue						
3020	16" x 16"	LF	0.800	43.75	72.00		120
3080	20" x 20"	"	1.000	61.00	90.00		150

CONCRETE UNIT MASONRY

ID Code	Description — Component Descriptions	Output — Unit of Meas.	Output — Manhr / Unit	Unit Costs — Material Cost	Unit Costs — Labor Cost	Unit Costs — Equipment Cost	Unit Costs — Total Cost
04 - 22001	**CONCRETE MASONRY UNITS**						**04 - 22001**
0110	Hollow, load bearing						
0120	4"	SF	0.059	1.62	5.32		6.94
0160	8"	"	0.067	2.73	5.99		8.72
0280	Solid, load bearing						
0300	4"	SF	0.059	2.55	5.32		7.87
0340	8"	"	0.067	3.91	5.99		9.90
0480	Back-up block, 8" x 16"						
0540	4"	SF	0.047	1.87	4.23		6.10
0580	8"	"	0.053	3.14	4.79		7.93
8000	Steel angles and plates						
8010	Minimum	LB	0.011	1.46	1.02		2.48
8020	Maximum	"	0.020	2.15	1.79		3.94
8200	Various size angle lintels						
8205	1/4" stock						
8210	3" x 3"	LF	0.050	7.54	4.49		12.00
8220	3" x 3-1/2"	"	0.050	8.30	4.49		12.75
8225	3/8" stock						
8230	3" x 4"	LF	0.050	13.00	4.49		17.50
8250	4" x 4"	"	0.050	15.00	4.49		19.50
8265	1/2" stock						
8280	6" x 4"	LF	0.050	19.75	4.49		24.25

REFRACTORIES

ID Code	Description — Component Descriptions	Output — Unit of Meas.	Output — Manhr / Unit	Unit Costs — Material Cost	Unit Costs — Labor Cost	Unit Costs — Equipment Cost	Unit Costs — Total Cost
04 - 51001	**FLUE LINERS**						**04 - 51001**
1000	Flue liners						
1020	Rectangular						
1040	8" x 12"	LF	0.133	12.75	12.00		24.75
1060	12" x 12"	"	0.145	15.75	13.00		28.75
1200	Round						
1220	18" dia.	LF	0.190	57.00	17.00		74.00
1240	24" dia.	"	0.229	110	20.50		130

DIVISION 05
METALS

METAL FASTENINGS

ID Code	Description — Component Descriptions	Unit of Meas.	Manhr / Unit	Material Cost	Labor Cost	Equipment Cost	Total Cost
05 - 05231	**STRUCTURAL WELDING**						**05 - 05231**
0080	Welding						
0100	Single pass						
0120	1/8"	LF	0.040	0.39	4.16		4.55
0140	3/16"	"	0.053	0.66	5.54		6.20
0160	1/4"	"	0.067	0.92	6.93		7.85
0180	Miscellaneous steel shapes						
0190	Plain	LB	0.002	1.44	0.16		1.60
0200	Galvanized	"	0.003	1.80	0.27		2.07
0210	Plates						
0220	Plain	LB	0.002	1.29	0.20		1.49
0240	Galvanized	"	0.003	1.66	0.33		1.99
05 - 05239	**METAL FASTENINGS**						**05 - 05239**
1000	Anchor bolts, material only						
1020	3/8" x						
1040	8" long	EA					1.66
1060	10" long	"					1.81
1080	12" long	"					1.97
1090	1/2" x						
1100	8" long	EA					2.48
1120	10" long	"					2.64
1140	12" long	"					2.90
1160	18" long	"					3.15
8020	Add 25% for galvanized anchor bolts						
05 - 05240	**METAL LINTELS**						**05 - 05240**
0080	Lintels, steel						
0100	Plain	LB	0.020	1.98	2.08		4.06
0120	Galvanized	"	0.020	2.39	2.08		4.47

COLD FORMED FRAMING

ID Code	Description — Component Descriptions	Unit of Meas.	Manhr / Unit	Material Cost	Labor Cost	Equipment Cost	Total Cost
05 - 41001	**METAL FRAMING**						**05 - 41001**
0100	Furring channel, galvanized						
0110	Beams and columns, 3/4"						
0120	12" o.c.	SF	0.080	0.48	8.32		8.80
0140	16" o.c.	"	0.073	0.37	7.56		7.93
0150	Walls, 3/4"						
0160	12" o.c.	SF	0.040	0.48	4.16		4.64

COLD FORMED FRAMING

ID Code	Component Descriptions	Unit of Meas.	Manhr / Unit	Material Cost	Labor Cost	Equipment Cost	Total Cost
	Description	**Output**		**Unit Costs**			
05 - 41001	**METAL FRAMING, Cont'd...**					**05 - 41001**	
0170	16" o.c.	SF	0.033	0.37	3.46		3.83
0177	Stud, load bearing						
0178	16" o.c.						
0179	16 ga.						
0180	2-1/2"	SF	0.036	1.33	3.69		5.02
0190	3-5/8"	"	0.036	1.57	3.69		5.26
0460	24" o.c.						
0470	16 ga.						
0480	2-1/2"	SF	0.031	0.91	3.20		4.11
0510	3-5/8"	"	0.031	1.08	3.20		4.28

DIVISION 06
WOOD AND PLASTICS

FASTENERS AND ADHESIVES

ID Code	Component Descriptions	Unit of Meas.	Manhr / Unit	Material Cost	Labor Cost	Equipment Cost	Total Cost
		Description		**Output**		**Unit Costs**	

ID Code	Component Descriptions	Unit of Meas.	Manhr / Unit	Material Cost	Labor Cost	Equipment Cost	Total Cost
06 - 05231	**ACCESSORIES**						**06 - 05231**
1000	Anchors						
1020	Bolts, threaded two ends, with nuts and washers						
1030	1/2" dia.						
1040	4" long	EA	0.050	3.75	4.71		8.46
1060	7-1/2" long	"	0.050	4.37	4.71		9.08
1070	3/4" dia.						
1080	7-1/2" long	EA	0.050	6.91	4.71		11.50
1100	15" long	"	0.050	10.50	4.71		15.25
1210	Bolts, carriage						
1212	1/4 x 4	EA	0.080	0.81	7.54		8.35
1214	5/16 x 6	"	0.084	1.84	7.93		9.77
1216	3/8 x 6	"	0.084	3.72	7.93		11.75
1218	1/2 x 6	"	0.084	5.19	7.93		13.00
1240	Joist and beam hangers						
1250	18 ga.						
1260	2 x 4	EA	0.080	1.50	7.54		9.04
1280	2 x 6	"	0.080	1.80	7.54		9.34
1282	2 x 8	"	0.080	2.10	7.54		9.64
1284	2 x 10	"	0.089	2.26	8.37		10.75
1286	2 x 12	"	0.100	3.81	9.42		13.25
1900	Strap ties, 14 ga., 1-3/8" wide						
1920	12" long	EA	0.067	3.04	6.28		9.32
1940	18" long	"	0.073	3.27	6.85		10.00
1960	24" long	"	0.080	4.86	7.54		12.50
1980	36" long	"	0.089	6.69	8.37		15.00

ROUGH CARPENTRY

ID Code	Component Descriptions	Unit of Meas.	Manhr / Unit	Material Cost	Labor Cost	Equipment Cost	Total Cost
06 - 11001	**BLOCKING**						**06 - 11001**
1100	Steel construction						
1105	Walls						
1110	2x4	LF	0.053	0.94	5.02		5.96
1120	2x6	"	0.062	1.44	5.80		7.24
1130	2x8	"	0.067	1.89	6.28		8.17
1140	2x10	"	0.073	2.52	6.85		9.37
1150	2x12	"	0.080	3.25	7.54		10.75
1160	Ceilings						
1170	2x4	LF	0.062	0.94	5.80		6.74

ROUGH CARPENTRY

ID Code	Description Component Descriptions	Output Unit of Meas.	Output Manhr / Unit	Unit Costs Material Cost	Unit Costs Labor Cost	Unit Costs Equipment Cost	Unit Costs Total Cost
06 - 11001	**BLOCKING, Cont'd...**						**06 - 11001**
1180	2x6	LF	0.073	1.44	6.85		8.29
1190	2x8	"	0.080	1.89	7.54		9.43
1200	2x10	"	0.089	2.52	8.37		11.00
1210	2x12	"	0.100	3.25	9.42		12.75
1215	Wood construction						
1220	Walls						
1230	2x4	LF	0.044	1.05	4.18		5.23
1240	2x6	"	0.050	1.60	4.71		6.31
1250	2x8	"	0.053	2.11	5.02		7.13
1260	2x10	"	0.057	2.80	5.38		8.18
1270	2x12	"	0.062	3.62	5.80		9.42
1280	Ceilings						
1290	2x4	LF	0.050	1.05	4.71		5.76
1300	2x6	"	0.057	1.60	5.38		6.98
1310	2x8	"	0.062	2.11	5.80		7.91
1320	2x10	"	0.067	2.80	6.28		9.08
1330	2x12	"	0.073	3.62	6.85		10.50
06 - 11002	**CEILING FRAMING**						**06 - 11002**
1000	Ceiling joists						
1010	12" o.c.						
1020	2x4	SF	0.019	1.55	1.79		3.34
1040	2x8	"	0.021	3.31	1.98		5.29
1070	16" o.c.						
1080	2x4	SF	0.015	1.27	1.45		2.72
1100	2x8	"	0.017	2.68	1.57		4.25
1300	Sister joists for ceilings						
1310	2x4	LF	0.057	1.05	5.38		6.43
1320	2x6	"	0.067	1.60	6.28		7.88
1330	2x8	"	0.080	2.11	7.54		9.65
1340	2x10	"	0.100	2.80	9.42		12.25
1350	2x12	"	0.133	3.45	12.50		16.00
06 - 11003	**FLOOR FRAMING**						**06 - 11003**
1000	Floor joists						
1010	12" o.c.						
1020	2x6	SF	0.016	1.72	1.50		3.22
1030	2x8	"	0.016	2.54	1.53		4.07
1040	2x10	"	0.017	3.52	1.57		5.09

ROUGH CARPENTRY

ID Code	Component Descriptions	Unit of Meas.	Manhr / Unit	Material Cost	Labor Cost	Equipment Cost	Total Cost
	Description	**Output**		**Unit Costs**			

06 - 11003 — FLOOR FRAMING, Cont'd... — 06 - 11003

ID Code	Component Descriptions	Unit of Meas.	Manhr / Unit	Material Cost	Labor Cost	Equipment Cost	Total Cost
1050	2x12	SF	0.017	5.16	1.63		6.79
1180	16" o.c.						
1190	2x6	SF	0.013	1.65	1.25		2.90
1200	2x8	"	0.014	2.32	1.27		3.59
1220	2x10	"	0.014	2.83	1.30		4.13
1230	2x12	"	0.014	3.52	1.34		4.86
2000	Sister joists for floors						
2010	2x4	LF	0.050	1.05	4.71		5.76
2020	2x6	"	0.057	1.60	5.38		6.98
2030	2x8	"	0.067	2.11	6.28		8.39
2040	2x10	"	0.080	2.80	7.54		10.25
2050	2x12	"	0.100	3.62	9.42		13.00
2060	3x6	"	0.080	5.27	7.54		12.75
2070	3x8	"	0.089	6.47	8.37		14.75
2080	3x10	"	0.100	8.59	9.42		18.00
2090	3x12	"	0.114	10.50	10.75		21.25
2100	4x6	"	0.080	6.79	7.54		14.25
2110	4x8	"	0.089	9.04	8.37		17.50
2120	4x10	"	0.100	11.75	9.42		21.25
2130	4x12	"	0.114	13.00	10.75		23.75

06 - 11004 — FURRING — 06 - 11004

ID Code	Component Descriptions	Unit of Meas.	Manhr / Unit	Material Cost	Labor Cost	Equipment Cost	Total Cost
1100	Furring, wood strips						
1102	Walls						
1105	On masonry or concrete walls						
1107	1x2 furring						
1110	12" o.c.	SF	0.025	0.84	2.35		3.19
1120	16" o.c.	"	0.023	0.71	2.15		2.86
1130	24" o.c.	"	0.021	0.69	1.98		2.67
1135	1x3 furring						
1140	12" o.c.	SF	0.025	1.05	2.35		3.40
1150	16" o.c.	"	0.023	0.96	2.15		3.11
1160	24" o.c.	"	0.021	0.74	1.98		2.72
1165	On wood walls						
1167	1x2 furring						
1170	12" o.c.	SF	0.018	0.84	1.67		2.51
1180	16" o.c.	"	0.016	0.71	1.50		2.21
1190	24" o.c.	"	0.015	0.67	1.37		2.04

ROUGH CARPENTRY

ID Code	Component Descriptions	Unit of Meas.	Manhr / Unit	Material Cost	Labor Cost	Equipment Cost	Total Cost
06 - 11004	**FURRING, Cont'd...**					**06 - 11004**	
1195	1x3 furring						
1200	12" o.c.	SF	0.018	1.08	1.67		2.75
1210	16" o.c.	"	0.016	0.91	1.50		2.41
1220	24" o.c.	"	0.015	0.74	1.37		2.11
1224	Ceilings						
1226	On masonry or concrete ceilings						
1228	1x2 furring						
1230	12" o.c.	SF	0.044	0.84	4.18		5.02
1240	16" o.c.	"	0.040	0.71	3.77		4.48
1250	24" o.c.	"	0.036	0.67	3.42		4.09
1254	1x3 furring						
1260	12" o.c.	SF	0.044	1.05	4.18		5.23
1270	16" o.c.	"	0.040	0.91	3.77		4.68
1280	24" o.c.	"	0.036	0.74	3.42		4.16
1286	On wood ceilings						
1288	1x2 furring						
1290	12" o.c.	SF	0.030	0.84	2.79		3.63
1300	16" o.c.	"	0.027	0.71	2.51		3.22
1310	24" o.c.	"	0.024	0.67	2.28		2.95
1316	1x3						
1320	12" o.c.	SF	0.030	1.05	2.79		3.84
1330	16" o.c.	"	0.027	0.91	2.51		3.42
1340	24" o.c.	"	0.024	0.74	2.28		3.02
06 - 11005	**ROOF FRAMING**					**06 - 11005**	
1000	Roof framing						
1005	Rafters, gable end						
1008	0-2 pitch (flat to 2-in-12)						
1010	12" o.c.						
1020	2x4	SF	0.017	1.51	1.57		3.08
1030	2x6	"	0.017	2.11	1.63		3.74
1040	2x8	"	0.018	3.02	1.71		4.73
1070	16" o.c.						
1080	2x6	SF	0.014	1.89	1.34		3.23
1090	2x8	"	0.015	2.66	1.39		4.05
4980	Sister rafters						
5000	2x4	LF	0.057	1.05	5.38		6.43
5010	2x6	"	0.067	1.60	6.28		7.88

ROUGH CARPENTRY

ID Code	Component Descriptions	Unit of Meas.	Manhr / Unit	Material Cost	Labor Cost	Equipment Cost	Total Cost
06 - 11005	**ROOF FRAMING, Cont'd...**						**06 - 11005**
5020	2x8	LF	0.080	2.11	7.54		9.65
5030	2x10	"	0.100	2.80	9.42		12.25
5040	2x12	"	0.133	3.62	12.50		16.00
06 - 11006	**SLEEPERS**						**06 - 11006**
0960	Sleepers, over concrete						
0980	12" o.c.						
1000	1x2	SF	0.018	0.50	1.71		2.21
1020	1x3	"	0.019	0.76	1.79		2.55
1060	2x4	"	0.022	1.64	2.09		3.73
1080	2x6	"	0.024	2.41	2.21		4.62
1090	16" o.c.						
1100	1x2	SF	0.016	0.46	1.50		1.96
1120	1x3	"	0.016	0.65	1.50		2.15
1140	2x4	"	0.019	1.37	1.79		3.16
1160	2x6	"	0.020	2.02	1.88		3.90
06 - 11007	**SOFFITS**						**06 - 11007**
0980	Soffit framing						
1000	2x3	LF	0.057	0.71	5.38		6.09
1020	2x4	"	0.062	0.88	5.80		6.68
1030	2x6	"	0.067	1.31	6.28		7.59
1040	2x8	"	0.073	1.83	6.85		8.68
06 - 11008	**WALL FRAMING**						**06 - 11008**
0960	Framing wall, studs						
0980	12" o.c.						
1000	2x3	SF	0.015	0.93	1.39		2.32
1040	2x4	"	0.015	1.31	1.39		2.70
1110	16" o.c.						
1120	2x3	SF	0.013	0.76	1.17		1.93
1140	2x4	"	0.013	1.05	1.17		2.22
1165	24" o.c.						
1170	2x3	SF	0.011	0.59	1.01		1.60
1180	2x4	"	0.011	0.80	1.01		1.81
1480	Plates, top or bottom						
1500	2x3	LF	0.024	0.71	2.21		2.92
1510	2x4	"	0.025	0.88	2.35		3.23
2000	Headers, door or window						

ROUGH CARPENTRY

ID Code	Component Descriptions	Unit of Meas.	Manhr / Unit	Material Cost	Labor Cost	Equipment Cost	Total Cost
	Description	**Output**		**Unit Costs**			
06 - 11008	**WALL FRAMING, Cont'd...**						**06 - 11008**
2005	2x6						
2008	Single						
2010	3' long	EA	0.400	4.27	37.75		42.00
2020	6' long	"	0.500	8.54	47.00		56.00
2025	Double						
2030	3' long	EA	0.444	8.56	42.00		51.00
2040	6' long	"	0.571	17.25	54.00		71.00
2044	2x8						
2046	Single						
2050	4' long	EA	0.500	7.82	47.00		55.00
2060	8' long	"	0.615	15.50	58.00		74.00
2065	Double						
2070	4' long	EA	0.571	15.50	54.00		70.00
2080	8' long	"	0.727	31.25	69.00		100
2085	2x10						
2088	Single						
2090	5' long	EA	0.615	13.00	58.00		71.00
2100	10' long	"	0.800	26.25	75.00		100
2110	Double						
2120	5' long	EA	0.667	26.25	63.00		89.00
2130	10' long	"	0.800	52.00	75.00		130

SHEATHING

ID Code	Component Descriptions	Unit of Meas.	Manhr / Unit	Material Cost	Labor Cost	Equipment Cost	Total Cost
06 - 16001	**FLOOR SHEATHING**						**06 - 16001**
1980	Sub-flooring, plywood, CDX						
2000	1/2" thick	SF	0.010	1.05	0.94		1.99
2020	5/8" thick	"	0.011	1.52	1.07		2.59
2080	3/4" thick	"	0.013	2.80	1.25		4.05
2090	Structural plywood						
2100	1/2" thick	SF	0.010	1.67	0.94		2.61
3100	Board type sub-flooring						
3105	1x6						
3110	Minimum	SF	0.018	2.54	1.67		4.21
3115	Maximum	"	0.020	3.22	1.88		5.10
3117	1x8						
3120	Minimum	SF	0.017	2.80	1.58		4.38
3140	Maximum	"	0.019	3.29	1.77		5.06

SHEATHING

ID Code	Description / Component Descriptions	Unit of Meas.	Manhr / Unit	Material Cost	Labor Cost	Equipment Cost	Total Cost
06 - 16001	**FLOOR SHEATHING, Cont'd...**						**06 - 16001**
3150	1x10						
3160	Minimum	SF	0.016	3.93	1.50		5.43
3180	Maximum	"	0.018	4.22	1.67		5.89
5990	Underlayment						
6000	Hardboard, 1/4" tempered	SF	0.010	1.56	0.94		2.50
6010	Plywood, CDX						
6020	3/8" thick	SF	0.010	1.63	0.94		2.57
6040	1/2" thick	"	0.011	1.94	1.00		2.94
6060	5/8" thick	"	0.011	2.25	1.07		3.32
6080	3/4" thick	"	0.012	2.80	1.16		3.96
06 - 16002	**ROOF SHEATHING**						**06 - 16002**
0080	Sheathing						
0090	Plywood, CDX						
1000	3/8" thick	SF	0.010	1.63	0.97		2.60
1020	1/2" thick	"	0.011	1.94	1.00		2.94
1040	5/8" thick	"	0.011	2.25	1.07		3.32
1060	3/4" thick	"	0.012	2.80	1.16		3.96
06 - 16003	**WALL SHEATHING**						**06 - 16003**
0980	Sheathing						
0990	Plywood, CDX						
1000	3/8" thick	SF	0.012	1.63	1.11		2.74
1020	1/2" thick	"	0.012	1.94	1.16		3.10
1040	5/8" thick	"	0.013	2.25	1.25		3.50
1060	3/4" thick	"	0.015	2.80	1.37		4.17

FINISH CARPENTRY

ID Code	Description / Component Descriptions	Unit of Meas.	Manhr / Unit	Material Cost	Labor Cost	Equipment Cost	Total Cost
06 - 22001	**MILLWORK**						**06 - 22001**
0070	Countertop, laminated plastic						
0080	25" x 7/8" thick						
0099	Minimum	LF	0.200	18.50	18.75		37.25
0100	Average	"	0.267	34.75	25.25		60.00
0110	Maximum	"	0.320	51.00	30.25		81.00
0115	25" x 1-1/4" thick						
0120	Minimum	LF	0.267	22.25	25.25		47.50
0130	Average	"	0.320	44.50	30.25		75.00
0140	Maximum	"	0.400	67.00	37.75		100

FINISH CARPENTRY

ID Code	Description — Component Descriptions	Output — Unit of Meas.	Output — Manhr / Unit	Unit Costs — Material Cost	Unit Costs — Labor Cost	Unit Costs — Equipment Cost	Unit Costs — Total Cost
06 - 22001	**MILLWORK, Cont'd...**						**06 - 22001**
0160	Add for cutouts	EA	0.500		47.00		47.00
0165	Backsplash, 4" high, 7/8" thick	LF	0.160	24.50	15.00		39.50
2500	Base cabinet, 34-1/2" high, 24" deep, hardwood						
2540	Minimum	LF	0.320	240	30.25		270
2560	Average	"	0.400	270	37.75		310
2580	Maximum	"	0.533	310	50.00		360
2600	Wall cabinets						
2640	Minimum	LF	0.267	73.00	25.25		98.00
2660	Average	"	0.320	99.00	30.25		130
2680	Maximum	"	0.400	120	37.75		160

DIVISION 07
THERMAL AND MOISTURE

MOISTURE PROTECTION

ID Code	Component Descriptions	Unit of Meas.	Manhr / Unit	Material Cost	Labor Cost	Equipment Cost	Total Cost
	Description	**Output**		**Unit Costs**			
07 - 11001	**DAMPPROOFING**						**07 - 11001**
1000	Silicone dampproofing, sprayed on						
1020	Concrete surface						
1040	1 coat	SF	0.004	0.68	0.32		1.00
1060	2 coats	"	0.006	1.12	0.45		1.57
1070	Concrete block						
1080	1 coat	SF	0.005	0.68	0.39		1.07
1100	2 coats	"	0.007	1.12	0.53		1.65
1110	Brick						
1120	1 coat	SF	0.006	0.78	0.45		1.23
1140	2 coats	"	0.008	1.21	0.59		1.80
07 - 11131	**BITUMINOUS DAMPPROOFING**						**07 - 11131**
0100	Building paper, asphalt felt						
0120	15 lb	SF	0.032	0.19	2.36		2.55
0140	30 lb	"	0.033	0.37	2.46		2.83
07 - 11161	**PARGING / MASONRY PLASTER**						**07 - 11161**
0080	Parging						
0100	1/2" thick	SF	0.053	0.44	4.79		5.23
0200	3/4" thick	"	0.067	0.48	5.99		6.47

INSULATION

ID Code	Component Descriptions	Unit of Meas.	Manhr / Unit	Material Cost	Labor Cost	Equipment Cost	Total Cost
07 - 21131	**BOARD INSULATION**						**07 - 21131**
1000	Insulation, rigid						
1010	Fiberglass, roof						
1020	0.75" thick, R2.78	SF	0.007	0.58	0.53		1.11
1040	1.06" thick, R4.17	"	0.008	0.88	0.56		1.44
1060	1.31" thick, R5.26	"	0.008	1.18	0.59		1.77
1080	1.63" thick, R6.67	"	0.008	1.46	0.62		2.08
1100	2.25" thick, R8.33	"	0.009	1.61	0.65		2.26
2200	Perlite board, roof						
2220	1.00" thick, R2.78	SF	0.007	0.60	0.49		1.09
2240	1.50" thick, R4.17	"	0.007	0.94	0.51		1.45
2260	2.00" thick, R5.92	"	0.007	1.16	0.53		1.69
2280	2.50" thick, R6.67	"	0.008	1.41	0.56		1.97
2290	3.00" thick, R8.33	"	0.008	1.78	0.59		2.37
2300	4.00" thick, R10.00	"	0.008	1.97	0.62		2.59
2320	5.25" thick, R14.29	"	0.009	2.18	0.65		2.83

INSULATION

ID Code	Component Descriptions	Unit of Meas.	Manhr / Unit	Material Cost	Labor Cost	Equipment Cost	Total Cost
07 - 21161	**BATT INSULATION**						**07 - 21161**
0980	Ceiling, fiberglass, unfaced						
1000	3-1/2" thick, R11	SF	0.009	0.45	0.69		1.14
1020	6" thick, R19	"	0.011	0.58	0.78		1.36
1030	9" thick, R30	"	0.012	1.15	0.90		2.05
1035	Suspended ceiling, unfaced						
1040	3-1/2" thick, R11	SF	0.009	0.45	0.65		1.10
1060	6" thick, R19	"	0.010	0.58	0.73		1.31
1070	9" thick, R30	"	0.011	1.15	0.84		1.99
1075	Crawl space, unfaced						
1080	3-1/2" thick, R11	SF	0.012	0.45	0.90		1.35
1100	6" thick, R19	"	0.013	0.58	0.98		1.56
1120	9" thick, R30	"	0.015	1.15	1.07		2.22
2000	Wall, fiberglass						
2010	Paper backed						
2020	2" thick, R7	SF	0.008	0.36	0.62		0.98
2040	3" thick, R8	"	0.009	0.40	0.65		1.05
2060	4" thick, R11	"	0.009	0.65	0.69		1.34
2080	6" thick, R19	"	0.010	0.97	0.73		1.70
2090	Foil backed, 1 side						
2100	2" thick, R7	SF	0.008	0.70	0.62		1.32
2120	3" thick, R11	"	0.009	0.75	0.65		1.40
2140	4" thick, R14	"	0.009	0.78	0.69		1.47
2160	6" thick, R21	"	0.010	1.03	0.73		1.76
2170	Foil backed, 2 sides						
2180	2" thick, R7	SF	0.009	0.80	0.69		1.49
2200	3" thick, R11	"	0.010	1.01	0.73		1.74
2220	4" thick, R14	"	0.011	1.20	0.78		1.98
2240	6" thick, R21	"	0.011	1.29	0.84		2.13
2250	Unfaced						
2260	2" thick, R7	SF	0.008	0.44	0.62		1.06
2280	3" thick, R9	"	0.009	0.50	0.65		1.15
2300	4" thick, R11	"	0.009	0.54	0.69		1.23
2320	6" thick, R19	"	0.010	0.70	0.73		1.43
2400	Mineral wool batts						
2410	Paper backed						
2420	2" thick, R6	SF	0.008	0.48	0.62		1.10
2440	4" thick, R12	"	0.009	1.09	0.65		1.74
2460	6" thick, R19	"	0.010	1.36	0.73		2.09

INSULATION

ID Code	Description / Component Descriptions	Output Unit of Meas.	Manhr / Unit	Material Cost	Labor Cost	Equipment Cost	Total Cost
07 - 21161	**BATT INSULATION, Cont'd...**						**07 - 21161**
8980	Fasteners, self adhering, attached to ceiling deck						
9000	2-1/2" long	EA	0.013	0.27	0.98		1.25
9020	4-1/2" long	"	0.015	0.30	1.07		1.37
9060	Capped, self-locking washers	"	0.008	0.27	0.59		0.86
07 - 21231	**LOOSE FILL INSULATION**						**07 - 21231**
1000	Blown-in type						
1010	Fiberglass						
1020	5" thick, R11	SF	0.007	0.41	0.49		0.90
1040	6" thick, R13	"	0.008	0.48	0.59		1.07
1060	9" thick, R19	"	0.011	0.58	0.84		1.42
2000	Rockwool, attic application						
2040	6" thick, R13	SF	0.008	0.38	0.59		0.97
2060	8" thick, R19	"	0.010	0.45	0.73		1.18
2080	10" thick, R22	"	0.012	0.53	0.90		1.43
2100	12" thick, R26	"	0.013	0.68	0.98		1.66
2120	15" thick, R30	"	0.016	0.82	1.18		2.00
6200	Poured type						
6210	Fiberglass						
6220	1" thick, R4	SF	0.005	0.45	0.36		0.81
6222	2" thick, R8	"	0.006	0.84	0.42		1.26
6224	3" thick, R12	"	0.007	1.24	0.49		1.73
6226	4" thick, R16	"	0.008	1.63	0.59		2.22
6230	Mineral wool						
6240	1" thick, R3	SF	0.005	0.55	0.36		0.91
6242	2" thick, R6	"	0.006	1.01	0.42		1.43
6244	3" thick, R9	"	0.007	1.54	0.49		2.03
6246	4" thick, R12	"	0.008	1.80	0.59		2.39
6300	Vermiculite or perlite						
6310	2" thick, R4.8	SF	0.006	0.98	0.42		1.40
6320	3" thick, R7.2	"	0.007	1.39	0.49		1.88
6330	4" thick, R9.6	"	0.008	1.81	0.59		2.40
8000	Masonry, poured vermiculite or perlite						
8020	4" block	SF	0.004	0.59	0.29		0.88
8040	6" block	"	0.005	0.90	0.36		1.26
8060	8" block	"	0.006	1.31	0.42		1.73
8100	10" block	"	0.006	1.73	0.45		2.18
8120	12" block	"	0.007	2.16	0.49		2.65

INSULATION

ID Code	Description Component Descriptions	Output Unit of Meas.	Output Manhr / Unit	Unit Costs Material Cost	Unit Costs Labor Cost	Unit Costs Equipment Cost	Unit Costs Total Cost
07 - 21291	**SPRAYED INSULATION**						**07 - 21291**
1000	Foam, sprayed on						
1010	Polystyrene						
1020	1" thick, R4	SF	0.008	0.76	0.59		1.35
1040	2" thick, R8	"	0.011	1.48	0.78		2.26
1050	Urethane						
1060	1" thick, R4	SF	0.008	0.65	0.59		1.24
1080	2" thick, R8	"	0.011	1.24	0.78		2.02
07 - 26001	**VAPOR BARRIERS**						**07 - 26001**
0980	Vapor barrier, polyethylene						
1000	2 mil	SF	0.004	0.02	0.29		0.31
1010	6 mil	"	0.004	0.07	0.29		0.36
1020	8 mil	"	0.004	0.08	0.32		0.40
1040	10 mil	"	0.004	0.09	0.32		0.41

SHINGLES AND TILES

ID Code	Description Component Descriptions	Output Unit of Meas.	Output Manhr / Unit	Unit Costs Material Cost	Unit Costs Labor Cost	Unit Costs Equipment Cost	Unit Costs Total Cost
07 - 31131	**ASPHALT SHINGLES**						**07 - 31131**
1000	Standard asphalt shingles, strip shingles						
1020	210 lb/square	SQ	0.800	110	72.00		180
1040	235 lb/square	"	0.889	110	80.00		190
1060	240 lb/square	"	1.000	120	90.00		210
1080	260 lb/square	"	1.143	170	100		270
1100	300 lb/square	"	1.333	190	120		310
1120	385 lb/square	"	1.600	260	140		400
5980	Roll roofing, mineral surface						
6000	90 lb	SQ	0.571	58.00	51.00		110
6020	110 lb	"	0.667	96.00	60.00		160
6040	140 lb	"	0.800	99.00	72.00		170
07 - 31161	**METAL SHINGLES**						**07 - 31161**
0980	Aluminum, .020" thick						
1000	Plain	SQ	1.600	290	140		430
1020	Colors	"	1.600	320	140		460
1960	Steel, galvanized						
1980	26 ga.						
2000	Plain	SQ	1.600	360	140		500
2020	Colors	"	1.600	460	140		600
2960	Porcelain enamel, 22 ga.						

SHINGLES AND TILES

ID Code	Description — Component Descriptions	Output — Unit of Meas.	Output — Manhr / Unit	Unit Costs — Material Cost	Unit Costs — Labor Cost	Unit Costs — Equipment Cost	Unit Costs — Total Cost
07 - 31161	**METAL SHINGLES, Cont'd...**						**07 - 31161**
3000	Minimum	SQ	2.000	870	180		1,050
3020	Average	"	2.000	1,000	180		1,180
3040	Maximum	"	2.000	1,120	180		1,300
1980	Replacement shingles						
2000	Small jobs	EA	0.267	12.75	24.00		36.75
2020	Large jobs	SF	0.133	9.94	12.00		22.00
07 - 31291	**WOOD SHINGLES**						**07 - 31291**
1000	Wood shingles, on roofs						
1010	White cedar, #1 shingles						
1020	4" exposure	SQ	2.667	270	240		510
1040	5" exposure	"	2.000	240	180		420
1140	On walls						
1150	White cedar, #1 shingles						
1160	4" exposure	SQ	4.000	270	360		630
1180	5" exposure	"	3.200	240	290		530
07 - 31292	**WOOD SHAKES**						**07 - 31292**
2010	Shakes, hand split, 24" red cedar, on roofs						
2020	5" exposure	SQ	4.000	530	360		890
2060	9" exposure	"	2.667	460	240		700

MEMBRANE ROOFING

ID Code	Description — Component Descriptions	Output — Unit of Meas.	Output — Manhr / Unit	Unit Costs — Material Cost	Unit Costs — Labor Cost	Unit Costs — Equipment Cost	Unit Costs — Total Cost
07 - 51131	**BUILT-UP ASPHALT ROOFING**						**07 - 51131**
0980	Built-up roofing, asphalt felt, including gravel						
1000	2 ply	SQ	2.000	97.00	180		280
1500	3 ply	"	2.667	130	240		370
2000	4 ply	"	3.200	190	290		480
07 - 53001	**SINGLE-PLY ROOFING**						**07 - 53001**
2000	Elastic sheet roofing						
2060	Neoprene, 1/16" thick	SF	0.010	2.83	0.90		3.73
2080	EPDM rubber						
2100	45 mil	SF	0.010	1.78	0.90		2.68
2110	60 mil	"	0.010	2.44	0.90		3.34
2115	PVC						
2120	45 mil	SF	0.010	2.34	0.90		3.24
2140	60 mil	"	0.010	2.78	0.90		3.68

MEMBRANE ROOFING

ID Code	Component Descriptions	Unit of Meas.	Manhr / Unit	Material Cost	Labor Cost	Equipment Cost	Total Cost
07 - 53001	**SINGLE-PLY ROOFING, Cont'd...**						**07 - 53001**
2200	Flashing						
2220	Pipe flashing, 90 mil thick						
2260	1" pipe	EA	0.200	34.00	18.00		52.00
2280	2" pipe	"	0.200	36.50	18.00		55.00
2290	3" pipe	"	0.211	36.75	19.00		56.00
2300	4" pipe	"	0.211	40.00	19.00		59.00
2310	5" pipe	"	0.222	42.75	20.00		63.00
2320	6" pipe	"	0.222	46.50	20.00		67.00
2360	Neoprene flashing, 60 mil thick strip						
2380	6" wide	LF	0.067	1.72	6.00		7.72
2390	12" wide	"	0.100	3.38	9.00		12.50
2400	18" wide	"	0.133	4.98	12.00		17.00
2420	24" wide	"	0.200	6.55	18.00		24.50
2500	Adhesives						
2520	Mastic sealer, applied at joints only						
2540	1/4" bead	LF	0.004	0.19	0.36		0.55
8000	Ballast, 3/4" through 1-1/2" gravel, 100lb/sf	SF	0.008	0.52	0.72		1.24
8100	Walkway for membrane roofs, 1/2" thick	"	0.027	2.69	2.40		5.09

FLASHING AND SHEET METAL

ID Code	Component Descriptions	Unit of Meas.	Manhr / Unit	Material Cost	Labor Cost	Equipment Cost	Total Cost
07 - 61001	**METAL ROOFING**						**07 - 61001**
1000	Sheet metal roofing, copper, 16 oz, batten seam	SQ	5.333	1,800	480		2,280
1020	Standing seam	"	5.000	1,760	450		2,210
2000	Aluminum roofing, natural finish						
2005	Corrugated, on steel frame						
2010	.0175" thick	SQ	2.286	140	210		350
2080	.032" thick	"	2.286	260	210		470
2500	Corrugated galvanized steel roofing, on steel frame						
2520	28 ga.	SQ	2.286	230	210		440
2560	22 ga.	"	2.286	330	210		540
07 - 62001	**FLASHING AND TRIM**						**07 - 62001**
0050	Counter flashing						
0060	Aluminum, .032"	SF	0.080	2.09	7.20		9.29
0100	Stainless steel, .015"	"	0.080	6.69	7.20		14.00
0105	Copper						
0110	16 oz.	SF	0.080	7.48	7.20		14.75
0380	Base flashing						

FLASHING AND SHEET METAL

	Description	Output		Unit Costs			
ID Code	Component Descriptions	Unit of Meas.	Manhr / Unit	Material Cost	Labor Cost	Equipment Cost	Total Cost
07 - 62001	**FLASHING AND TRIM, Cont'd...**					**07 - 62001**	
0400	Aluminum, .040"	SF	0.067	2.60	6.00		8.60
0410	Stainless steel, .018"	"	0.067	6.65	6.00		12.75
0415	Copper						
0420	16 oz.	SF	0.067	7.48	6.00		13.50
0424	24 oz.	"	0.067	10.75	6.00		16.75
2500	Scupper outlets						
2520	10" x 10" x 4"	EA	0.200	58.00	18.00		76.00
2530	22" x 4" x 4"	"	0.200	62.00	18.00		80.00
2540	8" x 8" x 5"	"	0.200	57.00	18.00		75.00
3400	Drainage boots, roof, cast iron						
3420	2 x 3	LF	0.100	140	9.00		150
3430	3 x 4	"	0.100	170	9.00		180
3440	4 x 5	"	0.107	190	9.60		200
3450	4 x 6	"	0.107	210	9.60		220
3460	5 x 7	"	0.114	230	10.25		240
7980	Pitch pocket, copper, 16 oz.						
8120	4 x 4	EA	0.200	130	18.00		150
8140	6 x 6	"	0.200	140	18.00		160
8160	8 x 8	"	0.200	180	18.00		200
8180	8 x 10	"	0.200	210	18.00		230
8200	8 x 12	"	0.200	240	18.00		260
8400	Reglets, copper 10 oz.	LF	0.053	7.76	4.80		12.50
8420	Stainless steel, .020"	"	0.053	3.51	4.80		8.31
07 - 71001	**MANUFACTURED SPECIALTIES**					**07 - 71001**	
3520	Ceiling access doors						
3540	Swing up model, metal frame						
3550	Steel door						
3560	2'6" x 2'6"	EA	0.800	690	75.00		770
3580	2'6" x 3'0"	"	0.800	740	75.00		820
3590	Aluminum door						
3600	2'6" x 2'6"	EA	0.800	850	75.00		930
3620	2'6" x 3'0"	"	0.800	920	75.00		990
3640	Swing down model, metal frame						
3650	Steel door						
3660	2'6" x 2'6"	EA	0.800	1,040	75.00		1,120
3670	2'6" x 3'0"	"	0.800	1,120	75.00		1,190
3680	Aluminum door						

FLASHING AND SHEET METAL

ID Code	Description — Component Descriptions	Output — Unit of Meas.	Output — Manhr / Unit	Unit Costs — Material Cost	Unit Costs — Labor Cost	Unit Costs — Equipment Cost	Unit Costs — Total Cost
07 - 71001	**MANUFACTURED SPECIALTIES, Cont'd...**						**07 - 71001**
3690	2'6" x 2'6"	EA	0.800	1,240	75.00		1,320
3700	2'6" x 3'0"	"	0.800	1,330	75.00		1,410
3800	Gravity ventilators, with curb, base, damper and screen						
3820	Stationary siphon						
3830	6" dia.	EA	0.533	54.00	48.00		100
3840	12" dia.	"	0.533	93.00	48.00		140
3850	24" dia.	"	0.800	340	72.00		410
3860	36" dia.	"	0.800	720	72.00		790
3900	Wind driven spinner						
3920	6" dia.	EA	0.533	82.00	48.00		130
3940	12" dia.	"	0.533	110	48.00		160
3960	24" dia.	"	0.800	410	72.00		480
3980	36" dia.	"	0.800	840	72.00		910
4000	Stationary mushroom						
4020	16" dia.	EA	0.800	670	72.00		740
4060	30" dia.	"	1.000	1,510	90.00		1,600
4080	36" dia.	"	1.333	1,940	120		2,060
4100	42" dia.	"	1.600	2,900	140		3,040

FIREPROOFING

ID Code	Description — Component Descriptions	Output — Unit of Meas.	Output — Manhr / Unit	Unit Costs — Material Cost	Unit Costs — Labor Cost	Unit Costs — Equipment Cost	Unit Costs — Total Cost
07 - 81001	**FIREPROOFING**						**07 - 81001**
0980	Sprayed on						
1000	1" thick						
1020	On beams	SF	0.018	0.89	1.31		2.20
1040	On columns	"	0.016	0.91	1.18		2.09
1050	On decks						
1060	Flat surface	SF	0.008	0.91	0.59		1.50
1080	Fluted surface	"	0.010	1.15	0.73		1.88
1100	1-1/2" thick						
1120	On beams	SF	0.023	1.60	1.68		3.28
1140	On columns	"	0.020	1.81	1.47		3.28
1150	On decks						
1160	Flat surface	SF	0.010	1.36	0.73		2.09
1170	Fluted surface	"	0.013	1.60	0.98		2.58

JOINT SEALANTS

ID Code	Component Descriptions	Unit of Meas.	Manhr / Unit	Material Cost	Labor Cost	Equipment Cost	Total Cost
	Description	**Output**		**Unit Costs**			
07 - 92001	**CAULKING**						**07 - 92001**
0100	Caulk exterior, two component						
0120	1/4 x 1/2	LF	0.040	0.43	3.77		4.20
0140	3/8 x 1/2	"	0.044	0.66	4.18		4.84
0160	1/2 x 1/2	"	0.050	0.90	4.71		5.61
0220	Caulk interior, single component						
0240	1/4 x 1/2	LF	0.038	0.29	3.59		3.88
0260	3/8 x 1/2	"	0.042	0.41	3.96		4.37
0280	1/2 x 1/2	"	0.047	0.54	4.43		4.97

DIVISION 09
FINISHES

ACCESS FLOORING

ID Code	Component Descriptions	Unit of Meas.	Manhr / Unit	Material Cost	Labor Cost	Equipment Cost	Total Cost
	Description	**Output**		**Unit Costs**			
09 - 69001	**ACCESS & PEDESTAL FLOOR**						**09 - 69001**
0980	Panels, no covering, 2'x2'						
1000	Plain	SF	0.010	13.25	0.94		14.25
1040	Perforated	"	0.400	18.50	37.75		56.00
1100	Pedestals						
1120	For 6" to 12" clearance	EA	0.080	10.75	7.54		18.25
1200	Stringers						
1220	2'	LF	0.038	3.44	3.59		7.03
1240	6'	"	0.027	3.44	2.51		5.95
1300	Accessories						
1320	Ramp assembly	SF	0.032	72.00	3.01		75.00
1330	Elevated floor assembly	"	0.030	110	2.79		110
1340	Handrail	LF	0.400	89.00	37.75		130
1360	Fascia plate	"	0.200	43.00	18.75		62.00
1400	For carpet tiles, add	SF					12.00
1420	For vinyl flooring, add	"					13.50
1500	RF shielding components, floor liner						
1520	Hot rolled steel sheet						
1540	14 ga.	SF	0.020	17.75	1.88		19.75
1560	11 ga.	"	0.062	24.75	5.80		30.50

DIVISION 10
SPECIALTIES

TELEPHONE SPECIALTIES

ID Code	Description / Component Descriptions	Unit of Meas.	Manhr / Unit	Material Cost	Labor Cost	Equipment Cost	Total Cost
10 - 17001	**TELEPHONE ENCLOSURES**						**10 - 17001**
1000	Enclosure, wall mounted, shelf, 28" x 30" x 15"	EA	2.000	2,590	190		2,780
1800	Directory shelf, stainless steel, 3 binders	"	1.333	2,220	130		2,350

FIRE PROTECTION SPECIALTIES

ID Code	Component Descriptions	Unit of Meas.	Manhr / Unit	Material Cost	Labor Cost	Equipment Cost	Total Cost
10 - 44001	**FIRE PROTECTION**						**10 - 44001**
1000	Portable fire extinguishers						
1020	Water pump tank type						
1030	2.5 gal.						
1040	Red enameled galvanized	EA	0.533	160	39.25		200
1060	Red enameled copper	"	0.533	240	39.25		280
1080	Polished copper	"	0.533	310	39.25		350
1200	Carbon dioxide type, red enamel steel						
1210	Squeeze grip with hose and horn						
1220	2.5 lb	EA	0.533	250	39.25		290
1240	5 lb	"	0.615	350	45.50		400
1260	10 lb	"	0.800	360	59.00		420
1280	15 lb	"	1.000	410	74.00		480
1300	20 lb	"	1.000	500	74.00		570
1310	Wheeled type						
1320	125 lb	EA	1.600	4,460	120		4,580
1340	250 lb	"	1.600	5,630	120		5,750
1360	500 lb	"	1.600	7,270	120		7,390
1400	Dry chemical, pressurized type						
1405	Red enameled steel						
1410	2.5 lb	EA	0.533	77.00	39.25		120
1430	5 lb	"	0.615	110	45.50		160
1440	10 lb	"	0.800	220	59.00		280
1450	20 lb	"	1.000	280	74.00		350
1460	30 lb	"	1.000	350	74.00		420
1480	Chrome plated steel, 2.5 lb	"	0.533	330	39.25		370
1500	Other type extinguishers						
1510	2.5 gal, stainless steel, pressurized water tanks	EA	0.533	250	39.25		290
1520	Soda and acid type	"	0.533	210	39.25		250
1530	Cartridge operated, water type	"	0.533	180	39.25		220
1540	Loaded stream, water type	"	0.533	220	39.25		260
1550	Foam type	"	0.533	300	39.25		340
1560	40 gal, wheeled foam type	"	1.600	6,650	120		6,770

FIRE PROTECTION SPECIALTIES

ID Code	Description — Component Descriptions	Output — Unit of Meas.	Output — Manhr / Unit	Unit Costs — Material Cost	Unit Costs — Labor Cost	Unit Costs — Equipment Cost	Unit Costs — Total Cost
10 - 44001	**FIRE PROTECTION, Cont'd...**						**10 - 44001**
1600	Fire extinguisher cabinets						
1605	Enameled steel						
1610	8" x 12" x 27"	EA	1.600	170	120		290
1620	8" x 16" x 38"	"	1.600	210	120		330
1625	Aluminum						
1630	8" x 12" x 27"	EA	1.600	260	120		380
1640	8" x 16" x 38"	"	1.600	310	120		430
1655	Stainless steel						
1660	8" x 16" x 38"	EA	1.600	300	120		420

PEST CONTROL DEVICES

ID Code	Description — Component Descriptions	Output — Unit of Meas.	Output — Manhr / Unit	Unit Costs — Material Cost	Unit Costs — Labor Cost	Unit Costs — Equipment Cost	Unit Costs — Total Cost
10 - 81001	**PEST CONTROL**						**10 - 81001**
1000	Termite control						
1010	Under slab spraying						
1020	Minimum	SF	0.002	1.31	0.14		1.45
1040	Average	"	0.004	1.31	0.29		1.60
1120	Maximum	"	0.008	1.87	0.59		2.46

DIVISION 11
EQUIPMENT

ARCHITECTURAL EQUIPMENT

	Description	Output		Unit Costs			
ID Code	Component Descriptions	Unit of Meas.	Manhr / Unit	Material Cost	Labor Cost	Equipment Cost	Total Cost
11 - 11001	**SPECIAL SYSTEMS**					**11 - 11001**	
1980	Air compressor, air cooled, two stage						
2000	5.0 cfm, 175 psi	EA	16.000	2,840	1,650		4,490
2020	10 cfm, 175 psi	"	17.778	3,470	1,840		5,310
2030	20 cfm, 175 psi	"	19.048	4,780	1,970		6,750
2040	50 cfm, 125 psi	"	21.053	6,970	2,180		9,150
2050	80 cfm, 125 psi	"	22.857	9,980	2,360		12,340
2055	Single stage, 125 psi						
2060	1.0 cfm	EA	11.429	2,740	1,180		3,920
2080	1.5 cfm	"	11.429	2,790	1,180		3,970
2090	2.0 cfm	"	11.429	2,860	1,180		4,040
8000	Automotive, hose reel, air and water, 50' hose	"	6.667	1,280	690		1,970
8010	Lube equipment, 3 reel, with pumps	"	32.000	6,680	3,310		9,990
8015	Tire changer						
8020	Truck	EA	11.429	15,150	1,180		16,330
8030	Passenger car	"	6.154	3,580	640		4,220
8040	Air hose reel, includes 50' hose	"	6.154	950	640		1,590
8050	Hose reel, 5 reel, motor oil, gear oil, lube, air & water	"	32.000	8,840	3,310		12,150
8100	Water hose reel, 50' hose	"	6.154	950	640		1,590
8120	Pump, for motor or gear oil, fits 55 gal drum	"	0.800	1,240	83.00		1,320
8140	For chassis lube	"	0.800	2,020	83.00		2,100
8490	Fuel dispensing pump, lighted dial, one product						
8500	One hose	EA	6.667	4,410	690		5,100
8520	Two hose	"	6.667	7,780	690		8,470
8530	Two products, two hose	"	6.667	8,200	690		8,890
11 - 13001	**LOADING DOCK EQUIPMENT**					**11 - 13001**	
0080	Dock leveler, 10 ton capacity						
0100	6' x 8'	EA	8.000	6,330	750		7,080
0120	7' x 8'	"	8.000	7,270	750		8,020
0360	Door seal, 12" x 12", vinyl covered	LF	0.200	63.00	18.75		82.00
1000	Dock boards, heavy duty, 5' x 5'						
1010	5000 lb						
1020	Minimum	EA	6.667	1,490	630		2,120
1040	Maximum	"	6.667	1,650	630		2,280
1050	9000 lb						
1060	Minimum	EA	6.667	1,720	630		2,350
1070	Maximum	"	7.273	2,060	690		2,750
1200	Truck shelters						

ARCHITECTURAL EQUIPMENT

ID Code	Description — Component Descriptions	Output — Unit of Meas.	Output — Manhr / Unit	Unit Costs — Material Cost	Unit Costs — Labor Cost	Unit Costs — Equipment Cost	Unit Costs — Total Cost
11 - 13001	**LOADING DOCK EQUIPMENT, Cont'd...**						**11 - 13001**
1220	Minimum	EA	6.154	1,240	580		1,820
1240	Maximum	"	11.429	2,070	1,080		3,150
11 - 15001	**SECURITY EQUIPMENT**						**11 - 15001**
1000	Bulletproof teller window						
1020	4' x 4'	EA	13.333	2,900	1,260		4,160
1040	5' x 4'	"	16.000	3,750	1,510		5,260
1045	Bulletproof partitions						
1050	Up to 12' high, 2.5" thick	SF	0.053	250	5.02		260
1060	Counter for banks						
1080	Minimum	LF	1.600	1,020	150		1,170
1100	Maximum	"	2.667	4,590	250		4,840
1280	Drive-up window						
1300	Minimum	EA	11.429	6,380	1,080		7,460
1310	Maximum	"	26.667	6,910	2,510		9,420
1400	Night depository						
1420	Minimum	EA	11.429	11,960	1,080		13,040
1440	Maximum	"	26.667	17,010	2,510		19,520
1450	Office safes, 30" x 20" x 20", 1 hr rating	"	2.000	4,650	190		4,840
1460	30" x 16" x 15", 2 hr rating	"	1.600	2,380	150		2,530
1470	30" x 28" x 20", H&G rating	"	1.000	5,510	94.00		5,600
1600	Service windows, pass through painted steel						
1620	24" x 36"	EA	8.000	4,010	750		4,760
1640	48" x 40"	"	10.000	6,250	940		7,190
1660	72" x 40"	"	16.000	7,840	1,510		9,350
1670	Special doors and windows						
1680	3' x 7' bulletproof door with frame	EA	11.429	7,970	1,080		9,050
1690	12" x 12" vision panel	"	5.714	4,620	540		5,160
1700	Surveillance system						
1720	Minimum	EA	16.000	7,900	1,510		9,410
1740	Maximum	"	80.000	14,290	7,540		21,830
2040	Vault door, 3' wide, 6'6" high						
2060	3-1/2" thick	EA	100.000	5,140	9,430		14,570
2070	7" thick	"	133.333	8,240	12,570		20,810
2080	10" thick	"	160.000	10,290	15,080		25,370
2160	Insulated vault door						
2170	2 hr rating						
2180	32" wide	EA	8.000	5,140	750		5,890

ARCHITECTURAL EQUIPMENT

ID Code	Component Descriptions	Unit of Meas.	Manhr / Unit	Material Cost	Labor Cost	Equipment Cost	Total Cost
11 - 15001	**SECURITY EQUIPMENT, Cont'd...**						**11 - 15001**
2200	40" wide	EA	8.421	5,720	790		6,510
2210	4 hr rating						
2220	32" wide	EA	8.889	5,690	840		6,530
2240	40" wide	"	10.000	6,630	940		7,570
2250	6 hr rating						
2260	32" wide	EA	8.889	6,550	840		7,390
2280	40" wide	"	10.000	7,650	940		8,590
3100	Insulated file room door						
3110	1 hr rating						
3120	32" wide	EA	8.000	5,060	750		5,810
3140	40" wide	"	8.889	5,630	840		6,470
11 - 21330	**CHECKROOM EQUIPMENT**						**11 - 21330**
1000	Motorized checkroom equipment						
1020	No shelf system, 6'4" height						
1040	7'6" length	EA	8.000	5,790	750		6,540
1060	14'6" length	"	8.000	5,860	750		6,610
1080	28' length	"	8.000	7,060	750		7,810
1100	One shelf, 6'8" height						
1120	7'6" length	EA	8.000	7,100	750		7,850
1140	14'6" length	"	8.000	7,290	750		8,040
1160	28' length	"	8.000	8,760	750		9,510
1180	Two shelves, 7'5" height						
1200	7'6" length	EA	8.000	8,720	750		9,470
1220	14'6" length	"	8.000	11,020	750		11,770
1240	28' length	"	8.000	11,240	750		11,990
1300	Three shelves, 8' height						
1320	7'6" length	EA	16.000	8,920	1,510		10,430
1340	14'6" length	"	16.000	11,230	1,510		12,740
1360	28' length	"	16.000	11,330	1,510		12,840
1400	Four shelves, 8'7" height						
1420	7'6" length	EA	16.000	9,020	1,510		10,530
1440	14'6" length	"	16.000	11,510	1,510		13,020
1460	28' length	"	16.000	11,760	1,510		13,270

ARCHITECTURAL EQUIPMENT

ID Code	Description — Component Descriptions	Output — Unit of Meas.	Output — Manhr / Unit	Unit Costs — Material Cost	Unit Costs — Labor Cost	Unit Costs — Equipment Cost	Unit Costs — Total Cost
11 - 23001	**LAUNDRY EQUIPMENT**						**11 - 23001**
1000	High capacity, heavy duty						
1020	Washer extractors						
1030	135 lb						
1040	Standard	EA	6.667	38,600	630		39,230
1060	Pass through	"	6.667	44,350	630		44,980
1070	200 lb						
1080	Standard	EA	6.667	47,400	630		48,030
1100	Pass through	"	6.667	57,560	630		58,190
1120	110 lb dryer	"	6.667	14,520	630		15,150
1140	Hand operated presser	"	8.889	10,420	840		11,260
1160	Mushroom press	"	8.889	6,510	840		7,350
1200	Spreader feeders						
1220	2 station	EA	8.889	74,120	840		74,960
1240	4 station	"	16.000	87,590	1,510		89,100
1300	Delivery carts						
1320	12 bushel	EA	0.100	380	9.42		390
1340	16 bushel	"	0.107	470	10.00		480
1350	18 bushel	"	0.114	590	10.75		600
1360	30 bushel	"	0.133	850	12.50		860
1370	40 bushel	"	0.160	1,020	15.00		1,030
1500	Low capacity						
1520	Pressers						
1530	Air operated	EA	3.200	8,090	300		8,390
1540	Hand operated	"	3.200	6,400	300		6,700
1560	Extractor, low capacity	"	3.200	5,830	300		6,130
1570	Ironer, 48"	"	1.600	4,410	150		4,560
1600	Coin washers						
1610	10 lb capacity	EA	1.600	2,140	150		2,290
1620	20 lb capacity	"	1.600	5,050	150		5,200
1630	Coin dryer	"	1.000	1,020	94.00		1,110
1680	Coin dry cleaner, 20 lb	"	3.200	4,410	300		4,710
11 - 24001	**MAINTENANCE EQUIPMENT**						**11 - 24001**
1000	Vacuum cleaning system						
1010	3 valves						
1020	1.5 hp	EA	8.889	1,110	840		1,950
1030	2.5 hp	"	11.429	1,340	1,080		2,420
1040	5 valves	"	16.000	2,090	1,510		3,600

ARCHITECTURAL EQUIPMENT

ID Code	Description Component Descriptions	Output Unit of Meas.	Manhr / Unit	Unit Costs Material Cost	Labor Cost	Equipment Cost	Total Cost
11 - 24001	**MAINTENANCE EQUIPMENT, Cont'd...**					**11 - 24001**	
1060	7 valves	EA	20.000	2,790	1,890		4,680
11 - 26001	**FOOD SERVICE EQUIPMENT**					**11 - 26001**	
1000	Unit kitchens						
1020	30" compact kitchen						
1040	Refrigerator, with range, sink	EA	4.000	1,820	380		2,200
1060	Sink only	"	2.667	2,320	260		2,580
1080	Range only	"	2.000	1,880	190		2,070
1100	Cabinet for upper wall section	"	1.143	470	110		580
1120	Stainless shield, for rear wall	"	0.320	190	30.75		220
1140	Side wall	"	0.320	140	30.75		170
1200	42" compact kitchen						
1220	Refrigerator with range, sink	EA	4.444	2,220	430		2,650
1240	Sink only	"	4.000	1,210	380		1,590
1260	Cabinet for upper wall section	"	1.333	940	130		1,070
1280	Stainless shield, for rear wall	"	0.333	750	32.00		780
1290	Side wall	"	0.333	210	32.00		240
1300	54" compact kitchen						
1310	Refrigerator, oven, range, sink	EA	5.714	2,970	550		3,520
1320	Cabinet for upper wall section	"	1.600	1,210	150		1,360
1330	Stainless shield, for						
1340	Rear wall	EA	0.364	750	35.00		780
1350	Side wall	"	0.364	210	35.00		250
1400	60" compact kitchen						
1420	Refrigerator, oven, range, sink	EA	5.714	4,040	550		4,590
1440	Cabinet for upper wall section	"	1.600	270	150		420
1450	Stainless shield, for						
1460	Rear wall	EA	0.364	900	35.00		930
1480	Side wall	"	0.364	230	35.00		260
1490	72" compact kitchen						
1500	Refrigerator, oven, range, sink	EA	6.667	4,260	640		4,900
1510	Cabinet for upper wall section	"	1.600	290	150		440
1520	Stainless shield for						
1540	Rear wall	EA	0.400	970	38.50		1,010
1550	Side wall	"	0.400	230	38.50		270
1560	Bake oven						
1580	Single deck						
1620	Minimum	EA	1.000	4,290	96.00		4,390

ARCHITECTURAL EQUIPMENT

ID Code	Description / Component Descriptions	Output Unit of Meas.	Manhr / Unit	Material Cost	Labor Cost	Equipment Cost	Total Cost

11 - 26001 — FOOD SERVICE EQUIPMENT, Cont'd... — 11 - 26001

ID Code	Description / Component Descriptions	Unit of Meas.	Manhr / Unit	Material Cost	Labor Cost	Equipment Cost	Total Cost
1640	Maximum	EA	2.000	8,210	190		8,400
1650	Double deck						
1660	Minimum	EA	1.333	7,650	130		7,780
1670	Maximum	"	2.000	23,900	190		24,090
1680	Triple deck						
1690	Minimum	EA	1.333	27,130	130		27,260
1700	Maximum	"	2.667	48,390	260		48,650
1710	Convection type oven, electric, 40" x 45" x 57"						
1720	Minimum	EA	1.000	4,230	96.00		4,330
1740	Maximum	"	2.000	7,450	190		7,640
1800	Broiler, without oven, 69" x 26" x 39"						
1820	Minimum	EA	1.000	6,740	96.00		6,840
1840	Maximum	"	1.333	10,610	130		10,740
1900	Coffee urns, 10 gallons						
1920	Minimum	EA	2.667	5,130	260		5,390
1940	Maximum	"	4.000	5,810	380		6,190
2000	Fryer, with submerger						
2010	Single						
2020	Minimum	EA	1.600	2,100	150		2,250
2040	Maximum	"	2.667	5,780	260		6,040
2050	Double						
2060	Minimum	EA	2.000	3,490	190		3,680
2080	Maximum	"	2.667	18,700	260		18,960
2100	Griddle, counter						
2110	3' long						
2120	Minimum	EA	1.333	3,400	130		3,530
2140	Maximum	"	1.600	6,990	150		7,140
2150	5' long						
2160	Minimum	EA	2.000	7,360	190		7,550
2180	Maximum	"	2.667	16,960	260		17,220
2200	Kettles, steam, jacketed						
2210	20 gallons						
2220	Minimum	EA	2.000	16,190	190		16,380
2240	Maximum	"	4.000	17,700	380		18,080
2250	40 gallons						
2260	Minimum	EA	2.000	24,840	190		25,030
2270	Maximum	"	4.000	35,880	380		36,260
2280	60 gallons						

ARCHITECTURAL EQUIPMENT

ID Code	Component Descriptions	Unit of Meas.	Manhr / Unit	Material Cost	Labor Cost	Equipment Cost	Total Cost
		Output		**Unit Costs**			
11 - 26001	**FOOD SERVICE EQUIPMENT, Cont'd...**						**11 - 26001**
2290	Minimum	EA	2.000	27,310	190		27,500
2300	Maximum	"	4.000	38,640	380		39,020
2310	Range						
2320	Heavy duty, single oven, open top						
2330	Minimum	EA	1.000	10,440	96.00		10,540
2340	Maximum	"	2.667	21,980	260		22,240
2350	Fry top						
2360	Minimum	EA	1.000	10,670	96.00		10,770
2380	Maximum	"	2.667	14,940	260		15,200
2390	Steamers, electric						
2400	27 kw						
2420	Minimum	EA	2.000	18,950	190		19,140
2440	Maximum	"	2.667	34,640	260		34,900
2450	18 kw						
2460	Minimum	EA	2.000	10,430	190		10,620
2480	Maximum	"	2.667	24,470	260		24,730
2500	Dishwasher, rack type						
2520	Single tank, 190 racks/hr	EA	4.000	27,780	380		28,160
2530	Double tank						
2540	234 racks/hr	EA	4.444	57,370	430		57,800
2560	265 racks/hr	"	5.333	68,570	510		69,080
2580	Dishwasher, automatic 100 meals/hr	"	2.667	22,310	260		22,570
2590	Disposals						
2620	100 gal/hr	EA	2.667	1,860	260		2,120
2640	120 gal/hr	"	2.759	2,160	260		2,420
2660	250 gal/hr	"	2.857	2,550	270		2,820
2670	Exhaust hood for dishwasher, gutter 4 sides						
2680	4'x4'x2'	EA	2.963	4,510	280		4,790
2690	4'x7'x2'	"	3.200	6,120	310		6,430
2700	Food preparation machines						
2710	Vertical cutter mixers						
2720	25 quart	EA	2.667	18,400	260		18,660
2730	40 quart	"	2.667	23,760	260		24,020
2740	80 quart	"	4.000	30,360	380		30,740
2750	130 quart	"	6.667	40,480	640		41,120
2760	Choppers						
2770	5 lb	EA	2.000	5,140	190		5,330
2780	16 lb	"	2.667	8,090	260		8,350

ARCHITECTURAL EQUIPMENT

ID Code	Component Descriptions	Unit of Meas.	Manhr / Unit	Material Cost	Labor Cost	Equipment Cost	Total Cost
	Description	**Output**		**Unit Costs**			
11 - 26001	**FOOD SERVICE EQUIPMENT, Cont'd...**						**11 - 26001**
2790	40 lb	EA	4.000	10,180	380		10,560
2800	Mixers, floor models						
2820	20 quart	EA	1.000	5,660	96.00		5,760
2840	60 quart	"	1.000	28,150	96.00		28,250
2860	80 quart	"	1.143	45,990	110		46,100
2870	140 quart	"	1.600	54,870	150		55,020
2890	Ice cube maker						
2900	50 lb per day						
2920	Minimum	EA	8.000	3,490	770		4,260
2940	Maximum	"	8.000	5,150	770		5,920
2950	500 lb per day						
2960	Minimum	EA	13.333	8,280	1,280		9,560
2970	Maximum	"	13.333	9,790	1,280		11,070
3000	Ice flakers						
3020	300 lb per day	EA	8.000	6,070	770		6,840
3040	600 lb per day	"	13.333	9,950	1,280		11,230
3050	1000 lb per day	"	17.778	11,350	1,710		13,060
3060	2000 lb per day	"	20.000	21,890	1,920		23,810
3100	Refrigerated cases						
3120	Dairy products						
3140	Multi-deck type	LF	0.533	1,970	51.00		2,020
3160	For rear sliding doors, add	"					370
3180	Delicatessen case, service deli						
3190	Single deck	LF	4.000	1,400	380		1,780
3200	Multi-deck	"	5.000	1,600	480		2,080
3210	Meat case						
3230	Single deck	LF	4.706	1,210	450		1,660
3240	Multi-deck	"	5.000	1,410	480		1,890
3260	Produce case						
3270	Single deck	LF	4.706	1,390	450		1,840
3280	Multi-deck	"	5.000	1,500	480		1,980
3300	Bottle coolers						
3310	6' long						
3320	Minimum	EA	16.000	3,970	1,540		5,510
3330	Maximum	"	16.000	5,890	1,540		7,430
3340	10' long						
3350	Minimum	EA	26.667	5,150	2,560		7,710
3360	Maximum	"	26.667	10,120	2,560		12,680

ARCHITECTURAL EQUIPMENT

ID Code	Component Descriptions	Unit of Meas.	Manhr / Unit	Material Cost	Labor Cost	Equipment Cost	Total Cost
Description		**Output**		**Unit Costs**			
11 - 26001	**FOOD SERVICE EQUIPMENT, Cont'd...**					**11 - 26001**	
3420	Frozen food cases						
3440	Chest type	LF	4.706	1,070	450		1,520
3460	Reach-in, glass door	"	5.000	1,480	480		1,960
3470	Island case, single	"	4.706	1,330	450		1,780
3480	Multi-deck	"	5.000	2,100	480		2,580
3500	Ice storage bins						
3520	500 lb capacity	EA	11.429	2,220	1,100		3,320
3530	1000 lb capacity	"	22.857	3,310	2,190		5,500
11 - 27001	**DARKROOM EQUIPMENT**					**11 - 27001**	
0600	Dryers						
0620	36" x 25" x 68"	EA	4.000	15,420	410		15,830
0640	48" x 25" x 68"	"	4.000	15,940	410		16,350
0700	Processors, film						
0720	Black and white	EA	4.000	24,580	410		24,990
0740	Color negatives	"	4.000	27,840	410		28,250
0760	Prints	"	4.000	31,920	410		32,330
0780	Transparencies	"	4.000	35,020	410		35,430
1000	Sinks with cabinet and/or stand						
1020	5" sink with stand						
1040	24" x 48"	EA	2.000	1,370	210		1,580
1060	32" x 64"	"	2.667	2,110	280		2,390
1080	38" x 52"	"	2.667	2,970	280		3,250
1100	42" x 132"	"	4.000	4,430	410		4,840
1120	48" x 52"	"	4.000	3,400	410		3,810
1200	5" sink with cabinet						
1220	24" x 48"	EA	2.000	2,840	210		3,050
1240	32" x 64"	"	2.667	3,550	280		3,830
1260	38" x 52"	"	2.667	3,640	280		3,920
1280	42" x 132"	"	4.000	5,930	410		6,340
1290	48" x 52"	"	4.000	4,960	410		5,370
1300	10" sink with stand						
1320	24" x 48"	EA	2.000	2,340	210		2,550
1340	32" x 64"	"	2.667	2,480	280		2,760
1360	38" x 52"	"	2.667	3,370	280		3,650
1400	10" sink with cabinet						
1420	24" x 48"	EA	2.000	2,570	210		2,780
1460	38" x 52"	"	2.667	4,760	280		5,040

ARCHITECTURAL EQUIPMENT

ID Code	Description — Component Descriptions	Output — Unit of Meas.	Output — Manhr / Unit	Unit Costs — Material Cost	Unit Costs — Labor Cost	Unit Costs — Equipment Cost	Unit Costs — Total Cost
11 - 31001	**RESIDENTIAL EQUIPMENT**						**11 - 31001**
0300	Compactor, 4 to 1 compaction	EA	2.000	2,220	190		2,410
1300	Dishwasher, built-in						
1320	2 cycles	EA	4.000	1,090	380		1,470
1330	4 or more cycles	"	4.000	2,940	380		3,320
1340	Disposal						
1350	Garbage disposer	EA	2.667	300	260		560
1360	Heaters, electric, built-in						
1362	Ceiling type	EA	2.667	620	260		880
1363	Wall type						
1370	Minimum	EA	2.000	310	190		500
1380	Maximum	"	2.667	1,080	260		1,340
1390	Hood for range, 2-speed, vented						
1420	30" wide	EA	2.667	860	260		1,120
1440	42" wide	"	2.667	1,590	260		1,850
1460	Ice maker, automatic						
1480	30 lb per day	EA	1.143	2,900	110		3,010
1500	50 lb per day	"	4.000	3,680	380		4,060
1820	Folding access stairs, disappearing metal stair						
1840	8' long	EA	1.143	1,520	110		1,630
1850	11' long	"	1.143	1,580	110		1,690
1860	12' long	"	1.143	1,690	110		1,800
1940	Wood frame, wood stair						
1950	22" x 54" x 8'9" long	EA	0.800	290	77.00		370
1960	25" x 54" x 10' long	"	0.800	360	77.00		440
2020	Ranges, electric						
2040	Built-in, 30", 1 oven	EA	2.667	3,180	260		3,440
2050	2 oven	"	2.667	3,680	260		3,940
2060	Countertop, 4 burner, standard	"	2.000	1,840	190		2,030
2070	With grill	"	2.000	4,600	190		4,790
2198	Freestanding, 21", 1 oven	"	2.667	1,660	260		1,920
2200	30", 1 oven	"	1.600	3,220	150		3,370
2220	2 oven	"	1.600	5,240	150		5,390
3600	Water softener						
3620	30 grains per gallon	EA	2.667	1,800	260		2,060
3640	70 grains per gallon	"	4.000	2,270	380		2,650

ARCHITECTURAL EQUIPMENT

ID Code	Component Descriptions	Unit of Meas.	Manhr / Unit	Material Cost	Labor Cost	Equipment Cost	Total Cost
	Description	**Output**		**Unit Costs**			
11 - 53001	**LABORATORY EQUIPMENT**						**11 - 53001**
1000	Cabinets, base						
1020	Minimum	LF	0.667	520	63.00		580
1040	Maximum	"	0.667	950	63.00		1,010
1080	Full storage, 7' high						
1100	Minimum	LF	0.667	500	63.00		560
1140	Maximum	"	0.667	950	63.00		1,010
1150	Wall						
1160	Minimum	LF	0.800	190	75.00		260
1200	Maximum	"	0.800	320	75.00		400
1220	Countertops						
1240	Minimum	SF	0.100	77.00	9.42		86.00
1260	Average	"	0.114	91.00	10.75		100
1280	Maximum	"	0.133	110	12.50		120
1300	Tables						
1320	Open underneath	SF	0.400	170	37.75		210
1330	Doors underneath	"	0.500	560	47.00		610
2000	Medical laboratory equipment						
2010	Analyzer						
2020	Chloride	EA	0.400	6,220	38.50		6,260
2060	Blood	"	0.667	34,210	64.00		34,270
2070	Bath, water, utility, countertop unit	"	0.800	1,280	77.00		1,360
2080	Hot plate, lab, countertop	"	0.727	490	70.00		560
2100	Stirrer	"	0.727	590	70.00		660
2120	Incubator, anaerobic, 23x23x36"	"	4.000	10,550	380		10,930
2140	Dry heat bath	"	1.333	1,140	130		1,270
2160	Incinerator, for sterilizing	"	0.080	780	7.68		790
2170	Meter, serum protein	"	0.100	1,210	9.60		1,220
2180	pH analog, general purpose	"	0.114	1,300	11.00		1,310
2190	Refrigerator, blood bank	"	1.333	9,980	130		10,110
2200	5.4 cf, undercounter type	"	1.333	6,770	130		6,900
2210	Refrigerator/freezer, 4.4 cf, undercounter type	"	1.333	1,280	130		1,410
2220	Sealer, impulse, free standing, 20x12x4"	"	0.267	710	25.50		740
2240	Timer, electric, 1-60 minutes, bench or wall mounted	"	0.444	270	42.75		310
2260	Glassware washer-dryer, undercounter	"	10.000	11,810	960		12,770
2300	Balance, torsion suspension, tabletop, 4.5 lb capacity	"	0.444	1,550	42.75		1,590
2340	Binocular microscope, with in-base illuminator	"	0.308	4,740	29.50		4,770
2400	Centrifuge, table model, 19x16x13"	"	0.320	1,820	30.75		1,850
2420	Clinical model, with four place head	"	0.178	1,960	17.00		1,980

ARCHITECTURAL EQUIPMENT

ID Code	Description Component Descriptions	Output Unit of Meas.	Output Manhr / Unit	Unit Costs Material Cost	Unit Costs Labor Cost	Unit Costs Equipment Cost	Unit Costs Total Cost
11 - 57001	**INDUSTRIAL EQUIPMENT**						**11 - 57001**
1000	Vehicular paint spray booth, solid back, 14'4" x 9'6"						
1020	24' deep	EA	8.000	9,800	750		10,550
1040	26'6" deep	"	8.000	11,200	750		11,950
1060	28'6" deep	"	8.000	12,940	750		13,690
1100	Drive through, 14'9" x 9'6"						
1120	24' deep	EA	8.000	10,430	750		11,180
1140	26'6" deep	"	8.000	12,580	750		13,330
1160	28'6" deep	"	8.000	14,350	750		15,100
1180	Water wash, paint spray booth						
1190	5' x 11'2" x 10'8"	EA	8.000	7,080	750		7,830
1200	6' x 11'2" x 10'8"	"	8.000	7,400	750		8,150
1220	8' x 11'2" x 10'8"	"	8.000	8,000	750		8,750
1240	10' x 11'2" x 11'2"	"	8.000	9,080	750		9,830
1260	12' x 12'2" x 11'2"	"	8.000	10,550	750		11,300
1280	14' x 12'2" x 11'2"	"	8.000	11,940	750		12,690
1290	16' x 12'2" x 11'2"	"	8.000	14,540	750		15,290
1300	20' x 12'2" x 11'2"	"	8.000	17,640	750		18,390
1320	Dry type spray booth, with paint arrestors						
1340	5'4" x 7'2" x 6'8"	EA	8.000	4,280	750		5,030
1360	6'4" x 7'2" x 6'8"	"	8.000	5,560	750		6,310
1380	8'4" x 7'2" x 9'2"	"	8.000	6,300	750		7,050
1400	10'4" x 7'2" x 9'2"	"	8.000	7,410	750		8,160
1420	12'4" x 7'6" x 9'2"	"	8.000	7,380	750		8,130
1440	14'4" x 7'6" x 9'8"	"	8.000	9,980	750		10,730
1460	16'4" x 7'7" x 9'8"	"	8.000	11,400	750		12,150
1480	20'4" x 7'7" x 10'8"	"	8.000	12,970	750		13,720
1500	Air compressor, electric						
1510	1 hp						
1520	115 volt	EA	5.333	1,630	500		2,130
1535	7.5 hp						
1540	115 volt	EA	8.000	4,850	750		5,600
1550	230 volt	"	8.000	6,060	750		6,810
1600	Hydraulic lifts						
1620	8,000 lb capacity	EA	20.000	3,350	1,890		5,240
1640	11,000 lb capacity	"	32.000	5,990	3,020		9,010
1660	24,000 lb capacity	"	53.333	10,550	5,030		15,580
1680	Power tools						
1700	Band saws						

ARCHITECTURAL EQUIPMENT

ID Code	Component Descriptions	Unit of Meas.	Manhr / Unit	Material Cost	Labor Cost	Equipment Cost	Total Cost
	Description	**Output**		**Unit Costs**			
11 - 57001	**INDUSTRIAL EQUIPMENT, Cont'd...**						**11 - 57001**
1720	10"	EA	0.667	1,430	63.00		1,490
1740	14"	"	0.800	2,140	75.00		2,210
1760	Motorized shaper	"	0.615	1,140	58.00		1,200
1780	Motorized lathe	"	0.667	1,350	63.00		1,410
1800	Bench saws						
1820	9" saw	EA	0.533	3,560	50.00		3,610
1830	10" saw	"	0.571	4,280	54.00		4,330
1840	12" saw	"	0.667	5,270	63.00		5,330
1900	Electric grinders						
1910	1/3 hp	EA	0.320	550	30.25		580
2000	1/2 hp	"	0.348	570	32.75		600
2020	3/4 hp	"	0.348	960	32.75		990
11 - 61001	**THEATER EQUIPMENT**						**11 - 61001**
1000	Roll out stage, steel frame, wood floor						
1020	Manual	SF	0.050	60.00	4.71		65.00
1040	Electric	"	0.080	57.00	7.54		65.00
1100	Portable stages						
1120	8" high	SF	0.040	23.50	3.77		27.25
1140	18" high	"	0.044	27.25	4.18		31.50
1160	36" high	"	0.047	31.75	4.43		36.25
1180	48" high	"	0.050	35.50	4.71		40.25
1300	Band risers						
1320	Minimum	SF	0.040	61.00	3.77		65.00
1340	Maximum	"	0.040	120	3.77		120
1400	Chairs for risers						
1420	Minimum	EA	0.036	790	2.68		790
1440	Maximum	"	0.036	1,270	2.68		1,270
11 - 66001	**ATHLETIC EQUIPMENT**						**11 - 66001**
2400	Gym divider curtain						
2420	Minimum	SF	0.011	4.61	1.00		5.61
2440	Maximum	"	0.011	6.91	1.00		7.91
2460	Scoreboards, single face						
2480	Minimum	EA	8.000	9,800	750		10,550
2500	Maximum	"	40.000	53,190	3,770		56,960

ARCHITECTURAL EQUIPMENT

ID Code	Description / Component Descriptions	Output Unit of Meas.	Output Manhr / Unit	Unit Costs Material Cost	Unit Costs Labor Cost	Unit Costs Equipment Cost	Unit Costs Total Cost
11 - 66002	**POLICE EQUIPMENT**						**11 - 66002**
9000	Firing range equipment, rifle						
9040	3 position	EA	26.667	20,350	2,510		22,860
9060	4 position	"	40.000	26,070	3,770		29,840
9080	5 position	"	44.444	31,790	4,190		35,980
9100	6 position	"	47.059	38,150	4,440		42,590
11 - 72001	**MEDICAL EQUIPMENT**						**11 - 72001**
1000	Hospital equipment, lights						
1020	Examination, portable	EA	0.667	2,170	64.00		2,230
1200	Meters						
1220	Air flow meter	EA	0.444	120	42.75		160
1240	Oxygen flow meters	"	0.333	140	32.00		170
1300	Racks						
1320	40 chart, revolving open frame; mobile caddy	EA	0.667	1,500	64.00		1,560
1400	Scales						
1420	Clinical, metric with measure rod, 350 lb	EA	0.727	800	70.00		870
1900	Physical therapy						
1930	Chair, hydrotherapy	EA	0.133	840	12.50		850
1940	Diathermy, shortwave, portable, on casters	"	0.320	3,610	30.25		3,640
1950	Exercise bicycle, floor standing, 35" x 15"	"	0.267	3,510	25.25		3,540
1960	Hydrocollator, 4 pack, portable, 129 x 90 x 160"	"	0.114	630	10.75		640
1970	Lamp, infrared, mobile with variable heat control	"	0.615	850	58.00		910
1980	Ultraviolet, base mounted	"	0.615	780	58.00		840
1990	Mirror, posture training, 27" wide and 72" high	"	0.200	860	18.75		880
2000	Parallel bars, adjustable	"	1.000	3,750	94.00		3,840
2020	Platform mat 10'x6', 1" thick	"	0.200	1,210	18.75		1,230
2030	Pulley, duplex, wall mounted	"	2.667	2,050	250		2,300
2040	Rack, crutch, wall mounted, 66 x 16 x 13"	"	0.800	500	75.00		580
2070	Stimulator, galvanic-faradic, handheld	"	0.053	440	5.02		450
2080	Ultrasound stimulator, portable, 13x13x8"	"	0.067	3,760	6.28		3,770
2100	Sandbag set, velcro straps, saddle bag type	"	0.114	200	10.75		210
2120	Whirlpool, 85 gallon	"	4.000	7,210	380		7,590
2141	65 gallon capacity	"	4.000	6,500	380		6,880
2260	Radiology						
2280	Radiographic table, motor driven tilting table	EA	80.000	63,590	7,540		71,130
2290	Fluoroscope image/tv system	"	160.000	105,510	15,080		120,590
2300	Processor for washing and drying radiographs						
2310	Water filter unit, 30" x 48-1/2" x 37-1/2"	EA	13.333	140	1,280		1,420

ARCHITECTURAL EQUIPMENT

ID Code	Component Descriptions	Unit of Meas.	Manhr / Unit	Material Cost	Labor Cost	Equipment Cost	Total Cost
	Description	**Output**		**Unit Costs**			
11 - 72001	**MEDICAL EQUIPMENT, Cont'd...**						**11 - 72001**
2340	Base storage cabinets, sectional design						
2350	With backsplash, 24" deep and 35" high	LF	0.667	720	64.00		780
2360	Wall storage cabinets	"	1.000	280	96.00		380
2400	Steam sterilizers						
2410	For heat and moisture stable materials	EA	0.800	6,140	77.00		6,220
2420	For fast drying after sterilization	"	1.000	7,950	96.00		8,050
2430	Compact unit	"	1.000	2,570	96.00		2,670
2440	Semi-automatic	"	4.000	3,040	380		3,420
2450	Floor loading						
2460	Single door	EA	6.667	90,040	640		90,680
2480	Double door	"	8.000	98,640	770		99,410
2490	Utensil washer, sanitizer	"	6.154	19,800	590		20,390
2500	Automatic washer/sterilizer	"	16.000	21,680	1,540		23,220
2510	16 x 16 x 26", including accessories	"	26.667	24,920	2,560		27,480
2520	Steam generator, elec., 10 kw to 180 kw	"	16.000	41,190	1,540		42,730
2550	Surgical scrub						
2560	Minimum	EA	2.667	2,170	260		2,430
2580	Maximum	"	2.667	12,580	260		12,840
2610	Gas sterilizers						
2620	Automatic, freestanding, 21x19x29"	EA	8.000	7,730	770		8,500
2640	Surgical tables						
2660	Minimum	EA	11.429	26,790	1,100		27,890
2680	Maximum	"	16.000	32,590	1,540		34,130
2720	Surgical lights, ceiling mounted						
2740	Minimum	EA	13.333	10,700	1,280		11,980
2760	Maximum	"	16.000	21,820	1,540		23,360
2880	Water stills						
2900	4 liters/hr	EA	2.667	4,840	260		5,100
2920	8 liters/hr	"	2.667	7,730	260		7,990
2940	19 liters/hr	"	6.667	15,900	640		16,540
3040	X-ray equipment						
3060	Mobile unit						
3080	Minimum	EA	4.000	14,160	380		14,540
3100	Maximum	"	8.000	27,420	770		28,190
3110	Film viewers						
3120	Minimum	EA	1.333	350	130		480
3140	Maximum	"	2.667	1,230	260		1,490
3200	Autopsy table						

ARCHITECTURAL EQUIPMENT

ID Code	Component Descriptions	Unit of Meas.	Manhr / Unit	Material Cost	Labor Cost	Equipment Cost	Total Cost
	Description	**Output**		**Unit Costs**			

11 - 72001 — MEDICAL EQUIPMENT, Cont'd... — 11 - 72001

ID Code	Component Descriptions	Unit of Meas.	Manhr / Unit	Material Cost	Labor Cost	Equipment Cost	Total Cost
3220	Minimum	EA	8.000	19,750	770		20,520
3240	Maximum	"	8.000	27,910	770		28,680
3300	Incubators						
3320	15 cf	EA	4.000	9,860	380		10,240
3330	29 cf	"	6.667	13,370	640		14,010
3340	Infant transport, portable	"	4.211	7,420	400		7,820
3400	Beds						
3420	Stretcher, with pad, 30" x 78"	EA	2.000	5,510	190		5,700
3440	Transfer, for patient transport	"	2.000	6,490	190		6,680
3450	Headwall						
3460	Aluminum, with back frame and console	EA	4.000	5,560	380		5,940
6000	Hospital ground detection system						
6010	Power ground module	EA	2.286	1,740	220		1,960
6020	Ground slave module	"	1.739	740	170		910
6030	Master ground module	"	1.509	650	140		790
6040	Remote indicator	"	1.600	690	150		840
6050	X-ray indicator	"	1.739	1,940	170		2,110
6060	Micro ammeter	"	2.000	2,310	190		2,500
6070	Supervisory module	"	1.739	1,940	170		2,110
6080	Ground cords	"	0.296	180	28.50		210
6100	Hospital isolation monitors, 5 ma						
6110	120v	EA	3.478	3,690	330		4,020
6120	208v	"	3.478	3,690	330		4,020
6130	240v	"	3.478	3,990	330		4,320
6210	Digital clock-timers separate display	"	1.600	1,880	150		2,030
6220	One display	"	1.600	1,200	150		1,350
6230	Remote control	"	1.250	590	120		710
6240	Battery pack	"	1.250	140	120		260
6310	Surgical chronometer clock and 3 timers	"	2.500	3,510	240		3,750
6320	Auxiliary control	"	1.159	960	110		1,070

11 - 74001 — DENTAL EQUIPMENT — 11 - 74001

ID Code	Component Descriptions	Unit of Meas.	Manhr / Unit	Material Cost	Labor Cost	Equipment Cost	Total Cost
3500	Dental care equipment						
3520	Drill console with accessories	EA	13.333	6,730	1,280		8,010
3540	Amalgamator	"	0.400	720	38.50		760
3560	Lathe	"	0.267	1,660	25.50		1,690
3580	Finish polisher	"	0.533	2,190	51.00		2,240
3590	Model trimmer	"	0.364	1,340	35.00		1,380

ARCHITECTURAL EQUIPMENT

ID Code	Component Descriptions	Unit of Meas.	Manhr / Unit	Material Cost	Labor Cost	Equipment Cost	Total Cost
11 - 74001	**DENTAL EQUIPMENT, Cont'd...**						**11 - 74001**
3600	Motor, wall mounted	EA	0.364	1,520	35.00		1,560
3640	Cleaner, ultrasonic	"	0.800	4,050	77.00		4,130
3660	Curing unit, bench mounted	"	1.333	6,220	130		6,350
3680	Oral evacuation system, dual pump	"	1.000	7,580	96.00		7,680
3700	Sterilizer, table top, self contained	"	0.444	3,200	42.75		3,240
3720	Dental lights						
3740	Light, floor or ceiling mounted	EA	4.000	2,720	380		3,100
3780	X-ray unit						
3790	Portable	EA	2.000	6,730	190		6,920
3820	Wall mounted with remote control	"	6.667	9,430	640		10,070
3830	Illuminator, single panel	"	11.429	970	1,100		2,070
3840	X-ray film processor	"	6.667	11,760	640		12,400
3850	Shield, portable x-ray, lead lined	"	0.533	2,190	51.00		2,240
11 - 82001	**WASTE HANDLING**						**11 - 82001**
1000	Incinerator, electric						
1010	100 lb/hr						
1020	Minimum	EA	8.000	16,930	770		17,700
1040	Maximum	"	8.000	29,090	770		29,860
1050	400 lb/hr						
1060	Minimum	EA	16.000	42,630	1,540		44,170
1070	Maximum	"	16.000	53,280	1,540		54,820
1075	1000 lb/hr						
1080	Minimum	EA	24.242	100,300	2,330		102,630
1090	Maximum	"	24.242	150,450	2,330		152,780
1200	Incinerator, medical-waste						
1220	25 lb/hr, 2-7 x 4-0	EA	16.000	14,420	1,540		15,960
1230	50 lb/hr, 2-11 x 4-11	"	16.000	27,960	1,540		29,500
1240	75 lb/hr, 3-8 x 5-0	"	32.000	37,610	3,070		40,680
1250	100 lb/hr, 3-8 x 6-0	"	32.000	56,420	3,070		59,490
1500	Industrial compactor						
1520	1 c.y.	EA	8.889	14,790	850		15,640
1540	3 c.y.	"	11.429	23,070	1,100		24,170
1560	5 c.y.	"	16.000	43,880	1,540		45,420
2000	Trash chutes, steel, including sprinklers						
2020	18" dia.	LF	4.000	110	380		490
2030	24" dia.	"	4.211	140	400		540
2040	30" dia.	"	4.444	170	420		590

ARCHITECTURAL EQUIPMENT

ID Code	Description Component Descriptions	Output Unit of Meas.	Output Manhr / Unit	Unit Costs Material Cost	Unit Costs Labor Cost	Unit Costs Equipment Cost	Unit Costs Total Cost
11 - 82001	**WASTE HANDLING, Cont'd...**						**11 - 82001**
2050	36" dia.	LF	4.706	210	440		650
2060	Refuse bottom hopper	EA	4.444	1,880	420		2,300

DIVISION 12
FURNISHINGS

MANUFACTURED WOOD CASEWORK

ID Code	Description	Output		Unit Costs			
	Component Descriptions	Unit of Meas.	Manhr / Unit	Material Cost	Labor Cost	Equipment Cost	Total Cost
12 - 32001		**CASEWORK**					**12 - 32001**
0080	Kitchen base cabinet, standard, 24" deep, 35" high						
0100	12" wide	EA	0.800	250	75.00		330
0120	18" wide	"	0.800	290	75.00		360
0140	24" wide	"	0.889	370	84.00		450
0160	27" wide	"	0.889	420	84.00		500
0180	36" wide	"	1.000	500	94.00		590
0200	48" wide	"	1.000	600	94.00		690
0210	Drawer base, 24" deep, 35" high						
0220	15" wide	EA	0.800	310	75.00		390
0230	18" wide	"	0.800	330	75.00		400
0240	24" wide	"	0.889	540	84.00		620
0250	27" wide	"	0.889	610	84.00		690
0260	30" wide	"	0.889	710	84.00		790
0270	Sink-ready base cabinet						
0280	30" wide	EA	0.889	330	84.00		410
0290	36" wide	"	0.889	350	84.00		430
0300	42" wide	"	0.889	380	84.00		460
0310	60" wide	"	1.000	450	94.00		540
0320	Corner cabinet, 36" wide	"	1.000	630	94.00		720
4000	Wall cabinet, 12" deep, 12" high						
4020	30" wide	EA	0.800	320	75.00		400
4060	36" wide	"	0.800	330	75.00		400
4070	15" high						
4080	30" wide	EA	0.889	370	84.00		450
4100	36" wide	"	0.889	560	84.00		640
4110	24" high						
4120	30" wide	EA	0.889	410	84.00		490
4140	36" wide	"	0.889	430	84.00		510
4150	30" high						
4160	12" wide	EA	1.000	240	94.00		330
4180	18" wide	"	1.000	270	94.00		360
4200	24" wide	"	1.000	290	94.00		380
4300	27" wide	"	1.000	350	94.00		440
4320	30" wide	"	1.143	390	110		500
4340	36" wide	"	1.143	400	110		510
4350	Corner cabinet, 30" high						
4360	24" wide	EA	1.333	440	130		570
4380	30" wide	"	1.333	530	130		660

MANUFACTURED WOOD CASEWORK

ID Code	Component Descriptions	Unit of Meas.	Manhr / Unit	Material Cost	Labor Cost	Equipment Cost	Total Cost
	Description	**Output**		**Unit Costs**			
12 - 32001	**CASEWORK, Cont'd...**						**12 - 32001**
4390	36" wide	EA	1.333	580	130		710
5020	Wardrobe	"	2.000	1,160	190		1,350
6980	Vanity with top, laminated plastic						
7000	24" wide	EA	2.000	960	190		1,150
7020	30" wide	"	2.000	1,070	190		1,260
7040	36" wide	"	2.667	1,240	250		1,490
7060	48" wide	"	3.200	1,380	300		1,680

COUNTERTOPS

ID Code	Component Descriptions	Unit of Meas.	Manhr / Unit	Material Cost	Labor Cost	Equipment Cost	Total Cost
12 - 36001	**COUNTERTOPS**						**12 - 36001**
1020	Stainless steel, countertop, with backsplash	SF	0.200	290	18.75		310
2000	Acid-proof, kemrock surface	"	0.133	120	12.50		130

DIVISION 13
SPECIAL CONSTRUCTION

CONSTRUCTION

ID Code	Description — Component Descriptions	Output — Unit of Meas.	Output — Manhr / Unit	Unit Costs — Material Cost	Unit Costs — Labor Cost	Unit Costs — Equipment Cost	Unit Costs — Total Cost
13 - 11001	**SWIMMING POOL EQUIPMENT**						**13 - 11001**
1700	Lights, underwater						
1705	12 volt, with transformer, 100 watt						
1710	Incandescent	EA	2.000	290	150		440
1715	Halogen	"	2.000	250	150		400
1720	LED	"	2.000	790	150		940
1730	110 volt						
1740	Minimum	EA	2.000	1,330	150		1,480
1760	Maximum	"	2.000	3,220	150		3,370
1780	Ground fault interrupter for 110 volt, each light	"	0.667	290	49.25		340

DIVISION 14
CONVEYING

CONVEYING EQUIPMENT

ID Code	Component Descriptions	Unit of Meas.	Manhr / Unit	Material Cost	Labor Cost	Equipment Cost	Total Cost
	Description	**Output**		**Unit Costs**			
14 - 21001	**ELEVATORS**					**14 - 21001**	
0420	Passenger elevators, electric, geared						
0440	Based on a shaft of 6 stops and 6 openings						
0450	50 fpm, 2000 lb	EA	24.000	148,550	1,760	1,500	151,810
0510	100 fpm, 2000 lb	"	26.667	154,040	1,950	1,670	157,660
0520	150 fpm						
0530	2000 lb	EA	30.000	169,960	2,200	1,880	174,030
0540	3000 lb	"	34.286	214,070	2,510	2,140	218,730
0550	4000 lb	"	40.000	222,750	2,930	2,500	228,180
0790	Based on a shaft of 8 stops and 8 openings						
1010	300 fpm						
1020	3000 lb	EA	48.000	275,590	3,520	3,000	282,110
1040	3500 lb	"	48.000	279,940	3,520	3,000	286,460
1060	4000 lb	"	53.333	293,720	3,910	3,330	300,960
1070	5000 lb	"	57.143	326,360	4,190	3,570	334,120
2080	Freight elevators, electric						
2090	Based on a shaft of 6 stops and 6 openings						
2100	50 fpm						
2110	3500 lb	EA	26.667	265,360	1,950	1,670	268,980
2120	4000 lb	"	26.667	266,040	1,950	1,670	269,660
2130	5000 lb	"	30.000	270,770	2,200	1,880	274,840
2140	100 fpm						
2150	3500 lb	EA	30.000	277,540	2,200	1,880	281,610
2160	4000 lb	"	30.000	281,610	2,200	1,880	285,680
2170	5000 lb	"	34.286	286,550	2,510	2,140	291,210
2230	For variable voltage control, add 20%						
14 - 31001	**ESCALATORS**					**14 - 31001**	
1000	Escalators						
1020	32" wide, floor to floor						
1040	12' high	EA	40.000	183,750	2,930	2,500	189,180
1050	15' high	"	48.000	200,770	3,520	3,000	207,290
1060	18' high	"	60.000	216,230	4,400	3,750	224,380
1070	22' high	"	80.000	213,970	5,860	5,000	224,830
1080	25' high	"	96.000	243,150	7,040	6,000	256,190

LIFTS

ID Code	Description / Component Descriptions	Output / Unit of Meas.	Output / Manhr / Unit	Unit Costs / Material Cost	Unit Costs / Labor Cost	Unit Costs / Equipment Cost	Total Cost
14 - 41001	**PERSONNEL LIFTS**						**14 - 41001**
1000	Electrically operated, 1 or 2 person lift						
1001	With attached foot platforms						
1020	3 stops	EA					12,640
1040	5 stops	"					19,710
1060	7 stops	"					22,970
2000	For each additional stop, add	"					2,750
3020	Residential stair climber, per story	"	6.667	5,860	640		6,500
14 - 92001	**PNEUMATIC SYSTEMS**						**14 - 92001**
1000	Pneumatic message tube system						
1010	Average, 20 station job						
1020	3" round system	EA	72.727	45,520	6,980		52,500
1040	4" round system	"	80.000	57,510	7,680		65,190
1060	6" round system	"	88.889	98,670	8,530		107,200
1080	4" x 7" oval system	"	160.000	103,940	15,360		119,300
5000	Trash and linen tube system						
5020	10 stations	EA	120.000	30,670	8,800	7,500	46,970
5030	15 stations	"	160.000	38,440	11,730	10,000	60,170
5040	20 stations	"	184.615	51,230	13,530	11,540	76,300
5060	30 stations	"	218.182	66,640	15,990	13,640	96,270

DIVISION 21
FIRE SUPPRESSION

COMPONENTS

ID Code	Component Descriptions	Unit of Meas.	Manhr / Unit	Material Cost	Labor Cost	Equipment Cost	Total Cost
Description		**Output**		**Unit Costs**			
21 - 11160	**HYDRANTS**						**21 - 11160**
0980	Wall hydrant						
1000	8" thick	EA	1.333	400	140		540
1020	12" thick	"	1.600	470	170		640

DIVISION 22
PLUMBING

BASIC MATERIALS

ID Code	Component Descriptions	Unit of Meas.	Manhr / Unit	Material Cost	Labor Cost	Equipment Cost	Total Cost
22 - 05236		**VALVES**					**22 - 05236**
0600	Gate valve, 125 lb, bronze, soldered						
0800	1/2"	EA	0.200	43.75	20.75		65.00
1000	3/4"	"	0.200	52.00	20.75		73.00
1010	1"	"	0.267	64.00	27.50		92.00
1055	Threaded						
1058	1/4", 125 lb	EA	0.320	37.50	33.00		71.00
1059	1/2"						
1060	125 lb	EA	0.320	35.75	33.00		69.00
1075	300 lb	"	0.320	90.00	33.00		120
1078	3/4"						
1083	125 lb	EA	0.320	42.00	33.00		75.00
1088	300 lb	"	0.320	110	33.00		140
1089	1"						
1091	125 lb	EA	0.320	54.00	33.00		87.00
1098	300 lb	"	0.400	150	41.25		190
1099	1-1/2"						
1100	125 lb	EA	0.400	94.00	41.25		140
1115	300 lb	"	0.444	280	46.00		330
1117	2"						
1118	125 lb	EA	0.571	130	59.00		190
1122	300 lb	"	0.667	350	69.00		420
1123	Cast iron, flanged						
1124	2", 150 lb	EA	0.667	490	69.00		560
1125	2-1/2"						
1126	125 lb	EA	0.667	480	69.00		550
1128	250 lb	"	0.667	1,310	69.00		1,380
1130	3"						
1132	125 lb	EA	0.800	570	83.00		650
1134	250 lb	"	0.800	1,200	83.00		1,280
1136	4"						
1138	125 lb	EA	1.143	750	120		870
1139	150 lb	"	1.143	1,250	120		1,370
1140	250 lb	"	1.143	1,610	120		1,730
1144	6"						
1148	125 lb	EA	1.600	1,380	170		1,550
1150	250 lb	"	1.600	3,230	170		3,400
1151	8"						
1152	125 lb	EA	2.000	2,200	210		2,410

BASIC MATERIALS

ID Code	Component Descriptions	Unit of Meas.	Manhr / Unit	Material Cost	Labor Cost	Equipment Cost	Total Cost
	Description	**Output**		**Unit Costs**			

ID Code	Component Descriptions	Unit of Meas.	Manhr / Unit	Material Cost	Labor Cost	Equipment Cost	Total Cost
22 - 05236	**VALVES, Cont'd...**						**22 - 05236**
1154	250 lb	EA	2.000	6,260	210		6,470
1160	OS&Y, flanged						
1165	2"						
1170	125 lb	EA	0.667	450	69.00		520
1180	250 lb	"	0.667	1,190	69.00		1,260
1185	2-1/2"						
1190	125 lb	EA	0.667	470	69.00		540
1200	250 lb	"	0.800	1,470	83.00		1,550
1205	3"						
1210	125 lb	EA	0.800	520	83.00		600
1215	250 lb	"	0.800	1,530	83.00		1,610
1218	4"						
1220	125 lb	EA	1.333	690	140		830
1225	250 lb	"	1.333	2,340	140		2,480
1227	6"						
1228	125 lb	EA	1.600	1,160	170		1,330
1230	250 lb	"	1.600	3,700	170		3,870
3980	Ball valve, bronze, 250 lb, threaded						
4000	1/2"	EA	0.320	20.50	33.00		54.00
4010	3/4"	"	0.320	30.50	33.00		64.00
4020	1"	"	0.400	38.75	41.25		80.00
4030	1-1/4"	"	0.444	57.00	46.00		100
4040	1-1/2"	"	0.500	90.00	52.00		140
4050	2"	"	0.571	100	59.00		160
4980	Angle valve, bronze, 150 lb, threaded						
5000	1/2"	EA	0.286	100	29.50		130
5010	3/4"	"	0.320	140	33.00		170
5020	1"	"	0.320	210	33.00		240
5030	1-1/4"	"	0.400	270	41.25		310
5040	1-1/2"	"	0.444	350	46.00		400
5980	Balancing valve, meter connections, circuit setter						
6000	1/2"	EA	0.320	99.00	33.00		130
6010	3/4"	"	0.364	100	37.50		140
6020	1"	"	0.400	130	41.25		170
6030	1-1/4"	"	0.444	190	46.00		240
6040	1-1/2"	"	0.533	230	55.00		290
6050	2"	"	0.667	320	69.00		390
6060	2-1/2"	"	0.800	630	83.00		710

BASIC MATERIALS

ID Code	Description / Component Descriptions	Output / Unit of Meas.	Manhr / Unit	Material Cost	Labor Cost	Equipment Cost	Total Cost
22 - 05236	**VALVES, Cont'd...**						**22 - 05236**
6070	3"	EA	1.000	920	100		1,020
6080	4"	"	1.333	1,290	140		1,430
8100	Pressure reducing valve, bronze, threaded, 250 lb						
8120	1/2"	EA	0.500	200	52.00		250
8140	3/4"	"	0.500	200	52.00		250
8200	1"	"	0.500	310	52.00		360
8210	1-1/4"	"	0.571	450	59.00		510
8220	1-1/2"	"	0.667	520	69.00		590
8225	Pressure regulating valve, bronze, class 300						
8230	1"	EA	0.500	750	52.00		800
8240	1-1/2"	"	0.615	1,000	64.00		1,060
8250	2"	"	0.800	1,130	83.00		1,210
8260	3"	"	1.143	1,280	120		1,400
8270	4"	"	1.600	1,600	170		1,770
8480	Solar water temperature regulating valve						
8500	3/4"	EA	0.667	770	69.00		840
8510	1"	"	0.800	790	83.00		870
8520	1-1/4"	"	0.889	850	92.00		940
8530	1-1/2"	"	1.000	950	100		1,050
8540	2"	"	1.143	1,170	120		1,290
8550	2-1/2"	"	2.000	2,210	210		2,420
8980	Tempering valve, threaded						
9000	3/4"	EA	0.267	470	27.50		500
9010	1"	"	0.320	600	33.00		630
9020	1-1/4"	"	0.400	880	41.25		920
9030	1-1/2"	"	0.400	1,010	41.25		1,050
9040	2"	"	0.500	1,390	52.00		1,440
9050	2-1/2"	"	0.667	2,340	69.00		2,410
9060	3"	"	0.800	3,090	83.00		3,170
9070	4"	"	1.143	6,180	120		6,300
9180	Thermostatic mixing valve, threaded						
9200	1/2"	EA	0.286	130	29.50		160
9210	3/4"	"	0.320	130	33.00		160
9220	1"	"	0.348	480	36.00		520
9230	1-1/2"	"	0.400	540	41.25		580
9240	2"	"	0.500	680	52.00		730
9245	Sweat connection						
9250	1/2"	EA	0.286	110	29.50		140

BASIC MATERIALS

ID Code	Description — Component Descriptions	Output — Unit of Meas.	Output — Manhr / Unit	Unit Costs — Material Cost	Unit Costs — Labor Cost	Unit Costs — Equipment Cost	Unit Costs — Total Cost
22 - 05236	**VALVES, Cont'd...**						**22 - 05236**
9260	3/4"	EA	0.320	140	33.00		170
9265	Mixing valve, sweat connection						
9270	1/2"	EA	0.286	79.00	29.50		110
9280	3/4"	"	0.320	79.00	33.00		110
9480	Liquid level gauge, aluminum body						
9500	3/4"	EA	0.320	400	33.00		430
9505	125 psi, PVC body						
9510	3/4"	EA	0.320	470	33.00		500
9520	150 psi, CRS body						
9530	3/4"	EA	0.320	380	33.00		410
9540	1"	"	0.320	410	33.00		440
9560	175 psi, bronze body, 1/2"	"	0.286	760	29.50		790
22 - 06291	**PIPE HANGERS, LIGHT**						**22 - 06291**
0010	A band, black iron						
0020	1/2"	EA	0.057	1.03	5.90		6.93
0030	1"	"	0.059	1.11	6.12		7.23
0040	1-1/4"	"	0.062	1.23	6.36		7.59
0050	1-1/2"	"	0.067	1.28	6.89		8.17
0060	2"	"	0.073	1.36	7.51		8.87
0070	2-1/2"	"	0.080	2.03	8.26		10.25
0080	3"	"	0.089	2.48	9.18		11.75
0090	4"	"	0.100	3.26	10.25		13.50
0130	Copper						
0140	1/2"	EA	0.057	1.67	5.90		7.57
0150	3/4"	"	0.059	1.94	6.12		8.06
0160	1"	"	0.059	1.94	6.12		8.06
0170	1-1/4"	"	0.062	2.09	6.36		8.45
0180	1-1/2"	"	0.067	2.24	6.89		9.13
0190	2"	"	0.073	2.37	7.51		9.88
0200	2-1/2"	"	0.080	4.79	8.26		13.00
0210	3"	"	0.089	4.99	9.18		14.25
0220	4"	"	0.100	5.51	10.25		15.75
1000	2 hole clips, galvanized						
1030	3/4"	EA	0.053	0.27	5.51		5.78
1040	1"	"	0.055	0.33	5.70		6.03
1050	1-1/4"	"	0.057	0.43	5.90		6.33
1060	1-1/2"	"	0.059	0.53	6.12		6.65

BASIC MATERIALS

ID Code	Description / Component Descriptions	Unit of Meas.	Manhr / Unit	Material Cost	Labor Cost	Equipment Cost	Total Cost
22 - 06291	**PIPE HANGERS, LIGHT, Cont'd...**						**22 - 06291**
1070	2"	EA	0.062	0.70	6.36		7.06
1080	2-1/2"	"	0.064	1.25	6.61		7.86
1090	3"	"	0.067	1.66	6.89		8.55
1110	4"	"	0.073	3.56	7.51		11.00
1120	Perforated strap						
1130	3/4"						
1140	Galvanized, 20 ga.	LF	0.040	0.44	4.13		4.57
1150	Copper, 22 ga.	"	0.040	2.20	4.13		6.33
1740	J-Hooks						
1750	1/2"	EA	0.036	0.79	3.75		4.54
1760	3/4"	"	0.036	0.84	3.75		4.59
1770	1"	"	0.038	0.86	3.93		4.79
1780	1-1/4"	"	0.039	0.91	4.03		4.94
1790	1-1/2"	"	0.040	0.93	4.13		5.06
1800	2"	"	0.040	0.97	4.13		5.10
1810	3"	"	0.042	1.12	4.35		5.47
1820	4"	"	0.042	1.21	4.35		5.56
1830	PVC coated hangers, galvanized, 28 ga.						
1840	1-1/2" x 12"	EA	0.053	1.55	5.51		7.06
1850	2" x 12"	"	0.057	1.70	5.90		7.60
1860	3" x 12"	"	0.062	1.90	6.36		8.26
1870	4" x 12"	"	0.067	2.11	6.89		9.00
1880	Copper, 30 ga.						
1890	1-1/2" x 12"	EA	0.053	2.28	5.51		7.79
1900	2" x 12"	"	0.057	2.71	5.90		8.61
1910	3" x 12"	"	0.062	3.01	6.36		9.37
1920	4" x 12"	"	0.067	3.30	6.89		10.25
2090	Wire hook hangers						
2095	Black wire, 1/2" x						
2100	4"	EA	0.040	0.44	4.13		4.57
2110	6"	"	0.042	0.50	4.35		4.85
4000	Copper wire hooks						
4010	1/2" x						
4020	4"	EA	0.040	0.58	4.13		4.71
4030	6"	"	0.042	0.66	4.35		5.01

BASIC MATERIALS

ID Code	Description Component Descriptions	Output Unit of Meas.	Output Manhr / Unit	Unit Costs Material Cost	Unit Costs Labor Cost	Unit Costs Equipment Cost	Unit Costs Total Cost
22 - 06481	**VIBRATION CONTROL**						**22 - 06481**
0120	Vibration isolator, in-line, stainless connector						
0140	1/2"	EA	0.444	100	46.00		150
0160	3/4"	"	0.471	120	48.75		170
0180	1"	"	0.500	120	52.00		170
0200	1-1/4"	"	0.533	170	55.00		230
0260	1-1/2"	"	0.571	190	59.00		250
0280	2"	"	0.615	230	64.00		290
0290	2-1/2"	"	0.667	340	69.00		410
0300	3"	"	0.727	400	75.00		480
0320	4"	"	0.800	510	83.00		590
22 - 06931	**SPECIALTIES**						**22 - 06931**
1000	Wall penetration						
1010	Concrete wall, 6" thick						
1020	2" dia.	EA	0.267		19.75		19.75
1040	4" dia.	"	0.400		29.50		29.50
1060	8" dia.	"	0.571		42.25		42.25
1090	12" thick						
1100	2" dia.	EA	0.364		26.75		26.75
1120	4" dia.	"	0.571		42.25		42.25
1140	8" dia.	"	0.889		66.00		66.00

INSULATION

ID Code	Description Component Descriptions	Output Unit of Meas.	Output Manhr / Unit	Unit Costs Material Cost	Unit Costs Labor Cost	Unit Costs Equipment Cost	Unit Costs Total Cost
22 - 07191	**FIBERGLASS PIPE INSULATION**						**22 - 07191**
1030	Fiberglass insulation on 1/2" pipe						
1040	1" thick	LF	0.027	1.75	2.75		4.50
1060	1-1/2" thick	"	0.033	3.69	3.44		7.13
1070	3/4" pipe						
1080	1" thick	LF	0.027	2.14	2.75		4.89
1100	1-1/2" thick	"	0.033	3.89	3.44		7.33
1110	1" pipe						
1120	1" thick	LF	0.027	2.14	2.75		4.89
1140	1-1/2" thick	"	0.033	4.08	3.44		7.52
1170	1-1/4" pipe						
1180	1" thick	LF	0.033	2.41	3.44		5.85
1200	1-1/2" thick	"	0.036	4.46	3.75		8.21
1215	1-1/2" pipe						
1220	1" thick	LF	0.033	2.63	3.44		6.07

INSULATION

ID Code	Description		Output		Unit Costs			
	Component Descriptions		Unit of Meas.	Manhr / Unit	Material Cost	Labor Cost	Equipment Cost	Total Cost
22 - 07191	**FIBERGLASS PIPE INSULATION, Cont'd...**							**22 - 07191**
1240	1-1/2" thick		LF	0.036	4.57	3.75		8.32
1310	2" pipe							
1340	1" thick		LF	0.033	2.91	3.44		6.35
1360	1-1/2" thick		"	0.036	5.05	3.75		8.80
1430	2-1/2" pipe							
1440	1" thick		LF	0.033	3.09	3.44		6.53
1460	1-1/2" thick		"	0.036	5.43	3.75		9.18
1530	3" pipe							
1540	1" thick		LF	0.038	3.49	3.93		7.42
1560	1-1/2" thick		"	0.040	5.63	4.13		9.76
1640	4" pipe							
1660	1" thick		LF	0.038	4.46	3.93		8.39
1680	1-1/2" thick		"	0.040	6.40	4.13		10.50
22 - 07193	**EXTERIOR PIPE INSULATION**							**22 - 07193**
0090	Fiberglass insulation, aluminum jacket							
0110	1/2" pipe							
0120	1" thick		LF	0.062	2.16	6.36		8.52
0140	1-1/2" thick		"	0.067	4.08	6.89		11.00
0190	1" pipe							
0200	1" thick		LF	0.062	2.64	6.36		9.00
0220	1-1/2" thick		"	0.067	4.56	6.89		11.50
1030	3" pipe							
1040	1" thick		LF	0.080	4.33	8.26		12.50
1060	1-1/2" thick		"	0.084	6.40	8.70		15.00
1500	4" pipe							
1520	1" thick		LF	0.080	5.44	8.26		13.75
1540	1-1/2" thick		"	0.084	7.35	8.70		16.00

FACILITY WATER DISTRIBUTION

ID Code	Description		Output		Unit Costs			
22 - 11161	**COPPER PIPE**							**22 - 11161**
0600	Type K copper							
0890	1/4"		LF	0.024	2.12	2.43		4.55
0895	3/8"		"	0.024	3.25	2.43		5.68
0900	1/2"		"	0.025	3.77	2.58		6.35
1000	3/4"		"	0.027	7.03	2.75		9.78
1020	1"		"	0.029	9.20	2.95		12.25
1320	3-1/2"		"	0.043	61.00	4.46		65.00

FACILITY WATER DISTRIBUTION

ID Code	Component Descriptions	Unit of Meas.	Manhr / Unit	Material Cost	Labor Cost	Equipment Cost	Total Cost
	Description	**Output**		**Unit Costs**			
22 - 11161	**COPPER PIPE, Cont'd...**						**22 - 11161**
3000	DWV, copper						
3020	1-1/4"	LF	0.033	10.25	3.44		13.75
3030	1-1/2"	"	0.036	13.00	3.75		16.75
3040	2"	"	0.040	17.00	4.13		21.25
3070	3"	"	0.044	29.00	4.59		33.50
3080	4"	"	0.050	50.00	5.16		55.00
3090	6"	"	0.057	200	5.90		210
4000	Refrigeration tubing, copper, sealed						
4010	1/8"	LF	0.032	0.84	3.30		4.14
4020	3/16"	"	0.033	0.98	3.44		4.42
4030	1/4"	"	0.035	1.18	3.59		4.77
6000	Type L copper						
6090	1/4"	LF	0.024	1.52	2.43		3.95
6095	3/8"	"	0.024	2.33	2.43		4.76
6100	1/2"	"	0.025	2.71	2.58		5.29
6190	3/4"	"	0.027	4.33	2.75		7.08
6240	1"	"	0.029	6.50	2.95		9.45
6580	Type M copper						
6595	3/8"	LF	0.024	1.65	2.43		4.08
6600	1/2"	"	0.025	1.91	2.58		4.49
6620	3/4"	"	0.027	3.12	2.75		5.87
6630	1"	"	0.029	5.06	2.95		8.01
6655	1-1/2"	"	0.033	10.00	3.44		13.50
6685	3-1/2"	"	0.043	42.25	4.46		46.75
7000	Type K tube, coil, material only						
7020	1/4" x 60'	EA					110
7030	1/2" x 60'	"					240
7050	3/4" x 60'	"					440
7070	1" x 60'	"					570
7150	Type L tube, coil						
7170	1/4" x 60'	EA					120
7180	3/8" x 60'	"					190
7190	1/2" x 60'	"					250
7210	3/4" x 60'	"					400
7230	1" x 60'	"					580

FACILITY WATER DISTRIBUTION

ID Code	Description — Component Descriptions	Unit of Meas.	Manhr / Unit	Material Cost	Labor Cost	Equipment Cost	Total Cost
22 - 11162	**COPPER FITTINGS**						**22 - 11162**
8000	Copper pipe fittings						
8010	1/2"						
8020	90 deg ell	EA	0.178	1.62	18.25		19.75
8040	45 deg ell	"	0.178	2.04	18.25		20.25
8060	Tee	"	0.229	2.71	23.50		26.25
8100	Cap	"	0.089	1.10	9.18		10.25
8120	Coupling	"	0.178	1.18	18.25		19.50
8160	Union	"	0.200	8.21	20.75		29.00
8200	3/4"						
8220	90 deg ell	EA	0.200	3.54	20.75		24.25
8240	45 deg ell	"	0.200	4.13	20.75		25.00
8260	Tee	"	0.267	5.92	27.50		33.50
8290	Cap	"	0.094	2.15	9.72		11.75
8300	Coupling	"	0.200	2.40	20.75		23.25
8320	Union	"	0.229	12.00	23.50		35.50
8360	1"						
8380	90 deg ell	EA	0.267	8.21	27.50		35.75
8390	45 deg ell	"	0.267	10.75	27.50		38.25
8400	Tee	"	0.320	13.50	33.00		46.50
8420	Cap	"	0.133	4.00	13.75		17.75
8430	Coupling	"	0.267	5.92	27.50		33.50
8450	Union	"	0.267	15.75	27.50		43.25
8480	1-1/4"						
8500	90 deg ell	EA	0.229	11.25	23.50		34.75
8510	45 deg ell	"	0.229	13.75	23.50		37.25
8520	Tee	"	0.400	18.25	41.25		60.00
8540	Cap	"	0.133	3.19	13.75		17.00
8560	Union	"	0.286	26.25	29.50		56.00
8580	1-1/2"						
8600	90 deg ell	EA	0.286	14.50	29.50		44.00
8610	45 deg ell	"	0.286	17.25	29.50		46.75
8620	Tee	"	0.444	24.00	46.00		70.00
8640	Cap	"	0.133	3.19	13.75		17.00
8660	Coupling	"	0.267	10.75	27.50		38.25
8680	Union	"	0.364	39.75	37.50		77.00
8905	2"						
8910	90 deg ell	EA	0.320	28.50	33.00		62.00
8920	45 deg ell	"	0.500	26.25	52.00		78.00

FACILITY WATER DISTRIBUTION

ID Code	Component Descriptions	Unit of Meas.	Manhr / Unit	Material Cost	Labor Cost	Equipment Cost	Total Cost
22 - 11162	**COPPER FITTINGS, Cont'd...**						**22 - 11162**
8930	Tee	EA	0.500	41.00	52.00		93.00
8950	Cap	"	0.160	6.61	16.50		23.00
8960	Coupling	"	0.320	17.25	33.00		50.00
8980	Union	"	0.400	43.25	41.25		85.00
9000	2-1/2"						
9020	90 deg ell	EA	0.400	55.00	41.25		96.00
9030	45 deg ell	"	0.400	47.75	41.25		89.00
9040	Tee	"	0.571	55.00	59.00		110
9070	Cap	"	0.200	13.50	20.75		34.25
9080	Coupling	"	0.400	26.25	41.25		68.00
9100	Union	"	0.444	80.00	46.00		130
22 - 11168	**PVC/CPVC PIPE**						**22 - 11168**
0900	PVC schedule 40						
1000	1/2" pipe	LF	0.033	0.50	3.44		3.94
1020	3/4" pipe	"	0.036	0.69	3.75		4.44
1040	1" pipe	"	0.040	0.88	4.13		5.01
1060	1-1/4" pipe	"	0.044	1.13	4.59		5.72
1080	1-1/2" pipe	"	0.050	1.69	5.16		6.85
1100	2" pipe	"	0.057	2.14	5.90		8.04
1110	2-1/2" pipe	"	0.067	3.46	6.89		10.25
1120	3" pipe	"	0.080	4.41	8.26		12.75
1130	4" pipe	"	0.100	6.30	10.25		16.50
1140	6" pipe	"	0.200	11.25	20.75		32.00
1150	8" pipe	"	0.267	16.50	27.50		44.00
2965	PVC schedule 80 pipe						
3070	1-1/2" pipe	LF	0.050	2.80	5.16		7.96
3071	2" pipe	"	0.057	3.78	5.90		9.68
3100	3" pipe	"	0.080	7.80	8.26		16.00
3110	4" pipe	"	0.100	10.25	10.25		20.50
7000	Polypropylene, acid resistant, DWV pipe						
7010	Schedule 40						
7030	1-1/2" pipe	LF	0.057	9.18	5.90		15.00
7040	2" pipe	"	0.067	12.50	6.89		19.50
7050	3" pipe	"	0.080	25.25	8.26		33.50
7060	4" pipe	"	0.100	32.25	10.25		42.50
7070	6" pipe	"	0.200	64.00	20.75		85.00

FACILITY WATER DISTRIBUTION

ID Code	Component Descriptions	Unit of Meas.	Manhr / Unit	Material Cost	Labor Cost	Equipment Cost	Total Cost
	Description	**Output**		**Unit Costs**			
22 - 11169	**STEEL PIPE**					**22 - 11169**	
1000	Black steel, extra heavy pipe, threaded						
1030	1/2" pipe	LF	0.032	2.81	3.30		6.11
1100	3/4" pipe	"	0.032	3.64	3.30		6.94
1200	1" pipe	"	0.040	4.68	4.13		8.81
4000	Fittings, malleable iron, threaded, 1/2" pipe						
4010	90 deg ell	EA	0.267	3.37	27.50		30.75
4020	45 deg ell	"	0.267	4.56	27.50		32.00
4030	Tee	"	0.400	3.66	41.25		45.00
4040	Reducing tee	"	0.400	8.22	41.25		49.50
4045	Cap	"	0.160	2.85	16.50		19.25
4050	Coupling	"	0.320	3.81	33.00		36.75
4070	Union	"	0.267	16.00	27.50		43.50
4080	Nipple, 4" long	"	0.267	3.00	27.50		30.50
4085	3/4" pipe						
4090	90 deg ell	EA	0.267	3.95	27.50		31.50
4100	45 deg ell	"	0.400	6.26	41.25		47.50
4120	Tee	"	0.400	5.31	41.25		46.50
4140	Reducing tee	"	0.267	9.18	27.50		36.75
4150	Cap	"	0.160	3.81	16.50		20.25
4160	Coupling	"	0.267	4.50	27.50		32.00
4170	Union	"	0.267	18.00	27.50		45.50
4175	Nipple, 4" long	"	0.267	3.47	27.50		31.00
4178	1" pipe						
4180	90 deg ell	EA	0.320	6.12	33.00		39.00
4200	45 deg ell	"	0.320	8.09	33.00		41.00
4210	Tee	"	0.444	9.18	46.00		55.00
4220	Reducing tee	"	0.444	12.50	46.00		59.00
4230	Cap	"	0.160	5.16	16.50		21.75
4240	Coupling	"	0.320	6.66	33.00		39.75
4250	Union	"	0.320	21.75	33.00		55.00
4260	Nipple, 4" long	"	0.320	4.90	33.00		38.00
22 - 11170	**GALVANIZED STEEL PIPE**					**22 - 11170**	
1000	Galvanized pipe						
1020	1/2" pipe	LF	0.080	4.97	8.26		13.25
1040	3/4" pipe	"	0.100	6.47	10.25		16.75
1050	1" pipe	"	0.114	6.61	11.75		18.25
1200	90 degree ell, 150 lb malleable iron, galvanized						

FACILITY WATER DISTRIBUTION

ID Code	Component Descriptions	Unit of Meas.	Manhr / Unit	Material Cost	Labor Cost	Equipment Cost	Total Cost
	Description	**Output**		**Unit Costs**			
22 - 11170	**GALVANIZED STEEL PIPE, Cont'd...**					**22 - 11170**	
1210	1/2"	EA	0.160	2.22	16.50		18.75
1220	3/4"	"	0.200	2.95	20.75		23.75
1230	1"	"	0.211	4.82	21.75		26.50
1400	45 degree ell, 150 lb m.i., galv.						
1410	1/2"	EA	0.160	10.25	16.50		26.75
1420	3/4"	"	0.200	4.82	20.75		25.50
1430	1"	"	0.211	6.83	21.75		28.50
1520	Tees, straight, 150 lb m.i., galv.						
1530	1/2"	EA	0.200	2.95	20.75		23.75
1540	3/4"	"	0.229	4.92	23.50		28.50
1550	1"	"	0.267	7.24	27.50		34.75
1640	Tees, reducing, out, 150 lb m.i., galv.						
1650	1/2"	EA	0.200	5.10	20.75		25.75
1660	3/4"	"	0.229	5.91	23.50		29.50
1670	1"	"	0.267	8.73	27.50		36.25
1800	Couplings, straight, 150 lb m.i., galv.						
1810	1/2"	EA	0.160	2.73	16.50		19.25
1820	3/4"	"	0.178	3.27	18.25		21.50
1830	1"	"	0.200	5.60	20.75		26.25
1920	Couplings, reducing, 150 lb m.i., galv						
1930	1/2"	EA	0.160	3.18	16.50		19.75
1940	3/4"	"	0.178	3.55	18.25		21.75
1950	1"	"	0.200	6.50	20.75		27.25
2040	Caps, 150 lb m.i., galv.						
2050	1/2"	EA	0.080	2.27	8.26		10.50
2060	3/4"	"	0.084	3.00	8.70		11.75
2070	1"	"	0.089	4.08	9.18		13.25
2170	Unions, 150 lb m.i., galv.						
2180	1/2"	EA	0.200	12.75	20.75		33.50
2190	3/4"	"	0.229	14.25	23.50		37.75
2200	1"	"	0.267	17.00	27.50		44.50
2260	Nipples, galvanized steel, 4" long						
2270	1/2"	EA	0.100	3.27	10.25		13.50
2280	3/4"	"	0.107	4.36	11.00		15.25
2290	1"	"	0.114	6.01	11.75		17.75
2360	90 degree reducing ell, 150 lb m.i., galv.						
2370	3/4" x 1/2"	EA	0.160	3.55	16.50		20.00
2380	1" x 3/4"	"	0.178	4.82	18.25		23.00

FACILITY WATER DISTRIBUTION

ID Code	Component Descriptions	Unit of Meas.	Manhr / Unit	Material Cost	Labor Cost	Equipment Cost	Total Cost
22 - 11170	**GALVANIZED STEEL PIPE, Cont'd...**						**22 - 11170**
2550	Square head plug (C.I.)						
2560	1/2"	EA	0.089	2.25	9.18		11.50
2570	3/4"	"	0.100	5.00	10.25		15.25
2580	1"	"	0.107	5.25	11.00		16.25
22 - 11171	**STAINLESS STEEL PIPE**						**22 - 11171**
3900	Stainless steel, schedule 40, threaded						
4000	1/2" pipe	LF	0.114	10.50	11.75		22.25
4090	1" pipe	"	0.123	17.25	12.75		30.00
4100	1-1/2" pipe	"	0.133	23.50	13.75		37.25
4200	2" pipe	"	0.145	35.25	15.00		50.00
4220	2-1/2" pipe	"	0.160	49.50	16.50		66.00
4240	3" pipe	"	0.178	69.00	18.25		87.00
4260	4" pipe	"	0.200	89.00	20.75		110
22 - 11191	**BACKFLOW PREVENTERS**						**22 - 11191**
0080	Backflow preventer, flanged, cast iron, with valves						
0100	3" pipe	EA	4.000	4,030	410		4,440
0120	4" pipe	"	4.444	4,700	460		5,160
1900	Threaded						
2000	3/4" pipe	EA	0.500	640	52.00		690
2020	2" pipe	"	0.800	1,120	83.00		1,200
22 - 11196	**VACUUM BREAKERS**						**22 - 11196**
1000	Vacuum breaker, atmospheric, threaded connection						
1010	3/4"	EA	0.320	55.00	33.00		88.00
1015	1"	"	0.320	81.00	33.00		110
1018	Anti-siphon, brass						
1020	3/4"	EA	0.320	60.00	33.00		93.00
1030	1"	"	0.320	93.00	33.00		130
1040	1-1/4"	"	0.400	160	41.25		200
1050	1-1/2"	"	0.444	190	46.00		240
1060	2"	"	0.500	300	52.00		350

FACILITY WATER DISTRIBUTION

ID Code	Description — Component Descriptions	Output — Unit of Meas.	Output — Manhr / Unit	Unit Costs — Material Cost	Unit Costs — Labor Cost	Unit Costs — Equipment Cost	Total Cost
22 - 11230	**PUMPS**						**22 - 11230**
0900	In-line pump, bronze, centrifugal						
1000	5 gpm, 20' head	EA	0.500	680	52.00		730
1010	20 gpm, 40' head	"	0.500	1,220	52.00		1,270
1015	50 gpm						
1020	50' head	EA	1.000	1,390	100		1,490
1065	Cast iron, centrifugal						
1070	50 gpm, 200' head	EA	1.000	1,320	100		1,420
1075	100 gpm						
1080	100' head	EA	1.333	2,250	140		2,390
1500	Centrifugal, close coupled, C.I., single stage						
1520	50 gpm, 100' head	EA	1.000	1,580	100		1,680
1530	100 gpm, 100' head	"	1.333	1,920	140		2,060
1535	Base mounted						
1540	50 gpm, 100' head	EA	1.000	3,220	100		3,320
1550	100 gpm, 50' head	"	1.333	3,660	140		3,800
1560	200 gpm, 100' head	"	2.000	4,690	210		4,900
1570	300 gpm, 175' head	"	2.000	4,980	210		5,190
5980	Condensate pump, simplex						
6000	1000 sf EDR, 2 gpm	EA	6.667	1,650	690		2,340
6010	2000 sf EDR, 3 gpm	"	6.667	1,680	690		2,370
6020	4000 sf EDR, 6 gpm	"	7.273	1,690	750		2,440
6030	6000 sf EDR, 9 gpm	"	7.273	1,720	750		2,470
6035	Duplex, bronze						
6040	8000 sf EDR, 12 gpm	EA	7.273	2,360	750		3,110
6050	10,000 sf EDR, 15 gpm	"	10.000	2,450	1,030		3,480
6060	15,000 sf EDR, 23 gpm	"	11.429	2,940	1,180		4,120
6070	20,000 sf EDR, 30 gpm	"	16.000	3,420	1,650		5,070
6080	25,000 sf EDR, 38 gpm	"	16.000	3,540	1,650		5,190

FACILITY SANITARY SEWERAGE

ID Code	Description — Component Descriptions	Output — Unit of Meas.	Output — Manhr / Unit	Unit Costs — Material Cost	Unit Costs — Labor Cost	Unit Costs — Equipment Cost	Total Cost
22 - 13150	**SERVICE WEIGHT PIPE**						**22 - 13150**
0005	Service weight pipe, single hub						
0020	3" x 5'	EA	0.170	62.00	17.50		80.00
0030	4" x 5'	"	0.178	72.00	18.25		90.00
0050	6" x 5'	"	0.200	140	20.75		160
0395	1/8 bend						
0410	3"	EA	0.320	16.25	33.00		49.25

FACILITY SANITARY SEWERAGE

ID Code	Component Descriptions	Unit of Meas.	Manhr / Unit	Material Cost	Labor Cost	Equipment Cost	Total Cost
22 - 13150	**SERVICE WEIGHT PIPE, Cont'd...**						**22 - 13150**
0420	4"	EA	0.364	23.75	37.50		61.00
0440	6"	"	0.400	40.50	41.25		82.00
0525	1/4 bend						
0540	3"	EA	0.320	19.50	33.00		53.00
0550	4"	"	0.364	30.50	37.50		68.00
0570	6"	"	0.400	53.00	41.25		94.00
0635	Sweep						
0650	3"	EA	0.320	31.75	33.00		65.00
0660	4"	"	0.364	46.50	37.50		84.00
0680	6"	"	0.400	94.00	41.25		140
0765	Sanitary T						
0790	3"	EA	0.571	33.00	59.00		92.00
0820	4"	"	0.667	40.50	69.00		110
0840	6"	"	0.727	91.00	75.00		170
0845	Wye						
0870	3"	EA	0.444	34.50	46.00		81.00
0900	4"	"	0.471	46.25	48.75		95.00
0990	6"	"	0.571	110	59.00		170
22 - 13160	**C.I. PIPE, ABOVE GROUND**						**22 - 13160**
0980	No hub pipe						
1000	1-1/2" pipe	LF	0.057	12.25	5.90		18.25
1010	2" pipe	"	0.067	10.75	6.89		17.75
1100	3" pipe	"	0.080	15.00	8.26		23.25
1200	4" pipe	"	0.133	19.50	13.75		33.25
4980	No hub fittings, 1-1/2" pipe						
5000	1/4 bend	EA	0.267	11.25	27.50		38.75
5060	1/8 bend	"	0.267	9.48	27.50		37.00
5100	Sanitary tee	"	0.400	15.75	41.25		57.00
5200	Wye	"	0.400	19.75	41.25		61.00
5370	2" pipe						
5380	1/4 bend	EA	0.320	13.00	33.00		46.00
5440	1/8 bend	"	0.320	10.50	33.00		43.50
5480	Sanitary tee	"	0.533	18.00	55.00		73.00
5560	Coupling	"					19.75
5600	Wye	"	0.667	16.75	69.00		86.00
5980	3" pipe						
6000	1/4 bend	EA	0.400	18.00	41.25		59.00

FACILITY SANITARY SEWERAGE

ID Code	Description — Component Descriptions	Output — Unit of Meas.	Output — Manhr / Unit	Unit Costs — Material Cost	Unit Costs — Labor Cost	Unit Costs — Equipment Cost	Unit Costs — Total Cost
22 - 13160	**C.I. PIPE, ABOVE GROUND, Cont'd...**						**22 - 13160**
6080	1/8 bend	EA	0.400	15.00	41.25		56.00
6120	Sanitary tee	"	0.500	22.00	52.00		74.00
6260	Coupling	"					22.50
6280	Wye	"	0.667	23.75	69.00		93.00
6810	4" pipe						
6820	1/4 bend	EA	0.400	26.00	41.25		67.00
6900	1/8 bend	"	0.400	19.00	41.25		60.00
6940	Sanitary tee	"	0.667	34.00	69.00		100
7100	Coupling	"					22.00
7120	Wye	"	0.667	38.75	69.00		110
22 - 13163	**ABS DWV PIPE**						**22 - 13163**
1480	Schedule 40 ABS						
1500	1-1/2" pipe	LF	0.040	1.81	4.13		5.94
1520	2" pipe	"	0.044	2.42	4.59		7.01
1530	3" pipe	"	0.057	4.97	5.90		10.75
1540	4" pipe	"	0.080	7.04	8.26		15.25
1550	6" pipe	"	0.100	14.50	10.25		24.75
22 - 13165	**PLASTIC PIPE**						**22 - 13165**
1000	Fiberglass reinforced pipe						
1010	2" pipe	LF	0.062	4.38	6.36		10.75
1020	3" pipe	"	0.067	6.23	6.89		13.00
1030	4" pipe	"	0.073	8.14	7.51		15.75
1040	6" pipe	"	0.080	15.50	8.26		23.75
22 - 13167	**DRAINS, ROOF & FLOOR**						**22 - 13167**
1020	Floor drain, cast iron, with cast iron top						
1030	2"	EA	0.667	220	69.00		290
1040	3"	"	0.667	230	69.00		300
1050	4"	"	0.667	480	69.00		550
1090	Roof drain, cast iron						
1100	2"	EA	0.667	280	69.00		350
1110	3"	"	0.667	290	69.00		360
1120	4"	"	0.667	370	69.00		440

FACILITY SANITARY SEWERAGE

ID Code	Component Descriptions	Unit of Meas.	Manhr / Unit	Material Cost	Labor Cost	Equipment Cost	Total Cost
22 - 13168	**TRAPS**						**22 - 13168**
0980	Bucket trap, threaded						
1000	3/4"	EA	0.500	230	52.00		280
1010	1"	"	0.533	650	55.00		700
1080	Inverted bucket steam trap, threaded						
1100	3/4"	EA	0.500	280	52.00		330
1110	1"	"	0.500	550	52.00		600
1150	With stainless interior						
1180	3/4"	EA	0.500	200	52.00		250
1200	1"	"	0.500	410	52.00		460
1215	Brass interior						
1220	3/4"	EA	0.500	310	52.00		360
1230	1"	"	0.533	610	55.00		670
1245	Cast steel body, threaded, high temperature						
1250	3/4"	EA	0.500	800	52.00		850
1260	1"	"	0.571	1,080	59.00		1,140
1480	Float trap, 15 psi						
1500	3/4"	EA	0.500	200	52.00		250
1510	1"	"	0.533	310	55.00		360
1980	Float and thermostatic trap, 15 psi						
2000	3/4"	EA	0.500	210	52.00		260
2010	1"	"	0.533	240	55.00		300
2135	Steam trap, cast iron body, threaded, 125 psi						
2140	3/4"	EA	0.500	250	52.00		300
2150	1"	"	0.533	290	55.00		350
2175	Thermostatic trap, low pressure, angle type, 25 psi						
2190	3/4"	EA	0.500	130	52.00		180
2200	1"	"	0.533	170	55.00		230
2235	Cast iron body, threaded, 125 psi						
2240	3/4"	EA	0.500	170	52.00		220
2250	1"	"	0.571	210	59.00		270
22 - 13192	**CLEANOUTS**						**22 - 13192**
0980	Cleanout, wall						
1000	2"	EA	0.533	240	55.00		300
1020	3"	"	0.533	340	55.00		400
1040	4"	"	0.667	340	69.00		410
1050	Floor						
1060	2"	EA	0.667	220	69.00		290

FACILITY SANITARY SEWERAGE

ID Code	Description / Component Descriptions	Output / Unit of Meas.	Manhr / Unit	Unit Costs / Material Cost	Labor Cost	Equipment Cost	Total Cost
22 - 13192	**CLEANOUTS, Cont'd...**						**22 - 13192**
1080	3"	EA	0.667	290	69.00		360
1100	4"	"	0.800	300	83.00		380

PLUMBING EQUIPMENT

ID Code	Description / Component Descriptions	Output / Unit of Meas.	Manhr / Unit	Material Cost	Labor Cost	Equipment Cost	Total Cost
22 - 33001	**DOMESTIC WATER HEATERS**						**22 - 33001**
0900	Water heater, electric						
1000	6 gal	EA	1.333	450	140		590
1020	10 gal	"	1.333	460	140		600
1030	15 gal	"	1.333	450	140		590
1040	20 gal	"	1.600	630	170		800
1050	30 gal	"	1.600	820	170		990
1060	40 gal	"	1.600	890	170		1,060
1070	52 gal	"	2.000	1,200	210		1,410
1080	66 gal	"	2.000	1,450	210		1,660
1090	80 gal	"	2.000	1,580	210		1,790
1100	100 gal	"	2.667	1,950	280		2,230
1120	120 gal	"	2.667	2,500	280		2,780
2980	Oil fired						
3000	20 gal	EA	4.000	1,430	410		1,840
3020	50 gal	"	5.714	2,230	590		2,820

PLUMBING FIXTURES

ID Code	Description / Component Descriptions	Output / Unit of Meas.	Manhr / Unit	Material Cost	Labor Cost	Equipment Cost	Total Cost
22 - 42260	**DISPOSALS & ACCESSORIES**						**22 - 42260**
0040	Disposal, continuous feed						
0050	Minimum	EA	1.600	83.00	170		250
0060	Average	"	2.000	230	210		440
0070	Maximum	"	2.667	440	280		720
0200	Batch feed, 1/2 hp						
0220	Minimum	EA	1.600	320	170		490
0230	Average	"	2.000	640	210		850
0240	Maximum	"	2.667	1,090	280		1,370
1100	Hot water dispenser						
1110	Minimum	EA	1.600	230	170		400
1120	Average	"	2.000	370	210		580
1130	Maximum	"	2.667	580	280		860
1140	Epoxy finish faucet	"	1.600	330	170		500

PLUMBING FIXTURES

ID Code	Description / Component Descriptions	Output Unit of Meas.	Output Manhr / Unit	Material Cost	Labor Cost	Equipment Cost	Total Cost
22 - 42260	**DISPOSALS & ACCESSORIES, Cont'd...**						**22 - 42260**
1160	Lock stop assembly	EA	1.000	70.00	100		170
1170	Mounting gasket	"	0.667	8.13	69.00		77.00
1180	Tailpipe gasket	"	0.667	1.19	69.00		70.00
1190	Stopper assembly	"	0.800	27.75	83.00		110
1200	Switch assembly, on/off	"	1.333	31.75	140		170
1210	Tailpipe gasket washer	"	0.400	1.27	41.25		42.50
1220	Stop gasket	"	0.444	2.79	46.00		48.75
1230	Tailpipe flange	"	0.400	0.31	41.25		41.50
1240	Tailpipe	"	0.500	3.62	52.00		56.00
22 - 42398	**HOSE BIBBS**						**22 - 42398**
0005	Hose bibb						
0010	1/2"	EA	0.267	10.50	27.50		38.00
0200	3/4"	"	0.267	11.00	27.50		38.50
22 - 47001	**MISCELLANEOUS FIXTURES**						**22 - 47001**
0900	Electric water cooler						
1000	Floor mounted	EA	2.667	1,110	280		1,390
1020	Wall mounted	"	2.667	1,040	280		1,320

GAS AND VACUUM SYSTEMS

ID Code	Description / Component Descriptions	Output Unit of Meas.	Output Manhr / Unit	Material Cost	Labor Cost	Equipment Cost	Total Cost
22 - 66001	**GLASS PIPE**						**22 - 66001**
0980	Glass pipe						
1000	1-1/2" dia.	LF	0.160	15.50	16.50		32.00
1020	2" dia.	"	0.178	21.00	18.25		39.25
1040	3" dia.	"	0.200	28.00	20.75		48.75
1060	4" dia.	"	0.229	51.00	23.50		75.00
1080	6" dia.	"	0.267	94.00	27.50		120

DIVISION 23
HVAC

INSULATION

ID Code	Description		Output		Unit Costs			
	Component Descriptions		Unit of Meas.	Manhr / Unit	Material Cost	Labor Cost	Equipment Cost	Total Cost
23 - 07131	**DUCTWORK INSULATION**						**23 - 07131**	
0980	Fiberglass duct insulation, plain blanket							
1000	1-1/2" thick		SF	0.010	0.32	1.03		1.35
1060	2" thick		"	0.013	0.42	1.37		1.79
1500	With vapor barrier							
1520	1-1/2" thick		SF	0.010	0.37	1.03		1.40
1540	2" thick		"	0.013	0.47	1.37		1.84
2000	Rigid with vapor barrier							
2020	2" thick		SF	0.027	2.03	2.75		4.78

CONTROLS

ID Code	Description		Output		Unit Costs			
23 - 09131	**HVAC CONTROLS**						**23 - 09131**	
1000	Pressure gauge, direct reading gauge cock and siphon		EA	0.500	130	52.00		180
1210	Control valve, 1", modulating							
1220	2-way		EA	0.667	1,010	69.00		1,080
1240	3-way		"	1.000	1,140	100		1,240
1260	Self contained control valve w/ sensing elmnt, 3/4"		"	0.500	190	52.00		240
1920	Inst air syst 2-1/2 hp comp, rcvr refrg dryer		"					8,490
2020	Thermostat primary control device		"					190
2040	Humidistat primary control device		"					150
2060	Timers primary control device, indoor/outdoor, 24 hour		"					300
2080	Thermometer, dir. reading, 3 dial		"					150
4380	Control dampers, round							
4400	6" dia.		EA	0.320	130	33.00		160
4401	8" dia		"	0.320	180	33.00		210
4402	10" dia		"	0.320	240	33.00		270
4403	12" dia		"	0.320	320	33.00		350
4404	16" dia		"	0.400	470	41.25		510
4405	18" dia		"	0.400	500	41.25		540
4406	20" dia		"	0.400	670	41.25		710
4407	Rectangular, parallel blade standard leakage							
4480	12" x 12"		EA	0.400	95.00	41.25		140
4525	16" x 16"		"	0.400	140	41.25		180
4530	20" x 20"		"	0.400	170	41.25		210
4532	28" x 28"		"	0.500	210	52.00		260
4536	32" x 32"		"	0.500	240	52.00		290
4541	36" x 36"		"	0.667	370	69.00		440
4542	40" x 40"		"	0.800	370	83.00		450

CONTROLS

ID Code	Component Descriptions	Unit of Meas.	Manhr / Unit	Material Cost	Labor Cost	Equipment Cost	Total Cost
	Description	**Output**		**Unit Costs**			

23 - 09131 — HVAC CONTROLS, Cont'd... — **23 - 09131**

ID Code	Component Descriptions	Unit of Meas.	Manhr / Unit	Material Cost	Labor Cost	Equipment Cost	Total Cost
4543	44" x 44"	EA	1.000	440	100		540
4600	48" x 48"	"	1.143	500	120		620
4620	48" x 52"	"	1.333	530	140		670
4640	48" x 56"	"	1.333	590	140		730
4680	48" x 60"	"	1.333	640	140		780
4681	48" x 64"	"	1.333	660	140		800
4682	48" x 68"	"	1.333	720	140		860
4700	48" x 72"	"	1.333	780	140		920
4980	Low leakage						
5000	12" x 12"	EA	0.400	180	41.25		220
5040	16" x 16"	"	0.400	230	41.25		270
5060	20" x 20"	"	0.400	300	41.25		340
5080	24" x 24"	"	0.400	410	41.25		450
5100	28" x 28"	"	0.500	470	52.00		520
5120	32" x 32"	"	0.571	530	59.00		590
5140	36" x 36"	"	0.667	590	69.00		660
5160	40" x 40"	"	0.800	960	83.00		1,040
5180	44" x 44"	"	1.000	1,010	100		1,110
5220	48" x 48"	"	1.143	1,070	120		1,190
5240	48" x 56"	"	1.333	1,230	140		1,370
5260	48" x 60"	"	1.333	1,300	140		1,440
5280	48" x 64"	"	1.333	1,340	140		1,480
5300	48" x 68"	"	1.333	1,400	140		1,540
5320	48" x 72"	"	1.333	1,660	140		1,800
5980	Rectangular, opposed horizontal blade						
6000	12" x 12"	EA	0.400	120	41.25		160
6040	16" x 16"	"	0.400	170	41.25		210
6060	20" x 20"	"	0.400	200	41.25		240
6080	24" x 24"	"	0.400	230	41.25		270
6100	28" x 28"	"	0.500	320	52.00		370
6120	32" x 32"	"	0.533	350	55.00		400
6140	36" x 36"	"	0.667	380	69.00		450
6160	40" x 40"	"	0.800	490	83.00		570
6180	44" x 44"	"	1.000	610	100		710
6200	48" x 48"	"	1.143	760	120		880
6220	48" x 52"	"	1.143	780	120		900
6240	48" x 56"	"	1.333	810	140		950
6260	48" x 60"	"	1.333	850	140		990

CONTROLS

ID Code	Description / Component Descriptions	Output / Unit of Meas.	Manhr / Unit	Unit Costs / Material Cost	Labor Cost	Equipment Cost	Total Cost
23 - 09131	**HVAC CONTROLS, Cont'd...**						**23 - 09131**
6280	48" x 64"	EA	1.333	900	140		1,040
6300	48" x 68"	"	1.333	990	140		1,130
6320	48" x 72"	"	1.333	1,070	140		1,210

HYDRONIC PIPING AND PUMPS

ID Code	Component Descriptions	Unit of Meas.	Manhr / Unit	Material Cost	Labor Cost	Equipment Cost	Total Cost
23 - 21137	**STRAINERS**						**23 - 21137**
0980	Strainer, Y pattern, 125 psi, cast iron body, threaded						
1000	3/4"	EA	0.286	13.75	29.50		43.25
1010	1"	"	0.320	17.75	33.00		51.00
1980	250 psi, brass body, threaded						
2000	3/4"	EA	0.320	36.00	33.00		69.00
2010	1"	"	0.320	50.00	33.00		83.00
2130	Cast iron body, threaded						
2140	3/4"	EA	0.320	21.00	33.00		54.00
2160	1"	"	0.320	26.75	33.00		60.00

AIR DISTRIBUTION

ID Code	Component Descriptions	Unit of Meas.	Manhr / Unit	Material Cost	Labor Cost	Equipment Cost	Total Cost
23 - 31130	**METAL DUCTWORK**						**23 - 31130**
0090	Rectangular duct						
0100	Galvanized steel						
1000	Minimum	LB	0.073	0.92	7.51		8.43
1010	Average	"	0.089	1.15	9.18		10.25
1020	Maximum	"	0.133	1.76	13.75		15.50
1080	Aluminum						
1100	Minimum	LB	0.160	2.41	16.50		19.00
1120	Average	"	0.200	3.21	20.75		24.00
1140	Maximum	"	0.267	3.98	27.50		31.50
1160	Fittings						
1180	Minimum	EA	0.267	7.62	27.50		35.00
1200	Average	"	0.400	11.50	41.25		53.00
1220	Maximum	"	0.800	16.75	83.00		100
1230	For work						
1240	10-20' high, add per pound	LB					0.66
1260	30-50', add per pound	"					1.10

AIR DISTRIBUTION

ID Code	Description — Component Descriptions	Output — Unit of Meas.	Output — Manhr / Unit	Unit Costs — Material Cost	Unit Costs — Labor Cost	Unit Costs — Equipment Cost	Unit Costs — Total Cost
23 - 33130		**DAMPERS**					**23 - 33130**
0980	Horizontal parallel aluminum backdraft damper						
1000	12" x 12"	EA	0.200	58.00	20.75		79.00
1010	16" x 16"	"	0.229	60.00	23.50		84.00
1020	20" x 20"	"	0.286	77.00	29.50		110
1030	24" x 24"	"	0.400	92.00	41.25		130
1040	28" x 28"	"	0.444	130	46.00		180
1050	32" x 32"	"	0.500	180	52.00		230
1060	36" x 36"	"	0.571	210	59.00		270
1070	40" x 40"	"	0.667	270	69.00		340
1080	44" x 44"	"	0.727	310	75.00		390
1100	48" x 48"	"	0.800	380	83.00		460
2000	"Up", parallel dampers						
2010	12" x 12"	EA	0.200	94.00	20.75		110
2020	16" x 16"	"	0.229	130	23.50		150
2030	20" x 20"	"	0.286	150	29.50		180
2040	24" x 24"	"	0.400	160	41.25		200
2050	28" x 28"	"	0.444	230	46.00		280
2060	32" x 32"	"	0.500	270	52.00		320
2070	36" x 36"	"	0.571	280	59.00		340
2080	40" x 40"	"	0.667	370	69.00		440
2090	44" x 44"	"	0.727	440	75.00		520
2100	48" x 48"	"	0.800	540	83.00		620
3000	"Down", parallel dampers						
3010	12" x 12"	EA	0.200	94.00	20.75		110
3020	16" x 16"	"	0.229	130	23.50		150
3030	20" x 20"	"	0.286	150	29.50		180
3040	24" x 24"	"	0.400	160	41.25		200
3050	28" x 28"	"	0.444	230	46.00		280
3060	32" x 32"	"	0.500	270	52.00		320
3070	36" x 36"	"	0.571	280	59.00		340
3080	40" x 40"	"	0.667	370	69.00		440
3090	44" x 44"	"	0.727	440	75.00		520
3100	48" x 48"	"	0.800	540	83.00		620
3980	Fire damper, 1.5 hr rating						
4000	12" x 12"	EA	0.400	38.25	41.25		80.00
4010	16" x 16"	"	0.400	61.00	41.25		100
4020	20" x 20"	"	0.400	66.00	41.25		110
4030	24" x 24"	"	0.400	77.00	41.25		120

AIR DISTRIBUTION

ID Code	Description / Component Descriptions	Unit of Meas.	Manhr / Unit	Material Cost	Labor Cost	Equipment Cost	Total Cost
23 - 33130	**DAMPERS, Cont'd...**						**23 - 33130**
4040	28" x 28"	EA	0.571	92.00	59.00		150
4050	32" x 32"	"	0.667	110	69.00		180
4060	36" x 36"	"	0.800	130	83.00		210
4070	40" x 40"	"	0.889	150	92.00		240
4080	44" x 44"	"	1.000	180	100		280
4090	48" x 48"	"	1.143	250	120		370
23 - 33460	**FLEXIBLE DUCTWORK**						**23 - 33460**
1010	Flexible duct, 1.25" fiberglass						
1020	5" dia.	LF	0.040	3.47	4.13		7.60
1040	6" dia.	"	0.044	3.86	4.59		8.45
1060	7" dia.	"	0.047	4.77	4.86		9.63
1080	8" dia.	"	0.050	5.00	5.16		10.25
1100	10" dia.	"	0.057	6.66	5.90		12.50
1120	12" dia.	"	0.062	7.27	6.36		13.75
1140	14" dia.	"	0.067	9.12	6.89		16.00
1160	16" dia.	"	0.073	13.75	7.51		21.25
9000	Flexible duct connector, 3" wide fabric	"	0.133	2.42	13.75		16.25
23 - 34001	**EXHAUST FANS**						**23 - 34001**
0160	Belt drive roof exhaust fans						
1020	640 cfm, 2618 fpm	EA	1.000	1,140	100		1,240
1030	940 cfm, 2604 fpm	"	1.000	1,480	100		1,580
1040	1050 cfm, 3325 fpm	"	1.000	1,320	100		1,420
1050	1170 cfm, 2373 fpm	"	1.000	1,920	100		2,020
1110	2440 cfm, 4501 fpm	"	1.000	1,500	100		1,600
1120	2760 cfm, 4950 fpm	"	1.000	1,660	100		1,760
1140	3890 cfm, 6769 fpm	"	1.000	1,890	100		1,990
1160	2380 cfm, 3382 fpm	"	1.000	2,100	100		2,200
1180	2880 cfm, 3859 fpm	"	1.000	2,200	100		2,300
1200	3200 cfm, 4173 fpm	"	1.333	2,220	140		2,360
1260	3660 cfm, 3437 fpm	"	1.333	2,260	140		2,400
1280	4070 cfm, 3694 fpm	"	1.333	2,860	140		3,000
1980	5030 cfm, 3251 fpm	"	1.333	2,000	140		2,140
2000	5830 cfm, 6932 fpm	"	1.600	2,780	170		2,950
2010	6380 cfm, 3817 fpm	"	1.600	2,780	170		2,950
2020	8460 cfm, 6721 fpm	"	1.600	2,680	170		2,850
2070	10,970 cfm, 5906 fpm	"	2.000	3,290	210		3,500
2080	12,470 cfm, 6620 fpm	"	2.667	3,750	280		4,030

AIR DISTRIBUTION

ID Code	Description / Component Descriptions	Unit of Meas.	Manhr / Unit	Material Cost	Labor Cost	Equipment Cost	Total Cost
23 - 34001	**EXHAUST FANS, Cont'd...**						**23 - 34001**
2100	7000 cfm, 3449 fpm	EA	2.000	2,680	210		2,890
2120	13,000 cfm, 5456 fpm	"	2.000	3,970	210		4,180
2140	11,250 cfm, 4854 fpm	"	2.000	3,630	210		3,840
2160	18,490 cfm, 7405 fpm	"	3.636	5,300	380		5,680
2180	11,300 cfm, 3232 fpm	"	3.478	3,440	360		3,800
2200	18,330 cfm, 4488 fpm	"	3.478	5,740	360		6,100
2220	21,720 cfm, 5131 fpm	"	3.478	6,030	360		6,390
2240	31,110 cfm, 6965 fpm	"	4.000	6,630	410		7,040
3020	Direct drive fans						
3040	60 to 390 cfm	EA	1.000	930	100		1,030
3060	145 to 590 cfm	"	1.000	1,130	100		1,230
3080	295 to 860 cfm	"	1.000	1,370	100		1,470
3100	235 to 1300 cfm	"	1.000	1,470	100		1,570
3120	415 to 1630 cfm	"	1.000	1,660	100		1,760
3160	590 to 2045 cfm	"	1.000	1,920	100		2,020
3180	805 cfm, 3235 fpm	"	1.000	1,260	100		1,360
3200	1455 cfm, 4360 fpm	"	1.000	1,360	100		1,460
3220	1385 cfm, 3655 fpm	"	1.000	1,390	100		1,490
3240	2260 cfm, 4930 fpm	"	1.000	1,470	100		1,570
3260	1720 cfm, 3870 fpm	"	1.000	1,600	100		1,700
3280	2700 cfm, 5220 fpm	"	1.000	1,490	100		1,590
9000	Terminal blenders and cooling						
9200	400 cfm	EA	1.600	510	170		680
9220	800 cfm	"	1.600	560	170		730
9240	1200 cfm	"	2.000	670	210		880
9260	2000 cfm	"	2.000	760	210		970
23 - 37131	**DIFFUSERS**						**23 - 37131**
1980	Ceiling diffusers, round, baked enamel finish						
2000	6" dia.	EA	0.267	40.25	27.50		68.00
2020	8" dia.	"	0.333	48.50	34.50		83.00
2040	10" dia.	"	0.333	54.00	34.50		89.00
2060	12" dia.	"	0.333	69.00	34.50		100
2080	14" dia.	"	0.364	84.00	37.50		120
2100	16" dia.	"	0.364	100	37.50		140
2120	18" dia.	"	0.400	120	41.25		160
2140	20" dia.	"	0.400	140	41.25		180
2480	Rectangular						

AIR DISTRIBUTION

ID Code	Component Descriptions	Unit of Meas.	Manhr / Unit	Material Cost	Labor Cost	Equipment Cost	Total Cost
	Description	**Output**		**Unit Costs**			

23 - 37131 — DIFFUSERS, Cont'd... — 23 - 37131

ID Code	Component Descriptions	Unit of Meas.	Manhr / Unit	Material Cost	Labor Cost	Equipment Cost	Total Cost
2500	6x6"	EA	0.267	43.00	27.50		71.00
2520	9x9"	"	0.400	52.00	41.25		93.00
2540	12x12"	"	0.400	76.00	41.25		120
2560	15x15"	"	0.400	95.00	41.25		140
2580	18x18"	"	0.400	120	41.25		160
2600	21x21"	"	0.500	150	52.00		200
2620	24x24"	"	0.500	170	52.00		220
3000	Lay in, flush mounted, perforated face, with grid						
3010	6x6/24x24	EA	0.320	62.00	33.00		95.00
3020	8x8/24x24	"	0.320	62.00	33.00		95.00
3040	9x9/24x24	"	0.320	62.00	33.00		95.00
3060	10x10/24x24	"	0.320	67.00	33.00		100
3080	12x12/24x24	"	0.320	72.00	33.00		110
3100	15x15/24x24	"	0.320	95.00	33.00		130
3120	18x6/24x24	"	0.320	72.00	33.00		110
3140	18x18/24x24	"	0.320	110	33.00		140
5000	Two-way slot diffuser with balancing damper, 4'	"	0.800	69.00	83.00		150

23 - 37134 — REGISTERS AND GRILLES — 23 - 37134

ID Code	Component Descriptions	Unit of Meas.	Manhr / Unit	Material Cost	Labor Cost	Equipment Cost	Total Cost
0980	Lay in flush mounted, perforated face, return						
1000	6x6/24x24	EA	0.320	54.00	33.00		87.00
1020	8x8/24x24	"	0.320	54.00	33.00		87.00
1040	9x9/24x24	"	0.320	58.00	33.00		91.00
1060	10x10/24x24	"	0.320	63.00	33.00		96.00
1080	12x12/24x24	"	0.320	63.00	33.00		96.00
3040	Rectangular, ceiling return, single deflection						
3060	10x10	EA	0.400	32.25	41.25		74.00
3080	12x12	"	0.400	37.50	41.25		79.00
3100	14x14	"	0.400	45.75	41.25		87.00
3120	16x8	"	0.400	37.50	41.25		79.00
3140	16x16	"	0.400	37.50	41.25		79.00
3160	18x8	"	0.400	43.00	41.25		84.00
3180	20x20	"	0.400	70.00	41.25		110
3220	24x12	"	0.400	100	41.25		140
3240	24x18	"	0.400	130	41.25		170
3260	36x24	"	0.444	250	46.00		300
3280	36x30	"	0.444	370	46.00		420
4980	Wall, return air register						

AIR DISTRIBUTION

ID Code	Description — Component Descriptions	Output — Unit of Meas.	Output — Manhr / Unit	Unit Costs — Material Cost	Unit Costs — Labor Cost	Unit Costs — Equipment Cost	Unit Costs — Total Cost
23 - 37134	**REGISTERS AND GRILLES, Cont'd...**						**23 - 37134**
5000	12x12	EA	0.200	53.00	20.75		74.00
5020	16x16	"	0.200	79.00	20.75		100
5040	18x18	"	0.200	93.00	20.75		110
5060	20x20	"	0.200	110	20.75		130
5080	24x24	"	0.200	150	20.75		170
5980	Ceiling, return air grille						
6000	6x6	EA	0.267	31.00	27.50		59.00
6020	8x8	"	0.320	38.50	33.00		72.00
6040	10x10	"	0.320	47.75	33.00		81.00
6980	Ceiling, exhaust grille, aluminum egg crate						
7000	6x6	EA	0.267	21.25	27.50		48.75
7020	8x8	"	0.320	21.25	33.00		54.00
7040	10x10	"	0.320	23.50	33.00		57.00
7060	12x12	"	0.400	29.00	41.25		70.00
7080	14x14	"	0.400	38.00	41.25		79.00
7100	16x16	"	0.400	44.75	41.25		86.00
7120	18x18	"	0.400	54.00	41.25		95.00
23 - 37232	**RELIEF VENTILATORS**						**23 - 37232**
0980	Intake ventilator, aluminum, with screen, no curbs						
1000	12" x 12"	EA	0.667	170	69.00		240
1020	16" x 16"	"	0.800	230	83.00		310
1040	20" x 20"	"	0.800	380	83.00		460
1080	30" x 30"	"	1.143	600	120		720
1100	36" x 36"	"	1.333	910	140		1,050
1120	42" x 42"	"	1.333	1,240	140		1,380
1140	48" x 48"	"	1.600	1,490	170		1,660

CENTRAL HEATING EQUIPMENT

ID Code	Description — Component Descriptions	Output — Unit of Meas.	Output — Manhr / Unit	Material Cost	Labor Cost	Equipment Cost	Total Cost
23 - 52230	**BOILERS**						**23 - 52230**
1980	Electric, hot water						
2000	115 mbh	EA	12.000	5,630	880	750	7,260
2020	175 mbh	"	12.000	6,230	880	750	7,860
2040	235 mbh	"	12.000	7,110	880	750	8,740
2060	940 mbh	"	24.000	17,290	1,760	1,500	20,550
2080	1600 mbh	"	48.000	24,490	3,520	3,000	31,010
2100	3000 mbh	"	60.000	36,580	4,400	3,750	44,730
2120	6000 mbh	"	80.000	42,430	5,860	5,000	53,290

CENTRAL HEATING EQUIPMENT

ID Code	Description		Output		Unit Costs			
	Component Descriptions	Unit of Meas.	Manhr / Unit	Material Cost	Labor Cost	Equipment Cost	Total Cost
23 - 54130		**FURNACES**					**23 - 54130**
0980	Electric, hot air						
1000	40 mbh	EA	4.000	850	410		1,260
1020	60 mbh	"	4.211	920	440		1,360
1040	80 mbh	"	4.444	1,000	460		1,460
1060	100 mbh	"	4.706	1,130	490		1,620
1080	125 mbh	"	4.848	1,380	500		1,880
1100	160 mbh	"	5.000	1,900	520		2,420
1120	200 mbh	"	5.161	2,760	530		3,290
1140	400 mbh	"	5.333	4,890	550		5,440

CENTRAL COOLING EQUIPMENT

ID Code	Description		Output		Unit Costs			
23 - 63001		**CONDENSING UNITS**					**23 - 63001**
0980	Air cooled condenser, single circuit						
1000	3 ton	EA	1.333	1,810	140		1,950
1030	5 ton	"	1.333	2,720	140		2,860
1040	7.5 ton	"	3.810	4,450	390		4,840
1050	20 ton	"	4.000	13,220	410		13,630
1060	25 ton	"	4.000	19,920	410		20,330
1070	30 ton	"	4.000	22,710	410		23,120
1080	40 ton	"	5.714	29,370	590		29,960
1090	50 ton	"	5.714	35,720	590		36,310
1100	60 ton	"	5.000	41,120	520		41,640
1480	With low ambient dampers						
1500	3 ton	EA	2.000	1,980	210		2,190
1530	5 ton	"	2.000	3,120	210		3,330
1550	7.5 ton	"	4.000	4,790	410		5,200
1570	20 ton	"	5.333	13,260	550		13,810
1590	25 ton	"	5.333	20,190	550		20,740
1610	30 ton	"	5.333	23,190	550		23,740
1630	40 ton	"	6.667	30,780	690		31,470
1650	50 ton	"	7.273	37,120	750		37,870
1670	60 ton	"	7.273	42,570	750		43,320
2980	Dual circuit						
3000	10 ton	EA	4.000	4,240	410		4,650
3010	15 ton	"	5.714	6,200	590		6,790
3030	20 ton	"	5.714	12,680	590		13,270
3040	25 ton	"	5.714	20,200	590		20,790

CENTRAL COOLING EQUIPMENT

ID Code	Description / Component Descriptions	Output / Unit of Meas.	Manhr / Unit	Unit Costs / Material Cost	Labor Cost	Equipment Cost	Total Cost
23 - 63001	**CONDENSING UNITS, Cont'd...**						**23 - 63001**
3050	30 ton	EA	5.714	23,500	590		24,090
3060	40 ton	"	6.667	33,600	690		34,290
3070	50 ton	"	6.667	37,120	690		37,810
3080	60 ton	"	6.667	38,530	690		39,220
3100	80 ton	"	8.889	48,180	920		49,100
3120	100 ton	"	8.889	56,400	920		57,320
3130	120 ton	"	8.889	66,970	920		67,890
4030	With low ambient dampers						
4050	15 ton	EA	5.714	6,910	590		7,500
4080	20 ton	"	5.714	13,480	590		14,070
4100	25 ton	"	5.714	21,380	590		21,970
4120	30 ton	"	5.714	24,110	590		24,700
4140	40 ton	"	6.667	34,770	690		35,460
4160	50 ton	"	6.667	38,310	690		39,000
4180	60 ton	"	6.667	39,710	690		40,400
4190	80 ton	"	8.889	50,510	920		51,430
4200	100 ton	"	8.889	58,750	920		59,670
4210	120 ton	"	8.889	70,030	920		70,950
23 - 64001	**CHILLERS**						**23 - 64001**
0980	Chiller, reciprocal						
1000	Air cooled, remote condenser, starter						
1020	20 ton	EA	8.000	31,460	590	500	32,550
1030	25 ton	"	8.000	35,500	590	500	36,590
1040	30 ton	"	8.000	37,540	590	500	38,630
1050	40 ton	"	12.000	41,630	880	750	43,260
1060	50 ton	"	13.333	46,210	980	830	48,020
1070	60 ton	"	14.118	51,920	1,030	880	53,840
1100	80 ton	"	21.818	64,060	1,600	1,360	67,020
1120	100 ton	"	24.000	75,860	1,760	1,500	79,120
1130	120 ton	"	26.667	85,980	1,950	1,670	89,600
1140	150 ton	"	30.000	109,570	2,200	1,880	113,640
1150	180 ton	"	34.286	125,590	2,510	2,140	130,250
1160	200 ton	"	40.000	138,230	2,930	2,500	143,660
2000	Water cooled, with starter						
2020	20 ton	EA	8.000	26,970	590	500	28,060
2030	25 ton	"	8.000	29,890	590	500	30,980
2040	30 ton	"	12.000	35,960	880	750	37,590

CENTRAL COOLING EQUIPMENT

ID Code	Description / Component Descriptions	Output / Unit of Meas.	Manhr / Unit	Unit Costs / Material Cost	Labor Cost	Equipment Cost	Total Cost
23 - 64001		**CHILLERS, Cont'd...**					**23 - 64001**
2050	40 ton	EA	12.000	49,450	880	750	51,080
2060	50 ton	"	13.333	53,940	980	830	55,750
2070	60 ton	"	14.118	58,430	1,030	880	60,350
2100	80 ton	"	21.818	67,430	1,600	1,360	70,390
2120	100 ton	"	24.000	78,670	1,760	1,500	81,930
2130	120 ton	"	26.667	92,160	1,950	1,670	95,780
2140	150 ton	"	30.000	114,630	2,200	1,880	118,700
2150	180 ton	"	34.286	119,130	2,510	2,140	123,790
2160	200 ton	"	40.000	125,860	2,930	2,500	131,290
2980	Packaged, air cooled, with starter						
3000	20 ton	EA	6.000	30,120	440	380	30,930
3010	25 ton	"	6.000	32,590	440	380	33,400
3020	30 ton	"	6.000	37,990	440	380	38,800
3030	40 ton	"	6.000	43,600	440	380	44,410
3040	50 ton	"	8.000	49,230	590	500	50,320
3050	60 ton	"	8.000	57,990	590	500	59,080
3070	80 ton	"	12.000	67,650	880	750	69,280
3080	100 ton	"	12.000	79,330	880	750	80,960
3090	120 ton	"	12.000	90,120	880	750	91,750
3980	Heat recovery, air cooled, with starter						
4000	40 ton	EA	12.000	50,570	880	750	52,200
4010	50 ton	"	12.000	58,430	880	750	60,060
4020	60 ton	"	16.000	64,060	1,170	1,000	66,230
4030	75 ton	"	24.000	73,270	1,760	1,500	76,530
4040	100 ton	"	24.000	83,610	1,760	1,500	86,870
4045	Water cooled, with starter						
4050	40 ton	EA	12.000	49,900	880	750	51,530
4060	50 ton	"	12.000	59,560	880	750	61,190
4070	60 ton	"	16.000	65,630	1,170	1,000	67,800
4080	75 ton	"	24.000	79,790	1,760	1,500	83,050
4090	100 ton	"	26.667	90,120	1,950	1,670	93,740
4980	Centrifugal, single bundle condenser, with starter						
5000	80 ton	EA	34.286	117,990	2,510	2,140	122,650
5010	130 ton	"	40.000	119,510	2,930	2,500	124,940
5020	160 ton	"	43.636	121,460	3,200	2,730	127,390
5030	180 ton	"	48.000	130,100	3,520	3,000	136,620
5040	230 ton	"	53.333	135,930	3,910	3,330	143,170
5050	280 ton	"	60.000	149,770	4,400	3,750	157,920

CENTRAL COOLING EQUIPMENT

ID Code	Description / Component Descriptions	Output / Unit of Meas.	Manhr / Unit	Unit Costs / Material Cost	Labor Cost	Equipment Cost	Total Cost
23 - 64001	**CHILLERS, Cont'd...**						**23 - 64001**
5060	360 ton	EA	60.000	167,060	4,400	3,750	175,210
5070	460 ton	"	80.000	205,960	5,860	5,000	216,820
5080	560 ton	"	85.714	225,340	6,280	5,360	236,980
5090	670 ton	"	96.000	274,660	7,040	6,000	287,700
23 - 65001	**COOLING TOWERS**						**23 - 65001**
5980	Cooling tower, propeller type						
6000	100 ton	EA	8.000	16,220	590	500	17,310
6010	200 ton	"	12.000	26,940	880	750	28,570
6020	300 ton	"	20.000	40,720	1,470	1,250	43,440
6030	400 ton	"	24.000	54,190	1,760	1,500	57,450
6040	600 ton	"	34.286	73,890	2,510	2,140	78,550
6050	800 ton	"	48.000	98,520	3,520	3,000	105,040
6060	1000 ton	"	60.000	118,280	4,400	3,750	126,430
6065	Centrifugal						
6070	100 ton	EA	8.000	22,500	590	500	23,590
6080	200 ton	"	12.000	34,440	880	750	36,070
6100	300 ton	"	20.000	48,670	1,470	1,250	51,390
6110	400 ton	"	24.000	64,900	1,760	1,500	68,160
6120	600 ton	"	34.286	92,900	2,510	2,140	97,560
6130	800 ton	"	48.000	123,830	3,520	3,000	130,350
6140	1000 ton	"	60.000	151,840	4,400	3,750	159,990

AIR HANDLING

ID Code	Description / Component Descriptions	Output / Unit of Meas.	Manhr / Unit	Unit Costs / Material Cost	Labor Cost	Equipment Cost	Total Cost
23 - 74001	**AIR HANDLING UNITS**						**23 - 74001**
0980	Air handling unit, medium pressure, single zone						
1000	1500 cfm	EA	5.000	4,840	520		5,360
1060	3000 cfm	"	8.889	6,360	920		7,280
1180	4000 cfm	"	10.000	8,150	1,030		9,180
2000	5000 cfm	"	10.667	10,270	1,100		11,370
2120	6000 cfm	"	11.429	13,200	1,180		14,380
2240	7000 cfm	"	12.308	15,150	1,270		16,420
3000	8500 cfm	"	13.333	18,630	1,380		20,010
3120	10,500 cfm	"	16.000	20,460	1,650		22,110
3240	12,500 cfm	"	17.778	23,550	1,840		25,390
4000	15,500 cfm	"	22.857	30,450	2,360		32,810
4120	17,500 cfm	"	26.667	33,780	2,760		36,540
4240	20,500 cfm	"	32.000	38,190	3,310		41,500

AIR HANDLING

ID Code	Description Component Descriptions	Output Unit of Meas.	Manhr / Unit	Unit Costs Material Cost	Labor Cost	Equipment Cost	Total Cost
23 - 74001	**AIR HANDLING UNITS, Cont'd...**						**23 - 74001**
6000	25,000 cfm	EA	40.000	43,280	4,130		47,410
6120	31,500 cfm	"	53.333	53,550	5,510		59,060
8980	Rooftop air handling units						
9000	4950 cfm	EA	8.889	13,910	920		14,830
9060	7370 cfm	"	11.429	17,640	1,180		18,820
9080	9790 cfm	"	13.333	18,770	1,380		20,150
9100	14,300 cfm	"	11.429	26,510	1,180		27,690
9120	21,725 cfm	"	11.429	37,550	1,180		38,730
9140	33,000 cfm	"	13.333	53,020	1,380		54,400
23 - 74009	**ROOF CURBS**						**23 - 74009**
0980	8" high, insulated, with liner and raised can						
1000	15" x 15"	EA	0.400	120	41.25		160
1020	17" x 17"	"	0.400	130	41.25		170
1040	19" x 19"	"	0.400	130	41.25		170
1060	21" x 21"	"	0.400	150	41.25		190
1100	25" x 25"	"	0.500	160	52.00		210
1120	28" x 28"	"	0.533	170	55.00		230
1140	32" x 32"	"	0.571	190	59.00		250
1160	36" x 36"	"	0.571	210	59.00		270
1180	40" x 40"	"	0.571	240	59.00		300
1190	44" x 44"	"	0.615	270	64.00		330
1200	48" x 48"	"	0.615	600	64.00		660

HVAC EQUIPMENT

ID Code	Description Component Descriptions	Output Unit of Meas.	Manhr / Unit	Unit Costs Material Cost	Labor Cost	Equipment Cost	Total Cost
23 - 81132	**ROOFTOP UNITS**						**23 - 81132**
0980	Packaged, single zone rooftop unit, with roof curb						
1000	2 ton	EA	8.000	4,130	830		4,960
1020	3 ton	"	8.000	4,340	830		5,170
1040	4 ton	"	10.000	4,740	1,030		5,770
1060	5 ton	"	13.333	5,140	1,380		6,520
1070	7.5 ton	"	16.000	7,470	1,650		9,120

HVAC EQUIPMENT

	Description	Output		Unit Costs			
ID Code	Component Descriptions	Unit of Meas.	Manhr / Unit	Material Cost	Labor Cost	Equipment Cost	Total Cost
23 - 81230	**COMPUTER ROOM A/C**					**23 - 81230**	
1010	Air cooled, alarm, high efficiency filter, elec. heat						
1020	3 ton	EA	6.154	20,500	640		21,140
1040	5 ton	"	6.667	21,890	690		22,580
1060	7.5 ton	"	8.000	39,660	830		40,490
1070	10 ton	"	10.000	41,450	1,030		42,480
1080	15 ton	"	11.429	45,550	1,180		46,730
1090	Steam heat						
1100	3 ton	EA	6.154	18,130	640		18,770
1120	5 ton	"	6.667	19,290	690		19,980
1140	7.5 ton	"	8.000	30,720	830		31,550
1160	10 ton	"	10.000	31,650	1,030		32,680
1180	15 ton	"	11.429	35,000	1,180		36,180
1190	Hot water heat						
1200	3 ton	EA	6.154	18,130	640		18,770
1220	5 ton	"	6.667	19,290	690		19,980
1240	7.5 ton	"	8.000	30,720	830		31,550
1260	10 ton	"	10.000	31,650	1,030		32,680
1300	15 ton	"	11.429	35,110	1,180		36,290
1310	Air cooled condenser, low ambient damper						
1320	3 ton	EA	1.600	1,810	170		1,980
1340	5 ton	"	2.000	2,850	210		3,060
1360	7.5 ton	"	4.000	4,380	410		4,790
1400	10 ton	"	5.714	6,400	590		6,990
1420	15 ton	"	4.706	7,070	490		7,560
3010	Water cooled, high efficiency filter, alarm, elec. heat						
3020	3 ton	EA	5.714	18,800	590		19,390
3040	5 ton	"	6.667	20,250	690		20,940
3060	7.5 ton	"	10.000	32,260	1,030		33,290
3080	10 ton	"	11.429	33,470	1,180		34,650
3100	15 ton	"	13.333	39,170	1,380		40,550
3110	Steam heat						
3120	3 ton	EA	5.714	21,470	590		22,060
3140	5 ton	"	6.667	24,500	690		25,190
3160	7.5 ton	"	10.000	34,560	1,030		35,590
3180	10 ton	"	11.429	35,780	1,180		36,960
3200	15 ton	"	13.333	41,600	1,380		42,980
3210	Hot water heat						
3220	3 ton	EA	5.714	21,470	590		22,060

HVAC EQUIPMENT

ID Code	Description / Component Descriptions	Unit of Meas.	Manhr / Unit	Material Cost	Labor Cost	Equipment Cost	Total Cost
23 - 81230	**COMPUTER ROOM A/C, Cont'd...**						**23 - 81230**
3240	5 ton	EA	6.667	22,920	690		23,610
3260	7.5 ton	"	10.000	34,560	1,030		35,590
3280	10 ton	"	11.429	35,780	1,180		36,960
3300	15 ton	"	13.333	41,600	1,380		42,980
5010	Chilled water, alarm, high eff. filter, elec. heat						
5020	7.5 ton	EA	7.273	15,040	750		15,790
5040	10 ton	"	8.889	15,770	920		16,690
5060	15 ton	"	10.000	17,950	1,030		18,980
5070	Steam heat						
5080	7.5 ton	EA	7.273	17,220	750		17,970
5100	10 ton	"	8.889	17,950	920		18,870
5120	15 ton	"	10.000	20,130	1,030		21,160
5130	Hot water heat						
5140	7.5 ton	EA	7.273	17,220	750		17,970
5160	10 ton	"	8.889	17,950	920		18,870
5180	15 ton	"	10.000	20,130	1,030		21,160

CONVECTION HEATING AND COOLING UNITS

ID Code	Description / Component Descriptions	Unit of Meas.	Manhr / Unit	Material Cost	Labor Cost	Equipment Cost	Total Cost
23 - 82190	**FAN COIL UNITS**						**23 - 82190**
0980	Fan coil unit, 2 pipe, complete						
1000	200 cfm ceiling hung	EA	2.667	1,220	280		1,500
1020	Floor mounted	"	2.000	1,150	210		1,360
1100	300 cfm, ceiling hung	"	3.200	1,290	330		1,620
1130	Floor mounted	"	2.667	1,230	280		1,510
1200	400 cfm, ceiling hung	"	3.810	1,360	390		1,750
1220	Floor mounted	"	2.667	1,310	280		1,590
1300	500 cfm, ceiling hung	"	4.000	1,580	410		1,990
1310	Floor mounted	"	3.077	1,520	320		1,840
1400	600 cfm, ceiling hung	"	4.420	2,000	460		2,460
1420	Floor mounted	"	3.636	1,860	380		2,240
2000	800 cfm, ceiling hung	"	5.000	2,340	520		2,860
2020	Floor mounted	"	3.810	1,860	390		2,250
2100	1000 cfm, ceiling hung	"	5.714	2,670	590		3,260
2120	Floor mounted	"	4.211	2,930	440		3,370
2200	1200 cfm ceiling hung	"	6.667	3,030	690		3,720
2220	Floor mounted	"	5.000	3,180	520		3,700

CONVECTION HEATING AND COOLING UNITS

ID Code	Component Descriptions	Unit of Meas.	Manhr / Unit	Material Cost	Labor Cost	Equipment Cost	Total Cost
23 - 82390	**UNIT HEATERS**						**23 - 82390**
0980	Steam unit heater, horizontal						
1000	12,500 btuh, 200 cfm	EA	1.333	560	140		700
1010	17,000 btuh, 300 cfm	"	1.333	740	140		880
1020	40,000 btuh, 500 cfm	"	1.333	900	140		1,040
1030	60,000 btuh, 700 cfm	"	1.333	940	140		1,080
1040	70,000 btuh, 1000 cfm	"	2.000	980	210		1,190
1045	Vertical						
1050	12,500 btuh, 200 cfm	EA	1.333	560	140		700
1060	17,000 btuh, 300 cfm	"	1.333	930	140		1,070
1070	40,000 btuh, 500 cfm	"	1.333	900	140		1,040
1080	60,000 btuh, 700 cfm	"	1.333	940	140		1,080
1090	70,000 btuh, 1000 cfm	"	1.333	980	140		1,120
1980	Gas unit heater, horizontal						
2000	27,400 btuh	EA	3.200	860	330		1,190
2010	38,000 btuh	"	3.200	900	330		1,230
2020	56,000 btuh	"	3.200	940	330		1,270
2030	82,200 btuh	"	3.200	980	330		1,310
2040	103,900 btuh	"	5.000	1,090	520		1,610
2060	125,700 btuh	"	5.000	1,280	520		1,800
2080	133,200 btuh	"	5.000	1,380	520		1,900
2090	149,000 btuh	"	5.000	1,630	520		2,150
2100	172,000 btuh	"	5.000	1,760	520		2,280
2120	190,000 btuh	"	5.000	1,850	520		2,370
2130	225,000 btuh	"	5.000	2,030	520		2,550
3980	Hot water unit heater, horizontal						
4000	12,500 btuh, 200 cfm	EA	1.333	450	140		590
4010	17,000 btuh, 300 cfm	"	1.333	500	140		640
4020	25,000 btuh, 500 cfm	"	1.333	580	140		720
4030	30,000 btuh, 700 cfm	"	1.333	680	140		820
4040	50,000 btuh, 1000 cfm	"	2.000	740	210		950
4050	60,000 btuh, 1300 cfm	"	2.000	780	210		990
4055	Vertical						
4060	12,500 btuh, 200 cfm	EA	1.333	660	140		800
4070	17,000 btuh, 300 cfm	"	1.333	660	140		800
4080	25,000 btuh, 500 cfm	"	1.333	660	140		800
4090	30,000 btuh, 700 cfm	"	1.333	660	140		800
4100	50,000 btuh, 1000 cfm	"	1.333	690	140		830
4120	60,000 btuh, 1300 cfm	"	1.333	850	140		990

CONVECTION HEATING AND COOLING UNITS

ID Code	Description — Component Descriptions	Output — Unit of Meas.	Output — Manhr / Unit	Unit Costs — Material Cost	Unit Costs — Labor Cost	Unit Costs — Equipment Cost	Unit Costs — Total Cost
23 - 82390	**UNIT HEATERS, Cont'd...**						**23 - 82390**
5000	Cabinet unit heaters, ceiling, exposed, hot water						
5010	200 cfm	EA	2.667	1,290	280		1,570
5030	300 cfm	"	3.200	1,380	330		1,710
5050	400 cfm	"	3.810	1,440	390		1,830
5070	600 cfm	"	4.211	1,480	440		1,920
5090	800 cfm	"	5.000	1,850	520		2,370
5120	1000 cfm	"	5.714	2,410	590		3,000
5140	1200 cfm	"	6.667	2,590	690		3,280
5160	2000 cfm	"	8.889	4,040	920		4,960

RESISTANCE HEATING

ID Code	Description — Component Descriptions	Output — Unit of Meas.	Output — Manhr / Unit	Unit Costs — Material Cost	Unit Costs — Labor Cost	Unit Costs — Equipment Cost	Unit Costs — Total Cost
23 - 83330	**ELECTRIC HEATING**						**23 - 83330**
1000	Baseboard heater						
1020	2', 375w	EA	1.000	51.00	96.00		150
1040	3', 500w	"	1.000	61.00	96.00		160
1060	4', 750w	"	1.143	68.00	110		180
1100	5', 935w	"	1.333	96.00	130		230
1120	6', 1125w	"	1.600	110	150		260
1140	7', 1310w	"	1.818	120	170		290
1160	8', 1500w	"	2.000	140	190		330
1180	9', 1680w	"	2.222	160	210		370
1200	10', 1875w	"	2.286	220	220		440
1210	Unit heater, wall mounted						
1215	750w	EA	1.600	200	150		350
1220	1500w	"	1.667	270	160		430
1225	2000w	"	1.739	280	170		450
1240	2500w	"	1.818	290	170		460
1250	3000w	"	2.000	350	190		540
1260	4000w	"	2.286	400	220		620
1270	Thermostat						
1280	Integral	EA	0.500	46.00	48.00		94.00
1300	Line voltage	"	0.500	47.25	48.00		95.00
1320	Electric heater connection	"	0.250	2.02	24.00		26.00
2000	Fittings						
2010	Inside corner	EA	0.400	29.75	38.50		68.00
2020	Outside corner	"	0.400	32.50	38.50		71.00
2030	Receptacle section	"	0.400	33.75	38.50		72.00

RESISTANCE HEATING

ID Code	Component Descriptions	Unit of Meas.	Manhr / Unit	Material Cost	Labor Cost	Equipment Cost	Total Cost
	Description	**Output**		**Unit Costs**			
23 - 83330	**ELECTRIC HEATING, Cont'd...**						**23 - 83330**
2040	Blank section	EA	0.400	42.00	38.50		81.00
2085	Infrared heaters						
2090	600w	EA	1.000	220	96.00		320
2100	2000w	"	1.194	240	110		350
2110	3000w	"	2.000	360	190		550
2120	4000w	"	2.500	500	240		740
2170	Controller	"	0.667	97.00	64.00		160
2180	Wall bracket	"	0.727	190	70.00		260
2185	Radiant ceiling heater panels						
2190	500w	EA	1.000	480	96.00		580
2200	750w	"	1.000	530	96.00		630
2210	Unit heaters, suspended, single phase						
2220	3.0 kw	EA	2.759	770	260		1,030
2230	5.0 kw	"	2.759	800	260		1,060
2240	7.5 kw	"	3.200	1,300	310		1,610
2250	10.0 kw	"	3.810	1,400	370		1,770
2255	Three phase						
2260	5 kw	EA	2.759	790	260		1,050
2270	7.5 kw	"	3.200	1,040	310		1,350
2280	10 kw	"	3.810	1,110	370		1,480
2290	15 kw	"	4.211	1,890	400		2,290
2300	20 kw	"	5.333	2,520	510		3,030
2310	25 kw	"	6.400	3,030	610		3,640
2320	30 kw	"	8.000	3,530	770		4,300
2330	35 kw	"	8.000	4,290	770		5,060
2340	Unit heater thermostat	"	0.533	74.00	51.00		130
2350	Mounting bracket	"	0.727	76.00	70.00		150
2360	Relay	"	0.615	97.00	59.00		160
2365	Duct heaters, three phase						
2370	10 kw	EA	3.810	1,380	370		1,750
2380	15 kw	"	3.810	1,650	370		2,020
2390	17.5 kw	"	4.000	1,740	380		2,120
2400	20 kw	"	6.154	1,870	590		2,460

HUMIDITY CONTROL EQUIPMENT

ID Code	Description / Component Descriptions	Output		Unit Costs			
		Unit of Meas.	Manhr / Unit	Material Cost	Labor Cost	Equipment Cost	Total Cost
23 - 84160	**DEHUMIDIFIERS**						**23 - 84160**
1000	Desiccant dehumidifier, 1125 cfm	EA					36,990

DIVISION 26
ELECTRICAL

CONDUCTORS, CONDUIT AND RACEWAYS

ID Code	Component Descriptions	Unit of Meas.	Manhr / Unit	Material Cost	Labor Cost	Equipment Cost	Total Cost
	Description	**Output**		**Unit Costs**			
26 - 05134	**COPPER CONDUCTORS**						**26 - 05134**
0980	Copper conductors, type THW, solid						
1000	#14	LF	0.004	0.22	0.38		0.60
1040	#12	"	0.005	0.34	0.48		0.82
1060	#10	"	0.006	0.52	0.57		1.09
1070	Stranded						
1080	#14	LF	0.004	0.24	0.38		0.62
1100	#12	"	0.005	0.30	0.48		0.78
1120	#10	"	0.006	0.46	0.57		1.03
1140	#8	"	0.008	0.76	0.76		1.52
1160	#6	"	0.009	1.23	0.86		2.09
1180	#4	"	0.010	1.92	0.96		2.88
1200	#3	"	0.010	2.42	0.96		3.38
1220	#2	"	0.012	3.05	1.15		4.20
1240	#1	"	0.014	3.86	1.34		5.20
1260	1/0	"	0.016	4.60	1.53		6.13
1280	2/0	"	0.020	5.78	1.92		7.70
1300	3/0	"	0.025	7.27	2.40		9.67
1520	4/0	"	0.028	9.09	2.69		11.75
1540	250 MCM	"	0.030	11.25	2.89		14.25
1560	300 MCM	"	0.033	13.25	3.20		16.50
1580	350 MCM	"	0.040	15.50	3.84		19.25
1600	400 MCM	"	0.044	17.50	4.26		21.75
1620	500 MCM	"	0.052	21.75	4.95		26.75
1640	600 MCM	"	0.059	28.75	5.68		34.50
1660	750 MCM	"	0.067	36.25	6.40		42.75
1680	1000 MCM	"	0.076	45.50	7.31		53.00
2010	THHN-THWN, solid						
2020	#14	LF	0.004	0.22	0.38		0.60
2040	#12	"	0.005	0.34	0.48		0.82
2060	#10	"	0.006	0.52	0.57		1.09
2070	Stranded						
2080	#14	LF	0.004	0.22	0.38		0.60
2100	#12	"	0.005	0.34	0.48		0.82
2120	#10	"	0.006	0.52	0.57		1.09
2140	#8	"	0.008	0.90	0.76		1.66
2160	#6	"	0.009	1.41	0.86		2.27
2180	#4	"	0.010	2.24	0.96		3.20
2200	#2	"	0.012	3.13	1.15		4.28

CONDUCTORS, CONDUIT AND RACEWAYS

ID Code	Description — Component Descriptions	Output — Unit of Meas.	Output — Manhr / Unit	Unit Costs — Material Cost	Unit Costs — Labor Cost	Unit Costs — Equipment Cost	Unit Costs — Total Cost
26 - 05134	**COPPER CONDUCTORS, Cont'd...**						**26 - 05134**
2220	#1	LF	0.014	3.96	1.34		5.30
2240	1/0	"	0.016	4.87	1.53		6.40
2260	2/0	"	0.020	6.02	1.92		7.94
2280	3/0	"	0.025	7.55	2.40		9.95
2300	4/0	"	0.028	9.45	2.69		12.25
2320	250 MCM	"	0.030	11.50	2.89		14.50
2340	350 MCM	"	0.040	13.75	3.84		17.50
2980	XHHW						
3000	#14	LF	0.004	0.38	0.38		0.76
3060	#10	"	0.006	0.84	0.57		1.41
3080	#8	"	0.008	1.23	0.76		1.99
3100	#6	"	0.009	1.92	0.86		2.78
3120	#4	"	0.009	2.99	0.86		3.85
3140	#2	"	0.011	4.64	1.05		5.69
3160	#1	"	0.014	5.94	1.34		7.28
3180	1/0	"	0.016	6.99	1.53		8.52
3200	2/0	"	0.019	8.77	1.82		10.50
3220	3/0	"	0.025	11.00	2.40		13.50
5000	XLP, 600v						
5020	#12	LF	0.005	0.58	0.48		1.06
5040	#10	"	0.006	0.84	0.57		1.41
5060	#8	"	0.008	1.11	0.76		1.87
5080	#6	"	0.009	1.69	0.86		2.55
5100	#4	"	0.010	2.62	0.96		3.58
5120	#3	"	0.011	3.27	1.05		4.32
5140	#2	"	0.012	4.06	1.15		5.21
5160	#1	"	0.014	5.21	1.34		6.55
5180	1/0	"	0.016	5.88	1.53		7.41
5200	2/0	"	0.020	7.33	1.92		9.25
5220	3/0	"	0.026	9.21	2.47		11.75
5240	4/0	"	0.028	11.50	2.69		14.25
5260	250 MCM	"	0.030	13.50	2.89		16.50
5280	300 MCM	"	0.033	16.00	3.20		19.25
5300	350 MCM	"	0.039	18.75	3.74		22.50
5320	400 MCM	"	0.044	21.25	4.26		25.50
5340	500 MCM	"	0.052	26.50	4.95		31.50
5360	600 MCM	"	0.059	31.75	5.68		37.50
5380	750 MCM	"	0.067	49.25	6.40		56.00

CONDUCTORS, CONDUIT AND RACEWAYS

ID Code	Component Descriptions	Unit of Meas.	Manhr / Unit	Material Cost	Labor Cost	Equipment Cost	Total Cost
	Description	**Output**		**Unit Costs**			
26 - 05134	**COPPER CONDUCTORS, Cont'd...**						**26 - 05134**
5390	1000 MCM	LF	0.076	65.00	7.31		72.00
6000	Bare solid wire						
6010	#14	LF	0.004	0.22	0.38		0.60
6020	#12	"	0.005	0.38	0.48		0.86
6030	#10	"	0.006	0.56	0.57		1.13
6040	#8	"	0.008	0.76	0.76		1.52
6050	#6	"	0.009	1.37	0.86		2.23
6060	#4	"	0.010	2.24	0.96		3.20
6070	#2	"	0.012	3.57	1.15		4.72
6075	Bare stranded wire						
6080	#8	LF	0.008	0.78	0.76		1.54
6090	#6	"	0.010	1.31	0.96		2.27
6100	#4	"	0.010	2.04	0.96		3.00
6110	#2	"	0.011	3.25	1.05		4.30
6120	#1	"	0.014	4.08	1.34		5.42
6130	1/0	"	0.018	4.81	1.72		6.53
6140	2/0	"	0.020	6.06	1.92		7.98
6150	3/0	"	0.025	7.64	2.40		10.00
6160	4/0	"	0.028	9.64	2.69		12.25
6170	250 MCM	"	0.030	11.50	2.89		14.50
6180	300 MCM	"	0.033	14.50	3.20		17.75
6190	350 MCM	"	0.040	16.00	3.84		19.75
6200	400 MCM	"	0.044	18.50	4.26		22.75
6210	500 MCM	"	0.052	22.75	4.95		27.75
6215	Type BX solid armored cable						
6220	#14/2	LF	0.025	1.49	2.40		3.89
6230	#14/3	"	0.028	2.36	2.69		5.05
6240	#14/4	"	0.031	3.31	2.95		6.26
6250	#12/2	"	0.028	1.53	2.69		4.22
6260	#12/3	"	0.031	2.46	2.95		5.41
6270	#12/4	"	0.035	3.41	3.33		6.74
6280	#10/2	"	0.031	2.85	2.95		5.80
6290	#10/3	"	0.035	4.08	3.33		7.41
6300	#10/4	"	0.040	6.34	3.84		10.25
6310	#8/2	"	0.035	5.67	3.33		9.00
6320	#8/3	"	0.040	7.98	3.84		11.75
6325	Steel type, metal clad cable, solid, with ground						
6330	#14/2	LF	0.018	1.23	1.72		2.95

CONDUCTORS, CONDUIT AND RACEWAYS

ID Code	Description — Component Descriptions	Output — Unit of Meas.	Output — Manhr / Unit	Unit Costs — Material Cost	Unit Costs — Labor Cost	Unit Costs — Equipment Cost	Unit Costs — Total Cost
26 - 05134	**COPPER CONDUCTORS, Cont'd...**						**26 - 05134**
6340	#14/3	LF	0.020	1.90	1.92		3.82
6350	#14/4	"	0.023	2.54	2.19		4.73
6360	#12/2	"	0.020	1.27	1.92		3.19
6370	#12/3	"	0.025	2.10	2.40		4.50
6380	#12/4	"	0.030	2.82	2.89		5.71
6390	#10/2	"	0.023	2.62	2.19		4.81
6400	#10/3	"	0.028	3.65	2.69		6.34
6410	#10/4	"	0.033	5.67	3.20		8.87
6415	Metal clad cable, stranded, with ground						
6420	#8/2	LF	0.028	4.60	2.69		7.29
6430	#8/3	"	0.035	6.60	3.33		9.93
6440	#8/4	"	0.042	8.63	4.04		12.75
6450	#6/2	"	0.030	6.28	2.89		9.17
6460	#6/3	"	0.038	7.57	3.65		11.25
6470	#6/4	"	0.044	9.03	4.26		13.25
6480	#4/2	"	0.040	8.20	3.84		12.00
6490	#4/3	"	0.044	9.29	4.26		13.50
6500	#4/4	"	0.055	10.50	5.29		15.75
6510	#3/3	"	0.050	10.75	4.80		15.50
6520	#3/4	"	0.059	12.00	5.68		17.75
6530	#2/3	"	0.057	8.67	5.48		14.25
6540	#2/4	"	0.067	14.75	6.40		21.25
6550	#1/3	"	0.076	15.25	7.31		22.50
6950	#1/4	"	0.084	18.00	8.08		26.00
26 - 05135	**SHEATHED CABLE**						**26 - 05135**
6700	Non-metallic sheathed cable						
6705	Type NM cable with ground						
6710	#14/2	LF	0.015	0.49	1.43		1.92
6720	#12/2	"	0.016	0.68	1.53		2.21
6730	#10/2	"	0.018	1.08	1.70		2.78
6740	#8/2	"	0.020	1.77	1.92		3.69
6750	#6/2	"	0.025	2.78	2.40		5.18
6760	#14/3	"	0.026	0.61	2.47		3.08
6770	#12/3	"	0.027	0.97	2.56		3.53
6780	#10/3	"	0.027	1.54	2.60		4.14
6790	#8/3	"	0.028	2.59	2.64		5.23
6800	#6/3	"	0.028	4.19	2.69		6.88

CONDUCTORS, CONDUIT AND RACEWAYS

ID Code	Component Descriptions	Unit of Meas.	Manhr / Unit	Material Cost	Labor Cost	Equipment Cost	Total Cost
	Description	**Output**		**Unit Costs**			

26 - 05135   SHEATHED CABLE, Cont'd...   26 - 05135

ID Code	Component Descriptions	Unit of Meas.	Manhr / Unit	Material Cost	Labor Cost	Equipment Cost	Total Cost
6810	#4/3	LF	0.032	8.67	3.07		11.75
6820	#2/3	"	0.035	13.00	3.33		16.25
6825	Type UF cable with ground						
6830	#14/2	LF	0.016	0.51	1.53		2.04
6840	#12/2	"	0.019	0.77	1.82		2.59
6850	#10/2	"	0.020	1.23	1.92		3.15
6860	#8/2	"	0.023	2.13	2.19		4.32
6870	#6/2	"	0.027	3.31	2.60		5.91
6880	#14/3	"	0.020	0.72	1.92		2.64
6890	#12/3	"	0.022	1.10	2.10		3.20
6900	#10/3	"	0.025	1.70	2.40		4.10
6910	#8/3	"	0.028	3.21	2.69		5.90
6920	#6/3	"	0.032	5.20	3.07		8.27
6925	Type SFU cable, 3 conductor						
6930	#8	LF	0.028	2.21	2.69		4.90
6940	#6	"	0.031	3.86	2.95		6.81
6960	#3	"	0.040	7.52	3.84		11.25
6970	#2	"	0.044	9.35	4.26		13.50
6980	#1	"	0.050	12.00	4.80		16.75
6990	#1/0	"	0.055	15.00	5.29		20.25
7000	#2/0	"	0.064	18.75	6.14		25.00
7010	#3/0	"	0.070	23.50	6.67		30.25
7020	#4/0	"	0.076	26.00	7.31		33.25
7025	Type SER cable, 4 conductor						
7030	#6	LF	0.036	5.53	3.49		9.02
7040	#4	"	0.039	7.75	3.74		11.50
7050	#3	"	0.044	10.50	4.26		14.75
7060	#2	"	0.048	12.25	4.65		17.00
7070	#1	"	0.055	15.25	5.29		20.50
7080	#1/0	"	0.064	19.00	6.14		25.25
7090	#2/0	"	0.067	24.00	6.40		30.50
7100	#3/0	"	0.076	30.00	7.31		37.25
7110	#4/0	"	0.084	37.75	8.08		45.75
7115	Flexible cord, type STO cord						
7120	#18/2	LF	0.004	0.94	0.38		1.32
7130	#18/3	"	0.005	1.10	0.48		1.58
7140	#18/4	"	0.006	1.52	0.57		2.09
7150	#16/2	"	0.004	1.08	0.38		1.46

CONDUCTORS, CONDUIT AND RACEWAYS

ID Code	Description		Output		Unit Costs			
	Component Descriptions		Unit of Meas.	Manhr / Unit	Material Cost	Labor Cost	Equipment Cost	Total Cost
26 - 05135	**SHEATHED CABLE, Cont'd...**						**26 - 05135**	
7160	#16/3		LF	0.004	0.91	0.42		1.33
7170	#16/4		"	0.005	1.27	0.48		1.75
7180	#14/2		"	0.005	1.70	0.48		2.18
7190	#14/3		"	0.006	1.54	0.59		2.13
7200	#14/4		"	0.007	1.90	0.67		2.57
7210	#12/2		"	0.006	2.14	0.57		2.71
7220	#12/3		"	0.007	1.61	0.64		2.25
7230	#12/4		"	0.008	2.33	0.76		3.09
7240	#10/2		"	0.007	2.66	0.67		3.33
7250	#10/3		"	0.008	2.56	0.76		3.32
7260	#10/4		"	0.009	3.95	0.86		4.81
7270	#8/2		"	0.008	4.43	0.76		5.19
7280	#8/3		"	0.009	4.93	0.85		5.78
7290	#8/4		"	0.010	6.90	0.96		7.86
26 - 05137	**ALUMINUM CONDUCTORS**						**26 - 05137**	
0080	Type XHHW, stranded aluminum, 600v							
0100	#8		LF	0.005	0.63	0.48		1.11
1000	#6		"	0.006	0.67	0.57		1.24
1020	#4		"	0.008	0.82	0.76		1.58
1040	#2		"	0.009	1.14	0.86		2.00
1060	1/0		"	0.011	1.79	1.05		2.84
1080	2/0		"	0.012	2.32	1.15		3.47
1081	3/0		"	0.014	2.90	1.34		4.24
1090	4/0		"	0.015	3.23	1.43		4.66
1220	300 MCM		"	0.020	5.42	1.92		7.34
1240	350 MCM		"	0.023	5.52	2.19		7.71
1260	400 MCM		"	0.028	6.45	2.69		9.14
1280	500 MCM		"	0.033	7.11	3.20		10.25
1300	600 MCM		"	0.040	8.99	3.84		12.75
1320	700 MCM		"	0.047	10.50	4.51		15.00
1340	750 MCM		"	0.052	10.50	4.95		15.50
1980	THW, stranded							
2000	#8		LF	0.005	0.63	0.48		1.11
2020	#6		"	0.006	0.67	0.57		1.24
2030	#4		"	0.008	0.82	0.76		1.58
2040	#3		"	0.009	1.05	0.85		1.90
2050	#1		"	0.010	1.79	0.96		2.75

CONDUCTORS, CONDUIT AND RACEWAYS

ID Code	Component Descriptions	Unit of Meas.	Manhr / Unit	Material Cost	Labor Cost	Equipment Cost	Total Cost
	Description	**Output**		**Unit Costs**			
26 - 05137	**ALUMINUM CONDUCTORS, Cont'd...**						**26 - 05137**
2060	1/0	LF	0.011	1.96	1.05		3.01
2070	2/0	"	0.012	2.32	1.13		3.45
2080	3/0	"	0.012	2.90	1.13		4.03
2090	4/0	"	0.015	3.23	1.43		4.66
2100	250 MCM	"	0.018	3.93	1.72		5.65
2120	300 MCM	"	0.020	5.42	1.92		7.34
2130	350 MCM	"	0.023	5.52	2.19		7.71
2140	400 MCM	"	0.028	6.45	2.69		9.14
2150	500 MCM	"	0.033	7.11	3.20		10.25
2160	600 MCM	"	0.040	8.99	3.84		12.75
2170	700 MCM	"	0.047	10.50	4.51		15.00
2180	750 MCM	"	0.052	10.50	4.95		15.50
4980	XLP, stranded						
5000	#6	LF	0.005	0.80	0.48		1.28
5020	#4	"	0.008	0.91	0.76		1.67
5040	#2	"	0.009	1.24	0.86		2.10
5060	#1	"	0.010	1.73	0.96		2.69
5080	1/0	"	0.011	2.11	1.05		3.16
5100	2/0	"	0.012	2.49	1.15		3.64
5120	3/0	"	0.014	2.98	1.34		4.32
5140	4/0	"	0.015	3.30	1.43		4.73
5160	250 MCM	"	0.016	4.42	1.55		5.97
5180	300 MCM	"	0.020	5.75	1.92		7.67
5200	350 MCM	"	0.023	5.88	2.19		8.07
5220	400 MCM	"	0.028	7.19	2.69		9.88
5240	500 MCM	"	0.033	7.89	3.20		11.00
5260	600 MCM	"	0.040	10.25	3.84		14.00
5280	700 MCM	"	0.047	11.75	4.51		16.25
5300	750 MCM	"	0.052	11.75	4.95		16.75
5320	1000 MCM	"	0.057	15.75	5.48		21.25
6000	Bare stranded aluminum wire						
6010	#4	LF	0.008	0.57	0.76		1.33
6020	#2	"	0.009	0.82	0.86		1.68
6030	1/0	"	0.011	1.10	1.05		2.15
6040	2/0	"	0.012	1.37	1.15		2.52
6050	3/0	"	0.014	1.71	1.34		3.05
6060	4/0	"	0.015	2.13	1.43		3.56
6065	Triplex XLP cable						

CONDUCTORS, CONDUIT AND RACEWAYS

ID Code	Description / Component Descriptions	Output / Unit of Meas.	Output / Manhr / Unit	Unit Costs / Material Cost	Unit Costs / Labor Cost	Unit Costs / Equipment Cost	Unit Costs / Total Cost
26 - 05137	**ALUMINUM CONDUCTORS, Cont'd...**						**26 - 05137**
6070	#4	LF	0.015	1.71	1.43		3.14
6080	#2	"	0.020	2.09	1.92		4.01
6090	1/0	"	0.030	3.34	2.89		6.23
6100	4/0	"	0.048	6.16	4.65		10.75
6105	Aluminum quadruplex XLP cable						
6110	#4	LF	0.018	2.27	1.72		3.99
6120	#2	"	0.023	2.94	2.19		5.13
6130	1/0	"	0.032	4.63	3.07		7.70
6140	2/0	"	0.042	5.56	4.04		9.60
6150	4/0	"	0.064	8.02	6.14		14.25
6155	Triplexed URD-XLP cable						
6160	#6	LF	0.011	1.38	1.05		2.43
6170	#4	"	0.014	1.96	1.34		3.30
6180	#2	"	0.018	2.54	1.72		4.26
6190	1/0	"	0.028	4.06	2.69		6.75
6200	2/0	"	0.033	4.63	3.20		7.83
6210	3/0	"	0.040	5.56	3.84		9.40
6220	4/0	"	0.047	6.50	4.51		11.00
6230	250 MCM	"	0.055	7.44	5.29		12.75
6240	350 MCM	"	0.057	9.64	5.48		15.00
6245	Type SEU cable						
6250	#8/3	LF	0.025	2.67	2.40		5.07
6260	#6/3	"	0.028	2.67	2.69		5.36
6270	#4/3	"	0.035	3.44	3.33		6.77
6280	#2/3	"	0.038	4.58	3.65		8.23
6290	#1/3	"	0.040	6.21	3.84		10.00
6300	1/0-3	"	0.042	6.98	4.04		11.00
6310	2/0-3	"	0.044	8.02	4.26		12.25
6320	3/0-3	"	0.052	11.25	4.95		16.25
6330	4/0-3	"	0.057	11.25	5.48		16.75
6335	Type SER cable with ground						
6340	#8/3	LF	0.028	3.23	2.69		5.92
6350	#6/3	"	0.035	3.65	3.33		6.98
6360	#4/3	"	0.038	4.10	3.65		7.75
6370	#2/3	"	0.040	6.02	3.84		9.86
6380	#1/3	"	0.044	7.85	4.26		12.00
6390	1/0-3	"	0.050	9.14	4.80		14.00
6400	2/0-3	"	0.055	10.75	5.29		16.00

CONDUCTORS, CONDUIT AND RACEWAYS

ID Code	Description — Component Descriptions	Output — Unit of Meas.	Output — Manhr / Unit	Unit Costs — Material Cost	Unit Costs — Labor Cost	Unit Costs — Equipment Cost	Unit Costs — Total Cost
26 - 05137	**ALUMINUM CONDUCTORS, Cont'd...**						**26 - 05137**
6410	3/0-3	LF	0.059	13.25	5.68		19.00
6420	4/0-3	"	0.067	15.25	6.40		21.75
6430	#6/4	"	0.038	6.21	3.65		9.86
6440	#4/4	"	0.044	7.00	4.26		11.25
6450	#2/4	"	0.044	10.25	4.26		14.50
6460	#1/4	"	0.050	13.25	4.80		18.00
6470	1/0-4	"	0.052	15.50	4.95		20.50
6480	2/0-4	"	0.057	18.25	5.48		23.75
6490	3/0-4	"	0.064	22.50	6.14		28.75
6500	4/0-4	"	0.076	26.25	7.31		33.50
26 - 05138	**FLAT CONDUCTOR CABLE**						**26 - 05138**
1000	Flat conductor cable, with shield, 3 conductor						
1010	#12 awg	LF	0.059	8.90	5.68		14.50
1020	#10 awg	"	0.059	10.50	5.68		16.25
1025	4 conductor						
1030	#12 awg	LF	0.080	12.00	7.68		19.75
1040	#10 awg	"	0.080	13.50	7.68		21.25
1045	Transition boxes						
1050	#12 awg	LF	0.089	15.00	8.53		23.50
1060	#10 awg	"	0.089	17.00	8.53		25.50
1065	Flat conductor cable, communication, with shield						
1070	10 conductor	LF	0.059	5.86	5.68		11.50
1080	16 conductor	"	0.070	6.77	6.67		13.50
1090	24 conductor	"	0.100	7.57	9.60		17.25
1100	Power and communication heads, duplex receptacle	EA	0.800	80.00	77.00		160
1110	Double duplex receptacle	"	0.952	89.00	91.00		180
1120	Telephone	"	0.800	51.00	77.00		130
1130	Receptacle and telephone	"	0.952	120	91.00		210
1140	Blank cover	"	0.145	12.75	14.00		26.75
1145	Transition boxes						
1150	Surface	EA	0.727	230	70.00		300
1160	Flush	"	1.000	130	96.00		230
1163	Flat conductor cable fittings						
1165	End caps	EA	0.145	2.45	14.00		16.50
1170	Insulators	"	0.296	26.50	28.50		55.00
1180	Splice connectors	"	0.444	1.83	42.75		44.50
1190	Tap connectors	"	0.444	2.01	42.75		44.75

CONDUCTORS, CONDUIT AND RACEWAYS

ID Code	Component Descriptions	Unit of Meas.	Manhr / Unit	Material Cost	Labor Cost	Equipment Cost	Total Cost
26 - 05138	**FLAT CONDUCTOR CABLE, Cont'd...**					**26 - 05138**	
1200	Cable connectors	EA	0.444	2.25	42.75		45.00
1210	Terminal blocks	"	0.615	15.50	59.00		75.00
1220	Tape	"					23.75
26 - 05234	**CONTROL CABLE**					**26 - 05234**	
0980	Control cable, 600v, #14 THWN, PVC jacket						
1000	2 wire	LF	0.008	0.53	0.76		1.29
1020	4 wire	"	0.010	0.89	0.96		1.85
1040	6 wire	"	0.131	1.56	12.50		14.00
1060	8 wire	"	0.145	1.93	14.00		16.00
1080	10 wire	"	0.160	2.19	15.25		17.50
1100	12 wire	"	0.182	2.66	17.50		20.25
1120	14 wire	"	0.211	3.11	20.25		23.25
1140	16 wire	"	0.222	3.39	21.25		24.75
1160	18 wire	"	0.242	3.66	23.25		27.00
1180	20 wire	"	0.250	4.30	24.00		28.25
1200	22 wire	"	0.286	4.30	27.50		31.75
2000	Audio cables, shielded, #24 gauge						
2010	3 conductor	LF	0.004	0.43	0.38		0.81
2020	4 conductor	"	0.006	0.53	0.57		1.10
2030	5 conductor	"	0.007	0.61	0.67		1.28
2040	6 conductor	"	0.009	0.69	0.86		1.55
2050	7 conductor	"	0.011	0.78	1.05		1.83
2060	8 conductor	"	0.012	0.88	1.15		2.03
2070	9 conductor	"	0.014	0.91	1.34		2.25
2080	10 conductor	"	0.015	1.03	1.43		2.46
2090	15 conductor	"	0.018	1.76	1.72		3.48
2100	20 conductor	"	0.023	2.34	2.19		4.53
2110	25 conductor	"	0.027	2.86	2.60		5.46
2120	30 conductor	"	0.030	3.51	2.89		6.40
2130	40 conductor	"	0.036	4.54	3.49		8.03
2140	50 conductor	"	0.042	5.70	4.04		9.74
2145	#22 gauge						
2150	3 conductor	LF	0.004	0.64	0.38		1.02
2160	4 conductor	"	0.006	0.83	0.57		1.40
2165	#20 gauge						
2170	3 conductor	LF	0.004	0.48	0.38		0.86
2180	10 conductor	"	0.015	1.53	1.43		2.96

CONDUCTORS, CONDUIT AND RACEWAYS

ID Code	Component Descriptions	Unit of Meas.	Manhr / Unit	Material Cost	Labor Cost	Equipment Cost	Total Cost
26 - 05234	**CONTROL CABLE, Cont'd...**						**26 - 05234**
2190	15 conductor	LF	0.018	1.97	1.72		3.69
2195	#18 gauge						
2200	3 conductor	LF	0.004	0.61	0.38		0.99
2210	4 conductor	"	0.006	0.91	0.57		1.48
2215	Microphone cables, #24 gauge						
2220	2 conductor	LF	0.004	0.61	0.38		0.99
2230	3 conductor	"	0.005	0.69	0.48		1.17
2235	#20 gauge						
2240	1 conductor	LF	0.004	0.57	0.38		0.95
2250	2 conductor	"	0.004	0.89	0.38		1.27
2270	3 conductor	"	0.006	1.23	0.57		1.80
2280	4 conductor	"	0.007	1.72	0.67		2.39
2290	5 conductor	"	0.009	2.10	0.86		2.96
2300	7 conductor	"	0.011	2.34	1.05		3.39
2310	8 conductor	"	0.012	2.59	1.15		3.74
2340	Computer cables, shielded, #24 gauge						
2350	1 pair	LF	0.004	0.33	0.38		0.71
2360	2 pair	"	0.004	0.46	0.38		0.84
2370	3 pair	"	0.006	0.57	0.57		1.14
2380	4 pair	"	0.007	0.66	0.67		1.33
2390	5 pair	"	0.009	0.87	0.86		1.73
2400	6 pair	"	0.011	1.03	1.05		2.08
2410	7 pair	"	0.012	1.07	1.15		2.22
2420	8 pair	"	0.014	1.24	1.34		2.58
2480	50 pair	"	0.039	7.14	3.74		11.00
2485	Coaxial cables						
2490	RG 6/u	LF	0.006	0.50	0.57		1.07
2500	RG 6a/u	"	0.006	0.78	0.57		1.35
2510	RG 8/u	"	0.006	0.91	0.57		1.48
2520	RG 8a/u	"	0.006	1.10	0.57		1.67
2530	RG 9/u	"	0.006	1.97	0.57		2.54
2540	RG 11/u	"	0.006	2.44	0.57		3.01
2550	RG 58/u	"	0.006	2.69	0.57		3.26
2560	RG 59/u	"	0.006	2.93	0.57		3.50
2570	RG 62/u	"	0.006	3.23	0.57		3.80
2580	RG 174/u	"	0.006	3.46	0.57		4.03
2590	RG 213/u	"	0.006	3.71	0.57		4.28
2600	MATV and CCTV camera cables						

CONDUCTORS, CONDUIT AND RACEWAYS

ID Code	Component Descriptions	Unit of Meas.	Manhr / Unit	Material Cost	Labor Cost	Equipment Cost	Total Cost
26 - 05234	**CONTROL CABLE, Cont'd...**						**26 - 05234**
2610	1 conductor	LF	0.004	0.55	0.38		0.93
2620	2 conductor	"	0.005	0.70	0.48		1.18
2630	4 conductor	"	0.006	1.51	0.57		2.08
2640	7 conductor	"	0.009	2.11	0.86		2.97
2650	12 conductor	"	0.015	3.24	1.43		4.67
2660	13 conductor	"	0.016	3.50	1.53		5.03
2670	14 conductor	"	0.018	3.57	1.72		5.29
2680	28 conductor	"	0.027	8.53	2.60		11.25
2685	Fire alarm cables, #22 gauge						
2690	6 conductor	LF	0.010	2.78	0.96		3.74
2700	9 conductor	"	0.015	3.57	1.43		5.00
2710	12 conductor	"	0.016	4.10	1.53		5.63
2715	#18 gauge						
2720	2 conductor	LF	0.005	2.78	0.48		3.26
2730	4 conductor	"	0.007	3.57	0.67		4.24
2735	#16 gauge						
2740	2 conductor	LF	0.007	2.78	0.67		3.45
2750	4 conductor	"	0.008	3.83	0.76		4.59
2755	#14 gauge						
2760	2 conductor	LF	0.008	4.10	0.76		4.86
2765	#12 gauge						
2770	2 conductor	LF	0.010	5.02	0.96		5.98
2775	Plastic jacketed thermostat cable						
2780	2 conductor	LF	0.004	0.18	0.38		0.56
2790	3 conductor	"	0.005	0.26	0.48		0.74
2800	4 conductor	"	0.006	0.34	0.57		0.91
2810	5 conductor	"	0.008	0.43	0.76		1.19
2820	6 conductor	"	0.009	0.51	0.86		1.37
2830	7 conductor	"	0.012	0.55	1.15		1.70
2840	8 conductor	"	0.013	0.82	1.24		2.06
26 - 05261	**GROUNDING**						**26 - 05261**
0400	Ground rods, copper clad, 1/2" x						
0510	6'	EA	0.667	21.75	64.00		86.00
0520	8'	"	0.727	30.00	70.00		100
0530	10'	"	1.000	37.50	96.00		130
0535	5/8" x						
0540	5'	EA	0.615	26.50	59.00		86.00

CONDUCTORS, CONDUIT AND RACEWAYS

ID Code	Component Descriptions	Unit of Meas.	Manhr / Unit	Material Cost	Labor Cost	Equipment Cost	Total Cost
	Description	**Output**		**Unit Costs**			
26 - 05261		**GROUNDING, Cont'd...**					**26 - 05261**
0550	6'	EA	0.727	28.50	70.00		99.00
0560	8'	"	1.000	36.75	96.00		130
0570	10'	"	1.250	45.50	120		170
0580	3/4" x						
0590	8'	EA	0.727	80.00	70.00		150
0600	10'	"	0.800	88.00	77.00		160
1060	Ground rod clamp						
1080	5/8"	EA	0.123	10.75	11.75		22.50
1100	3/4"	"	0.123	15.00	11.75		26.75
1120	Coupling, on threaded rods, 3/4"	"	0.050	31.25	4.80		36.00
1140	Ground receptacles	"	0.250	39.75	24.00		64.00
1160	Bus bar, copper, 2" x 1/4"	LF	0.145	11.50	14.00		25.50
1180	Copper braid, 1" x 1/8", for door ground	EA	0.100	8.93	9.60		18.50
2000	Brazed connection for						
2020	#6 wire	EA	0.500	37.25	48.00		85.00
2040	#2 wire	"	0.800	47.00	77.00		120
2060	#2/0 wire	"	1.000	63.00	96.00		160
2080	#4/0 wire	"	1.143	86.00	110		200
2580	Ground rod couplings						
2600	1/2"	EA	0.100	19.50	9.60		29.00
2610	5/8"	"	0.100	27.50	9.60		37.00
2615	Ground rod, driving stud						
2620	1/2"	EA	0.100	15.75	9.60		25.25
2630	5/8"	"	0.100	18.50	9.60		28.00
2640	3/4"	"	0.100	21.00	9.60		30.50
2645	Ground rod clamps, #8-2 to						
2650	1" pipe	EA	0.200	17.75	19.25		37.00
2660	2" pipe	"	0.250	22.00	24.00		46.00
2670	3" pipe	"	0.296	89.00	28.50		120
2680	5" pipe	"	0.348	140	33.50		170
2690	6" pipe	"	0.444	190	42.75		230
2695	#4-4/0 to						
2700	1" pipe	EA	0.200	42.00	19.25		61.00
2710	2" pipe	"	0.250	66.00	24.00		90.00
2720	3" pipe	"	0.296	100	28.50		130
2730	4" pipe	"	0.348	150	33.50		180
2740	6" pipe	"	0.444	220	42.75		260
2750	8" pipe	"	0.667	250	64.00		310

CONDUCTORS, CONDUIT AND RACEWAYS

ID Code	Description — Component Descriptions	Output — Unit of Meas.	Output — Manhr / Unit	Unit Costs — Material Cost	Unit Costs — Labor Cost	Unit Costs — Equipment Cost	Unit Costs — Total Cost
26 - 05261	**GROUNDING, Cont'd...**						**26 - 05261**
2760	10" pipe	EA	0.952	290	91.00		380
2770	12" pipe	"	1.290	340	120		460
26 - 05292	**CONDUIT SPECIALTIES**						**26 - 05292**
8005	Rod beam clamp, 1/2"	EA	0.050	9.35	4.80		14.25
8007	Hanger rod						
8010	3/8"	LF	0.040	1.97	3.84		5.81
8020	1/2"	"	0.050	4.92	4.80		9.72
8030	All thread rod						
8040	1/4"	LF	0.030	0.63	2.89		3.52
8060	3/8"	"	0.040	0.72	3.84		4.56
8080	1/2"	"	0.050	1.34	4.80		6.14
8100	5/8"	"	0.080	2.37	7.68		10.00
8120	Hanger channel, 1-1/2"						
8140	No holes	EA	0.030	6.22	2.89		9.11
8160	Holes	"	0.030	7.69	2.89		10.50
8170	Channel strap						
8180	1/2"	EA	0.050	1.94	4.80		6.74
8200	3/4"	"	0.050	2.60	4.80		7.40
8220	1"	"	0.050	3.33	4.80		8.13
8240	1-1/4"	"	0.080	2.68	7.68		10.25
8260	1-1/2"	"	0.080	3.23	7.68		11.00
8280	2"	"	0.080	3.49	7.68		11.25
8290	2-1/2"	"	0.123	6.69	11.75		18.50
8300	3"	"	0.123	7.29	11.75		19.00
8310	3-1/2"	"	0.123	9.08	11.75		20.75
8320	4"	"	0.145	10.25	14.00		24.25
8340	5"	"	0.145	16.75	14.00		30.75
8360	6"	"	0.145	18.75	14.00		32.75
8410	Conduit penetrations, roof and wall, 8" thick						
8420	1/2"	EA	0.615		59.00		59.00
8460	3/4"	"	0.615		59.00		59.00
8480	1"	"	0.800		77.00		77.00
8500	1-1/4"	"	0.800		77.00		77.00
8520	1-1/2"	"	0.800		77.00		77.00
8540	2"	"	1.600		150		150
8560	2-1/2"	"	1.600		150		150
8580	3"	"	1.600		150		150

CONDUCTORS, CONDUIT AND RACEWAYS

ID Code	Component Descriptions	Unit of Meas.	Manhr / Unit	Material Cost	Labor Cost	Equipment Cost	Total Cost
	Description	**Output**		**Unit Costs**			
26 - 05292	**CONDUIT SPECIALTIES, Cont'd...**					**26 - 05292**	
8590	3-1/2"	EA	2.000		190		190
8600	4"	"	2.000		190		190
8810	Plastic duct bank conduit spacer, 3" separation						
8820	2"	EA	0.050	2.47	4.80		7.27
8840	3"	"	0.050	2.73	4.80		7.53
8860	4"	"	0.050	3.08	4.80		7.88
8880	5"	"	0.050	3.34	4.80		8.14
8900	6"	"	0.050	5.41	4.80		10.25
8910	Intermediate, 3" separation						
8920	2"	EA	0.050	2.56	4.80		7.36
8940	3"	"	0.050	2.83	4.80		7.63
8960	4"	"	0.050	3.16	4.80		7.96
8980	5"	"	0.050	3.48	4.80		8.28
8990	6"	"	0.050	5.49	4.80		10.25
9005	Base with 1-1/2" separation						
9010	2"	EA	0.050	2.44	4.80		7.24
9020	3"	"	0.160	2.66	15.25		18.00
9040	4"	"	0.160	2.92	15.25		18.25
9050	5"	"	0.160	3.16	15.25		18.50
9060	6"	"	0.160	5.10	15.25		20.25
9065	Intermediate, 1-1/2" separation						
9070	2"	EA	0.160	2.60	15.25		17.75
9080	3"	"	0.160	2.80	15.25		18.00
9090	3-1/2"	"	0.160	3.03	15.25		18.25
9100	4"	"	0.160	3.09	15.25		18.25
9110	5"	"	0.160	3.27	16.50		19.75
9120	6"	"	0.160	5.17	15.25		20.50
9130	OD beam clamp, 1/4"	"	0.200	1.18	19.25		20.50
9140	Threaded rod couplings						
9150	1/4"	EA	0.050	2.26	4.80		7.06
9160	3/8"	"	0.050	2.39	4.80		7.19
9170	1/2"	"	0.050	2.70	4.80		7.50
9180	5/8"	"	0.050	4.15	4.80		8.95
9190	3/4"	"	0.050	4.54	4.80		9.34
9195	Hex nuts						
9200	Hex nuts, 1/4"	EA	0.050	0.18	4.80		4.98
9210	3/8"	"	0.050	0.28	4.80		5.08
9220	1/2"	"	0.050	0.61	4.80		5.41

CONDUCTORS, CONDUIT AND RACEWAYS

ID Code	Component Descriptions	Unit of Meas.	Manhr / Unit	Material Cost	Labor Cost	Equipment Cost	Total Cost
	Description	**Output**		**Unit Costs**			
26 - 05292	**CONDUIT SPECIALTIES, Cont'd...**					**26 - 05292**	
9230	5/8"	EA	0.050	1.31	4.80		6.11
9240	3/4"	"	0.050	1.73	4.80		6.53
9245	Square nuts						
9250	1/4"	EA	0.050	0.16	4.80		4.96
9260	3/8"	"	0.050	0.31	4.80		5.11
9270	1/2"	"	0.050	0.54	4.80		5.34
9280	5/8"	"	0.050	0.71	4.80		5.51
9290	3/4"	"	0.050	1.25	4.80		6.05
9295	Flat washers, material only						
9300	1/4"	EA					0.18
9310	3/8"	"					0.25
9320	1/2"	"					0.35
9330	5/8"	"					0.71
9340	3/4"	"					0.99
9345	Lockwashers						
9350	1/4"	EA					0.11
9360	3/8"	"					0.19
9370	1/2"	"					0.24
9380	5/8"	"					0.42
9390	3/4"	"					0.71
9400	Channel closure strip	LF	0.133	2.70	12.75		15.50
9410	Channel end cap	EA	0.133	1.24	12.75		14.00
9415	Li-channel trapeze hangers						
9420	12" long	EA	0.145	17.75	14.00		31.75
9430	18" long	"	0.145	20.00	14.00		34.00
9440	24" long	"	0.145	24.75	14.00		38.75
9450	30" long	"	0.250	28.25	24.00		52.00
9460	36" long	"	0.250	32.75	24.00		57.00
9470	42" long	"	0.296	39.25	28.50		68.00
9475	Channel spring nuts						
9480	1/4"	EA	0.059	2.21	5.68		7.89
9490	3/8"	"	0.080	3.02	7.68		10.75
9500	1/2"	"	0.100	3.22	9.60		12.75
9505	Fireproofing, for conduit penetrations						
9510	1/2"	EA	0.500	4.04	48.00		52.00
9520	3/4"	"	0.500	4.19	48.00		52.00
9530	1"	"	0.500	4.28	48.00		52.00
9540	1-1/4"	"	0.727	10.00	75.00		85.00

CONDUCTORS, CONDUIT AND RACEWAYS

ID Code	Description / Component Descriptions	Output Unit of Meas.	Output Manhr / Unit	Material Cost	Labor Cost	Equipment Cost	Total Cost
26 - 05292	**CONDUIT SPECIALTIES, Cont'd...**						**26 - 05292**
9550	1-1/2"	EA	0.727	5.95	70.00		76.00
9560	2"	"	0.727	6.10	70.00		76.00
9570	2-1/2"	"	0.899	11.75	86.00		98.00
9580	3"	"	0.899	12.00	93.00		110
9590	3-1/2"	"	1.250	14.00	120		130
9600	4"	"	1.509	17.00	140		160
26 - 05332	**EXPLOSION PROOF FITTINGS**						**26 - 05332**
1000	Flexible couplings with female unions						
1010	1/2" x 18"	EA	0.200	370	19.25		390
1020	3/4" x 18"	"	0.276	470	26.50		500
1030	1" x 18"	"	0.348	850	33.50		880
1040	1-1/4" x 18"	"	0.421	1,410	40.50		1,450
1050	1-1/2" x 18"	"	0.500	1,720	48.00		1,770
1060	2" x 18"	"	0.571	2,390	55.00		2,440
1070	1/2" x 24"	"	0.250	500	24.00		520
1080	3/4" x 24"	"	0.296	620	28.50		650
1090	1" x 24"	"	0.400	1,130	38.50		1,170
1100	1-1/4" x 24"	"	0.444	1,800	42.75		1,840
1110	1-1/2" x 24"	"	0.571	2,290	55.00		2,340
1120	2" x 24"	"	0.615	3,180	59.00		3,240
1125	Female seal-offs						
1130	1/2"	EA	0.571	29.50	55.00		85.00
1140	3/4"	"	0.667	34.75	64.00		99.00
1150	1"	"	0.727	44.75	70.00		110
1160	1-1/4"	"	0.851	54.00	82.00		140
1170	1-1/2"	"	1.000	82.00	96.00		180
1180	2"	"	1.159	110	110		220
1190	2-1/2"	"	1.739	160	170		330
1200	3"	"	2.162	220	210		430
1210	4"	"	2.667	820	260		1,080
1215	Conduit plugs						
1220	1/2"	EA	0.145	5.41	14.00		19.50
1230	3/4"	"	0.145	5.67	14.00		19.75
1240	1"	"	0.145	6.82	14.00		20.75
1250	1-1/4"	"	0.250	7.44	24.00		31.50
1260	1-1/2"	"	0.250	10.50	24.00		34.50
1270	2"	"	0.296	18.00	28.50		46.50

CONDUCTORS, CONDUIT AND RACEWAYS

ID Code	Description — Component Descriptions	Output — Unit of Meas.	Output — Manhr / Unit	Unit Costs — Material Cost	Unit Costs — Labor Cost	Unit Costs — Equipment Cost	Unit Costs — Total Cost
26 - 05332	**EXPLOSION PROOF FITTINGS, Cont'd...**						**26 - 05332**
1280	2-1/2"	EA	0.296	28.75	28.50		57.00
1290	3"	"	0.348	40.75	33.50		74.00
1300	4"	"	0.348	68.00	33.50		100
1305	Sealing cement						
1310	1 pound	EA					21.75
1320	5 pound	"					61.00
1325	Fiber						
1330	1 ounce	EA					11.25
1340	8 ounce	"					79.00
1345	Male unions						
1350	1/2"	EA	0.200	20.75	19.25		40.00
1360	3/4"	"	0.242	29.00	23.25		52.00
1370	1"	"	0.276	51.00	26.50		78.00
1380	1-1/4"	"	0.296	78.00	28.50		110
1390	1-1/2"	"	0.348	100	33.50		130
1400	2"	"	0.421	130	40.50		170
1410	2-1/2"	"	0.500	200	48.00		250
1420	3"	"	0.727	280	70.00		350
1430	4"	"	0.899	320	86.00		410
1435	Female unions						
1440	1/2"	EA	0.200	14.25	19.25		33.50
1450	3/4"	"	0.242	19.50	23.25		42.75
1460	1"	"	0.276	35.50	26.50		62.00
1470	1-1/4"	"	0.296	53.00	28.50		82.00
1480	1-1/2"	"	0.348	68.00	33.50		100
1490	2"	"	0.421	88.00	40.50		130
1500	2-1/2"	"	0.500	130	48.00		180
1510	3"	"	0.727	180	70.00		250
1520	4"	"	0.899	260	86.00		350
1525	Male elbows						
1530	1/2"	EA	0.250	24.25	24.00		48.25
1540	3/4"	"	0.296	27.00	28.50		56.00
1550	1"	"	0.348	40.50	33.50		74.00
1560	1-1/4"	"	0.444	46.50	42.75		89.00
1565	Female elbows						
1570	1/2"	EA	0.250	20.25	24.00		44.25
1580	3/4"	"	0.296	23.00	28.50		52.00
1590	1"	"	0.348	31.25	33.50		65.00

CONDUCTORS, CONDUIT AND RACEWAYS

ID Code	Description — Component Descriptions	Unit of Meas.	Manhr / Unit	Material Cost	Labor Cost	Equipment Cost	Total Cost
26 - 05332	**EXPLOSION PROOF FITTINGS, Cont'd...**						**26 - 05332**
1600	1-1/4"	EA	0.444	43.50	42.75		86.00
1605	Pulling elbows						
1610	1/2"	EA	0.348	110	33.50		140
1620	3/4"	"	0.444	120	42.75		160
1630	1"	"	0.500	290	48.00		340
1640	1-1/4"	"	0.615	350	59.00		410
1650	1-1/2"	"	0.727	460	70.00		530
1660	2"	"	1.905	480	180		660
1670	2-1/2"	"	2.500	1,070	240		1,310
1680	3"	"	2.963	1,020	280		1,300
1690	3-1/2"	"	3.478	1,930	330		2,260
1700	4"	"	4.211	1,960	400		2,360
1705	Male expansion couplings						
1710	1/2"	EA	0.250	30.25	24.00		54.00
1720	3/4"	"	0.296	39.75	28.50		68.00
1730	1"	"	0.444	74.00	42.75		120
1735	Female expansion couplings						
1740	1/2"	EA	0.250	27.25	24.00		51.00
1750	3/4"	"	0.296	42.75	28.50		71.00
1760	1"	"	0.444	74.00	42.75		120
26 - 05333	**FLEXIBLE WIRING SYSTEMS**						**26 - 05333**
1000	Single circuit cables						
1010	5'	EA	0.059	38.25	5.68		44.00
1020	10'	"	0.100	54.00	9.60		64.00
1030	15'	"	0.145	68.00	14.00		82.00
1040	20'	"	0.200	95.00	19.25		110
1060	25'	"	0.267	120	25.50		150
1070	30'	"	0.296	150	28.50		180
1080	40'	"	0.400	200	38.50		240
1085	Two circuit cables						
1090	5'	EA	0.059	52.00	5.68		58.00
1100	10'	"	0.100	61.00	9.60		71.00
1110	15'	"	0.145	77.00	14.00		91.00
1120	20'	"	0.200	110	19.25		130
1130	25'	"	0.267	140	25.50		170
1140	30'	"	0.296	170	28.50		200
1150	40'	"	0.400	240	38.50		280

CONDUCTORS, CONDUIT AND RACEWAYS

ID Code	Description — Component Descriptions	Output — Unit of Meas.	Manhr / Unit	Unit Costs — Material Cost	Labor Cost	Equipment Cost	Total Cost
26 - 05333	**FLEXIBLE WIRING SYSTEMS, Cont'd...**						**26 - 05333**
1155	Two wire switch and receptacle cables						
1160	5'	EA	0.059	21.50	5.68		27.25
1170	10'	"	0.100	43.50	9.60		53.00
1210	15'	"	0.145	56.00	14.00		70.00
1220	20'	"	0.200	96.00	19.25		120
1230	25'	"	0.267	130	25.50		160
1240	30'	"	0.296	170	28.50		200
1250	40'	"	0.348	220	33.50		250
1255	Three wire switch						
1260	5'	EA	0.059	24.75	5.68		30.50
1270	10'	"	0.100	49.75	9.60		59.00
1280	15'	"	0.145	65.00	14.00		79.00
1290	20'	"	0.200	96.00	19.25		120
1300	25'	"	0.267	130	25.50		160
1310	30'	"	0.296	170	28.50		200
1320	40'	"	0.348	220	33.50		250
1325	Distribution boxes						
1330	2 circuit	EA	0.533	37.00	51.00		88.00
1340	3 circuit	"	0.667	42.50	64.00		110
1350	4 circuit	"	0.800	64.00	77.00		140
1360	6 circuit	"	1.096	85.00	110		200
1370	12 circuit	"	2.222	130	210		340
1380	18 circuit	"	2.963	200	280		480
1385	Tap boxes						
1390	1 single pole switch	EA	0.400	74.00	38.50		110
1400	2 single pole switches	"	0.533	82.00	51.00		130
1410	1 3 way switch	"	0.444	87.00	42.75		130
1420	1 4 way switch	"	0.533	110	51.00		160
1430	1 receptacle	"	0.400	67.00	38.50		110
1440	2 receptacles	"	0.533	84.00	51.00		140
1450	4 receptacles	"	0.800	100	77.00		180
1460	1 clock	"	0.400	84.00	38.50		120
1470	2 clocks	"	0.533	97.00	51.00		150
1480	4 clocks	"	0.800	110	77.00		190
1500	Dust cap	"	0.100	11.00	9.60		20.50
1510	Cable coupler	"	0.200	26.00	19.25		45.25
1520	Reversing connector	"	0.296	37.00	28.50		66.00

CONDUCTORS, CONDUIT AND RACEWAYS

ID Code	Description / Component Descriptions	Output Unit of Meas.	Manhr / Unit	Material Cost	Labor Cost	Equipment Cost	Total Cost
26 - 05334	**SURFACE MOUNTED RACEWAY**						**26 - 05334**
0980	Single Raceway						
1000	3/4" x 17/32" Conduit	LF	0.040	2.87	3.84		6.71
1020	Mounting Strap	EA	0.053	0.77	5.12		5.89
1040	Connector	"	0.053	1.03	5.12		6.15
1060	Elbow						
2000	45 degree	EA	0.050	13.00	4.80		17.75
2020	90 degree	"	0.050	4.17	4.80		8.97
2040	internal	"	0.050	5.25	4.80		10.00
2050	external	"	0.050	4.85	4.80		9.65
2060	Switch	"	0.400	34.00	38.50		73.00
2100	Utility Box	"	0.400	22.75	38.50		61.00
2110	Receptacle	"	0.400	40.25	38.50		79.00
2140	3/4" x 21/32" Conduit	LF	0.040	3.26	3.84		7.10
2160	Mounting Strap	EA	0.053	1.20	5.12		6.32
2180	Connector	"	0.053	1.24	5.12		6.36
2200	Elbow						
2210	45 degree	EA	0.050	9.70	4.80		14.50
2220	90 degree	"	0.050	3.60	4.80		8.40
2240	internal	"	0.050	4.88	4.80		9.68
2260	external	"	0.050	4.88	4.80		9.68
3000	Switch	"	0.400	27.50	38.50		66.00
3010	Utility Box	"	0.400	18.50	38.50		57.00
3020	Receptacle	"	0.400	32.50	38.50		71.00
3040	1-1/4" x 7/8" Conduit	LF	0.040	6.26	3.84		10.00
3050	Mounting Strap	EA	0.053	1.05	5.12		6.17
3060	Connector	"	0.053	1.67	5.12		6.79
3070	Elbow						
3080	90 degree	EA	0.050	13.75	4.80		18.50
3085	internal	"	0.050	11.00	4.80		15.75
3090	external	"	0.050	21.50	4.80		26.25
3100	Switch Box	"	0.400	22.00	38.50		61.00
3110	Receptacle Box	"	0.400	22.00	38.50		61.00
3130	1-29/32" x 7/8" Conduit	LF	0.040	5.86	3.84		9.70
3140	Mounting Strap	EA	0.053	0.97	5.12		6.09
3150	Connector	"	0.053	3.45	5.12		8.57
3160	Elbow						
3165	90 degree	EA	0.050	12.75	4.80		17.50
3170	internal	"	0.050	31.00	4.80		35.75

CONDUCTORS, CONDUIT AND RACEWAYS

ID Code	Description / Component Descriptions	Output		Unit Costs			
		Unit of Meas.	Manhr / Unit	Material Cost	Labor Cost	Equipment Cost	Total Cost
26 - 05334	**SURFACE MOUNTED RACEWAY, Cont'd...**						**26 - 05334**
3180	external	EA	0.050	18.75	4.80		23.50
3190	Switch Box	"	0.400	19.00	38.50		58.00
3200	Receptacle Box	"	0.400	19.00	38.50		58.00
3220	2-3/4" x 1-15/32" Conduit	LF	0.040	9.37	3.84		13.25
3230	Mounting Strap	EA	0.053	2.96	5.12		8.08
3240	Connector	"	0.053	5.84	5.12		11.00
3245	Elbow						
3250	90 degree	EA	0.050	42.75	4.80		47.50
3260	internal	"	0.050	28.75	4.80		33.50
3270	external	"	0.050	36.50	4.80		41.25
3280	Switch Cover	"	0.400	18.00	38.50		57.00
3290	Receptacle Cover	"	0.400	17.00	38.50		56.00
3320	Double Raceway						
3325	5-1/2" x 2" Conduit	LF	0.044	24.25	4.26		28.50
3330	Mounting Strap	EA	0.067	4.44	6.40		10.75
3340	Connector	"	0.067	9.51	6.40		16.00
3350	Elbow						
3360	90 degree	EA	0.057	93.00	5.48		98.00
3370	internal	"	0.057	130	5.48		140
3375	external	"	0.057	140	5.48		150
3380	Receptacle Cover	"	0.400	22.25	38.50		61.00
26 - 05335	**PULL BOXES AND CABINETS**						**26 - 05335**
2500	Galvanized pull boxes, screw cover						
2520	4x4x4	EA	0.190	13.50	18.25		31.75
2540	4x6x4	"	0.190	16.00	18.25		34.25
2560	6x6x4	"	0.190	20.50	18.25		38.75
2580	6x8x4	"	0.190	24.25	18.25		42.50
2620	8x8x4	"	0.250	30.25	24.00		54.00
3020	8x10x4	"	0.242	34.75	23.25		58.00
3040	8x12x4	"	0.250	38.25	24.00		62.00
3050	Screw cover						
3060	10x10x4	EA	0.308	38.25	29.50		68.00
3200	12x12x6	"	0.444	56.00	42.75		99.00
3220	12x15x6	"	0.444	67.00	42.75		110
3240	12x18x6	"	0.500	75.00	48.00		120
3250	15x18x6	"	0.571	83.00	55.00		140
3260	18x24x6	"	0.615	150	59.00		210

CONDUCTORS, CONDUIT AND RACEWAYS

ID Code	Component Descriptions	Unit of Meas.	Manhr / Unit	Material Cost	Labor Cost	Equipment Cost	Total Cost
	Description	**Output**		**Unit Costs**			
26 - 05335	**PULL BOXES AND CABINETS, Cont'd...**						**26 - 05335**
3270	18x30x6	EA	0.727	170	70.00		240
3280	24x36x6	"	0.727	270	70.00		340
3500	Cast iron junction box, unflanged						
3510	6x6x4						
3520	3/4" tap	EA	0.500	130	48.00		180
3540	1" tap	"	0.500	140	48.00		190
3560	Two 1/2" taps	"	0.500	140	48.00		190
3580	3/4" taps	"	0.500	140	48.00		190
3620	6" adapter plate	"	0.348	55.00	33.50		89.00
3640	6" exterior collar	"	0.348	90.00	33.50		120
4000	Screw cover cabinet						
4021	12x12x4	EA	0.615	110	59.00		170
4040	12x16x4	"	0.615	140	59.00		200
4060	12x16x6	"	0.615	160	59.00		220
4080	12x18x4	"	0.667	150	64.00		210
4100	12x18x6	"	0.667	180	64.00		240
4120	18x18x4	"	1.000	180	96.00		280
4140	18x18x6	"	1.000	210	96.00		310
4160	18x24x6	"	1.143	290	110		400
4180	24x24x6	"	1.333	330	130		460
4200	24x36x6	"	1.667	480	160		640
4220	36x48x6	"	2.500	980	240		1,220
5010	NEMA 3R, rain tight screw cover enclosures						
5020	6x6x4	EA	0.211	37.75	20.25		58.00
5030	8x6x4	"	0.296	41.50	28.50		70.00
5040	8x8x4	"	0.296	47.75	28.50		76.00
5050	10x8x4	"	0.400	54.00	38.50		93.00
5060	10x10x4	"	0.400	61.00	38.50		100
5070	12x8x4	"	0.444	70.00	42.75		110
5080	12x12x4	"	0.444	77.00	42.75		120
5090	15x12x4	"	0.533	90.00	51.00		140
5100	8x8x6	"	0.400	56.00	38.50		95.00
5110	10x8x6	"	0.444	64.00	42.75		110
5120	10x10x6	"	0.444	72.00	42.75		110
5130	12x8x6	"	0.533	71.00	51.00		120
5200	12x10x6	"	0.548	80.00	53.00		130
5210	12x12x6	"	0.548	90.00	53.00		140
5220	18x12x6	"	0.702	130	67.00		200

CONDUCTORS, CONDUIT AND RACEWAYS

ID Code	Description — Component Descriptions	Output — Unit of Meas.	Manhr / Unit	Unit Costs — Material Cost	Labor Cost	Equipment Cost	Total Cost
26 - 05336	**WALL DUCT**						**26 - 05336**
1010	Lay-in wall duct, 10"	LF	0.059	170	5.68		180
1020	Horizontal elbow	EA	0.727	820	70.00		890
1030	Edgewise elbow	"	0.727	190	70.00		260
1040	Tee	"	0.899	660	86.00		750
1050	Cross	"	1.096	390	110		500
1060	Cabinet connector	"	1.600	390	150		540
1070	Reverse elbow	"	0.533	340	51.00		390
1080	Sweep elbow	"	0.727	590	70.00		660
1090	Partition	"	0.059	120	5.68		130
1100	Straight tunnel	"	0.276	170	26.50		200
1110	Elbow tunnel	"	0.348	250	33.50		280
1120	Tee kit	"	0.400	250	38.50		290
1130	Ceiling dropout	"	1.000	990	96.00		1,090
1140	Coupling device	"	0.145	110	14.00		120
1150	End cap	"	0.276	140	26.50		170
1160	Lay-in wall duct, 18"	LF	0.080	210	7.68		220
1170	Horizontal elbow	EA	0.800	300	77.00		380
1180	Edgeware elbow	"	1.000	660	96.00		760
1190	Tee	"	1.096	670	110		780
1200	Cross	"	1.250	820	120		940
1210	Reverse elbow	"	0.727	410	70.00		480
1220	Sweep elbow	"	0.800	880	77.00		960
1230	Partition	"	0.100	170	9.60		180
1240	Straight tunnel	"	0.400	210	38.50		250
1250	Elbow tunnel	"	0.500	380	48.00		430
1260	Tee kit	"	0.533	380	51.00		430
1270	Ceiling dropout	"	1.096	990	110		1,100
1280	Coupling device	"	0.200	110	19.25		130
1290	Reducer coupling	"	0.400	250	38.50		290
1300	Cabinet connector	"	2.000	270	190		460
1310	End cap	"	0.400	180	38.50		220
26 - 05337	**WIREWAYS**						**26 - 05337**
0960	Wireway, hinge cover type						
0980	2-1/2" x 2-1/2"						
1000	1' section	EA	0.154	30.75	14.75		45.50
1040	2'	"	0.190	43.75	18.25		62.00
1060	3'	"	0.250	59.00	24.00		83.00

CONDUCTORS, CONDUIT AND RACEWAYS

ID Code	Component Descriptions	Unit of Meas.	Manhr / Unit	Material Cost	Labor Cost	Equipment Cost	Total Cost
	Description	**Output**		**Unit Costs**			
26 - 05337		**WIREWAYS, Cont'd...**				**26 - 05337**	
1080	5'	EA	0.381	100	36.50		140
1100	10'	"	0.667	200	64.00		260
1110	4" x 4"						
1120	1'	EA	0.250	33.50	24.00		58.00
1140	2'	"	0.250	49.00	24.00		73.00
1160	3'	"	0.308	73.00	29.50		100
1180	4'	"	0.308	100	29.50		130
1200	10'	"	0.800	300	77.00		380
1210	6" x 6"						
1220	1'	EA	0.381	64.00	36.50		100
1240	2'	"	0.381	79.00	36.50		120
1260	3'	"	0.444	110	42.75		150
1280	4'	"	0.444	150	42.75		190
1300	5'	"	0.571	160	55.00		210
1320	10'	"	0.889	320	85.00		400
1340	8" x 8"						
1360	1'	EA	0.444	100	42.75		140
1380	2'	"	0.444	160	42.75		200
1400	3'	"	0.500	220	48.00		270
1420	4'	"	0.500	260	48.00		310
1440	5'	"	0.615	330	59.00		390
1450	12" x 12"						
1460	1'	EA	0.615	140	59.00		200
1480	2'	"	0.615	220	59.00		280
1500	3'	"	0.727	310	70.00		380
1600	4'	"	0.727	360	70.00		430
1620	5'	"	0.889	440	85.00		530
2000	Fittings						
2005	2-1/2" x 2-1/2"						
2010	Drop hanger	EA	0.123	38.75	11.75		51.00
2020	Bracket hanger	"	0.123	28.50	11.75		40.25
2030	Panel adapter	"	0.500	34.25	48.00		82.00
2040	End plate	"	0.123	12.75	11.75		24.50
2050	U-connector	"	0.123	24.00	11.75		35.75
2060	Tee	"	0.200	160	19.25		180
2070	Cross	"	0.250	100	24.00		120
2080	90 degree elbow	"	0.200	81.00	19.25		100
2090	Sweep elbow	"	0.200	140	19.25		160

CONDUCTORS, CONDUIT AND RACEWAYS

ID Code	Component Descriptions	Unit of Meas.	Manhr / Unit	Material Cost	Labor Cost	Equipment Cost	Total Cost
26 - 05337	**WIREWAYS, Cont'd...**						**26 - 05337**
2100	45 degree elbow	EA	0.200	80.00	19.25		99.00
2110	Lay-in adapter	"	0.145	34.00	14.00		48.00
2115	4" x 4"						
2120	Drop hanger	EA	0.145	25.50	14.00		39.50
2130	Bracket hanger	"	0.145	34.25	14.00		48.25
2140	Panel adapter	"	0.615	40.25	59.00		99.00
2150	End plate	"	0.145	12.25	14.00		26.25
2160	U-connector	"	0.145	22.75	14.00		36.75
2170	Tee	"	0.250	180	24.00		200
2180	Cross	"	0.348	120	33.50		150
2190	90 degree elbow	"	0.250	98.00	24.00		120
2200	Sweep elbow	"	0.250	160	24.00		180
2210	45 degree elbow	"	0.250	97.00	24.00		120
2220	Lay-in adapter	"	0.200	68.00	19.25		87.00
2225	6" x 6"						
2230	Drop hanger	EA	0.145	57.00	14.00		71.00
2240	Bracket hanger	"	0.145	48.75	14.00		63.00
2250	Reducing bushing	"	0.145	100	14.00		110
2260	Panel adapter	"	0.727	53.00	70.00		120
2270	End plate	"	0.145	15.00	14.00		29.00
2280	U-connector	"	0.145	32.50	14.00		46.50
2290	Tee	"	0.250	210	24.00		230
2300	Cross	"	0.348	150	33.50		180
2310	90 degree elbow	"	0.250	110	24.00		130
2320	Sweep elbow	"	0.250	230	24.00		250
2330	45 degree elbow	"	0.250	110	24.00		130
2340	Lay-in adapter	"	0.250	90.00	24.00		110
2345	8" x 8"						
2350	Drop hanger	EA	0.145	54.00	14.00		68.00
2360	Bracket hanger	"	0.145	47.00	14.00		61.00
2370	Reducing bushing	"	0.145	110	14.00		120
2380	Panel adapter	"	0.899	65.00	86.00		150
2390	End plate	"	0.145	9.04	14.00		23.00
2400	U-connector	"	0.145	26.00	14.00		40.00
2410	Tee	"	0.400	210	38.50		250
2420	Cross	"	0.444	220	42.75		260
2430	90 degree elbow	"	0.400	160	38.50		200
2440	Sweep elbow	"	0.400	130	38.50		170

CONDUCTORS, CONDUIT AND RACEWAYS

ID Code	Description / Component Descriptions	Output Unit of Meas.	Output Manhr / Unit	Material Cost	Labor Cost	Equipment Cost	Total Cost
26 - 05337	**WIREWAYS, Cont'd...**						**26 - 05337**
2450	45 degree elbow	EA	0.400	150	38.50		190
2460	Lay-in adapter	"	0.250	81.00	24.00		110
2465	10" x 10"						
2470	Drop hanger	EA	0.145	55.00	14.00		69.00
2480	Bracket hanger	"	0.145	56.00	14.00		70.00
2490	Reducing bushing	"	0.145	160	14.00		170
2500	Panel adapter	"	1.000	100	96.00		200
2510	End plate	"	0.145	14.50	14.00		28.50
2520	U-connector	"	0.145	66.00	14.00		80.00
2530	Tee	"	0.444	200	42.75		240
2540	Cross	"	0.500	290	48.00		340
2550	90 degree elbow	"	0.444	170	42.75		210
2560	Sweep elbow	"	0.444	160	42.75		200
2570	45 degree elbow	"	0.444	200	42.75		240
2580	Lay-in adapter	"	0.296	92.00	28.50		120
2585	12" x 12"						
2590	Drop hanger	EA	0.145	150	14.00		160
2600	Bracket hanger	"	0.145	110	14.00		120
2610	Reducing bushing	"	0.145	160	14.00		170
2620	Panel adapter	"	1.250	110	120		230
2630	End plate	"	0.145	20.25	14.00		34.25
2640	U-connector	"	0.615	98.00	59.00		160
2650	Tee	"	0.533	410	51.00		460
2660	Cross	"	0.615	410	59.00		470
2670	90 degree elbow	"	0.615	260	59.00		320
2680	Sweep elbow	"	0.615	380	59.00		440
2690	45 degree elbow	"	0.615	410	59.00		470
2700	Lay-in adapter	"	0.533	100	51.00		150
2705	Raintight wireway, 4" x 4"						
2710	1' section	EA	0.400	120	38.50		160
2720	5'	"	0.400	200	38.50		240
2730	10'	"	1.000	290	96.00		390
2735	Fittings						
2740	90 degree elbow	EA	0.400	120	38.50		160
2750	Tee	"	0.444	210	42.75		250
2760	Cross	"	0.500	210	48.00		260
2770	Panel adapter	"	0.727	54.00	70.00		120
2780	End plate	"	0.145	27.75	14.00		41.75

CONDUCTORS, CONDUIT AND RACEWAYS

	Description	Output		Unit Costs			
ID Code	Component Descriptions	Unit of Meas.	Manhr / Unit	Material Cost	Labor Cost	Equipment Cost	Total Cost
26 - 05337	**WIREWAYS, Cont'd...**						**26 - 05337**
2790	Gusset bracket	EA	0.145	20.50	14.00		34.50
2795	6" x 6"						
2800	1' section	EA	0.500	96.00	48.00		140
2810	5'	"	0.727	230	70.00		300
2820	10'	"	1.509	400	140		540
2825	Fittings						
2830	90 degree elbow	EA	0.500	140	48.00		190
2840	Tee	"	0.500	240	48.00		290
2850	Cross	"	0.727	250	70.00		320
2860	Panel adapter	"	1.000	71.00	96.00		170
2870	End plate	"	0.145	31.50	14.00		45.50
2880	Gusset bracket	"	0.145	26.00	14.00		40.00
26 - 05338	**BOXES**						**26 - 05338**
5000	Round cast box, type SEH						
5010	1/2"	EA	0.348	35.50	33.50		69.00
5020	3/4"	"	0.421	35.50	40.50		76.00
5025	SEHC						
5030	1/2"	EA	0.348	42.25	33.50		76.00
5040	3/4"	"	0.421	42.25	40.50		83.00
5045	SEHL						
5050	1/2"	EA	0.348	43.50	33.50		77.00
5060	3/4"	"	0.444	42.25	42.75		85.00
5065	SEHT						
5070	1/2"	EA	0.421	46.75	40.50		87.00
5080	3/4"	"	0.500	46.75	48.00		95.00
5085	SEHX						
5090	1/2"	EA	0.500	50.00	48.00		98.00
5100	3/4"	"	0.615	50.00	59.00		110
5110	Blank cover	"	0.145	8.58	14.00		22.50
5120	1/2", hub cover	"	0.145	8.20	14.00		22.25
5130	Cover with gasket	"	0.178	8.97	17.00		26.00
5135	Rectangle, type FS boxes						
5140	1/2"	EA	0.348	18.25	33.50		52.00
5150	3/4"	"	0.400	19.25	38.50		58.00
5160	1"	"	0.500	21.00	48.00		69.00
5165	FSA						
5170	1/2"	EA	0.348	32.75	33.50		66.00

CONDUCTORS, CONDUIT AND RACEWAYS

ID Code	Component Descriptions	Unit of Meas.	Manhr / Unit	Material Cost	Labor Cost	Equipment Cost	Total Cost
	Description	**Output**		**Unit Costs**			
26 - 05338		**BOXES, Cont'd...**				**26 - 05338**	
5180	3/4"	EA	0.400	30.50	38.50		69.00
5185	FSC						
5190	1/2"	EA	0.348	20.25	33.50		54.00
5200	3/4"	"	0.421	22.25	40.50		63.00
5210	1"	"	0.500	27.75	48.00		76.00
5215	FSL						
5220	1/2"	EA	0.348	32.50	33.50		66.00
5230	3/4"	"	0.400	32.50	38.50		71.00
5235	FSR						
5240	1/2"	EA	0.348	34.00	33.50		68.00
5250	3/4"	"	0.400	34.50	38.50		73.00
5255	FSS						
5260	1/2"	EA	0.348	20.25	33.50		54.00
5270	3/4"	"	0.400	22.25	38.50		61.00
5275	FSLA						
5280	1/2"	EA	0.348	14.00	33.50		47.50
5290	3/4"	"	0.400	15.75	38.50		54.00
5295	FSCA						
5300	1/2"	EA	0.348	41.00	33.50		75.00
5310	3/4"	"	0.400	39.50	38.50		78.00
5315	FSCC						
5320	1/2"	EA	0.400	24.75	38.50		63.00
5330	3/4"	"	0.500	37.00	48.00		85.00
5335	FSCT						
5340	1/2"	EA	0.400	24.75	38.50		63.00
5350	3/4"	"	0.500	31.00	48.00		79.00
5360	1"	"	0.571	25.25	55.00		80.00
5365	FST						
5370	1/2"	EA	0.500	36.50	48.00		85.00
5380	3/4"	"	0.571	36.50	55.00		92.00
5385	FSX						
5390	1/2"	EA	0.615	41.75	59.00		100
5400	3/4"	"	0.727	38.50	70.00		110
5405	FSCD boxes						
5410	1/2"	EA	0.615	34.75	59.00		94.00
5420	3/4"	"	0.727	36.50	70.00		110
5425	Rectangle, type FS, 2 gang boxes						
5430	1/2"	EA	0.348	39.00	33.50		73.00

CONDUCTORS, CONDUIT AND RACEWAYS

ID Code	Description Component Descriptions	Output Unit of Meas.	Manhr / Unit	Material Cost	Labor Cost	Equipment Cost	Total Cost
26 - 05338		**BOXES, Cont'd...**					**26 - 05338**
5440	3/4"	EA	0.400	40.00	38.50		79.00
5450	1"	"	0.500	42.25	48.00		90.00
5455	FSC, 2 gang boxes						
5460	1/2"	EA	0.348	41.25	33.50		75.00
5470	3/4"	"	0.400	45.75	38.50		84.00
5480	1"	"	0.500	55.00	48.00		100
5485	FSS, 2 gang boxes						
5490	3/4"	EA	0.400	43.25	38.50		82.00
5495	FS, tandem boxes						
5500	1/2"	EA	0.400	43.00	38.50		82.00
5510	3/4"	"	0.444	44.25	42.75		87.00
5515	FSC, tandem boxes						
5520	1/2"	EA	0.400	58.00	38.50		97.00
5530	3/4"	"	0.444	62.00	42.75		100
5535	FS, three gang boxes						
5540	3/4"	EA	0.444	63.00	42.75		110
5550	1"	"	0.500	70.00	48.00		120
5560	FSS, three gang boxes, 3/4"	"	0.500	82.00	48.00		130
5565	Weatherproof cast aluminum boxes, 1 gang, 3 outlets						
5570	1/2"	EA	0.400	11.50	38.50		50.00
5580	3/4"	"	0.500	12.50	48.00		61.00
5585	2 gang, 3 outlets						
5590	1/2"	EA	0.500	22.25	48.00		70.00
5600	3/4"	"	0.533	23.75	51.00		75.00
5605	1 gang, 4 outlets						
5610	1/2"	EA	0.615	20.25	59.00		79.00
5620	3/4"	"	0.727	22.25	70.00		92.00
5625	2 gang, 4 outlets						
5630	1/2"	EA	0.615	21.25	59.00		80.00
5640	3/4"	"	0.727	23.50	70.00		94.00
5645	1 gang, 5 outlets						
5650	1/2"	EA	0.727	16.75	70.00		87.00
5660	3/4"	"	0.800	20.00	77.00		97.00
5665	2 gang, 5 outlets						
5670	1/2"	EA	0.727	30.25	70.00		100
5680	3/4"	"	0.800	36.75	77.00		110
5685	2 gang, 6 outlets						
5690	1/2"	EA	0.851	34.25	82.00		120

CONDUCTORS, CONDUIT AND RACEWAYS

ID Code	Component Descriptions	Unit of Meas.	Manhr / Unit	Material Cost	Labor Cost	Equipment Cost	Total Cost
	Description	**Output**		**Unit Costs**			
26 - 05338		**BOXES, Cont'd...**					**26 - 05338**
5700	3/4"	EA	0.899	36.75	86.00		120
5705	2 gang, 7 outlets						
5710	1/2"	EA	1.000	36.50	96.00		130
5720	3/4"	"	1.096	45.25	110		160
5730	Weatherproof and type FS box covers, blank, 1 gang	"	0.145	5.28	14.00		19.25
5740	Tumbler switch, 1 gang	"	0.145	10.75	14.00		24.75
5750	1 gang, single recept	"	0.145	6.83	14.00		20.75
5760	Duplex recept	"	0.145	8.71	14.00		22.75
5770	Despard	"	0.145	8.75	14.00		22.75
5780	Red pilot light	"	0.145	41.25	14.00		55.00
5785	SW and						
5790	Single recept	EA	0.200	18.25	19.25		37.50
5800	Duplex recept	"	0.200	15.00	19.25		34.25
5805	2 gang						
5810	Blank	EA	0.182	5.49	17.50		23.00
5820	Tumbler switch	"	0.182	7.21	17.50		24.75
5830	Single recept	"	0.182	7.21	17.50		24.75
5840	Duplex recept	"	0.182	7.21	17.50		24.75
5845	3 gang						
5850	Blank	EA	0.200	12.50	19.25		31.75
5860	Tumbler switch	"	0.200	15.50	19.25		34.75
5865	4 gang						
5870	Tumbler switch	EA	0.250	19.75	24.00		43.75
5875	Explosion proof boxes type E						
5880	1/2"	EA	0.348	71.00	33.50		100
5890	3/4"	"	0.400	69.00	38.50		110
5900	1"	"	0.500	71.00	48.00		120
5910	1-1/4"	"	0.571	120	55.00		180
5920	1-1/2"	"	0.615	280	59.00		340
5925	Type LB						
5930	1/2"	EA	0.400	68.00	38.50		110
5940	3/4"	"	0.500	72.00	48.00		120
5950	1"	"	0.571	75.00	55.00		130
5960	1-1/4"	"	0.667	130	64.00		190
5970	1-1/2"	"	0.727	260	70.00		330
5980	2"	"	0.800	300	77.00		380
5985	Type C						
5990	1/2"	EA	0.400	61.00	38.50		100

CONDUCTORS, CONDUIT AND RACEWAYS

ID Code	Description — Component Descriptions	Output — Unit of Meas.	Manhr / Unit	Material Cost	Labor Cost	Equipment Cost	Total Cost
26 - 05338	**BOXES, Cont'd...**						**26 - 05338**
6000	3/4"	EA	0.500	64.00	48.00		110
6010	1"	"	0.571	66.00	55.00		120
6020	1-1/4"	"	0.667	110	64.00		170
6030	1-1/2"	"	0.727	230	70.00		300
6040	2"	"	0.800	250	77.00		330
6045	Type CA						
6050	1/2"	EA	0.571	59.00	55.00		110
6060	3/4"	"	0.727	67.00	70.00		140
6065	Type L						
6070	1/2"	EA	0.400	62.00	38.50		100
6080	3/4"	"	0.500	66.00	48.00		110
6090	1"	"	0.571	68.00	55.00		120
6100	1-1/4"	"	0.667	120	64.00		180
6110	1-1/2"	"	0.727	230	70.00		300
6120	2"	"	0.800	230	77.00		310
6125	Type N						
6130	1/2"	EA	0.400	63.00	38.50		100
6140	3/4"	"	0.500	65.00	48.00		110
6150	1"	"	0.615	67.00	59.00		130
6160	1-1/4"	"	0.667	120	64.00		180
6165	Type T						
6170	1/2"	EA	0.533	60.00	51.00		110
6180	3/4"	"	0.727	64.00	70.00		130
6190	1"	"	0.851	66.00	82.00		150
6200	1-1/4"	"	1.000	110	96.00		210
6210	1-1/2"	"	1.159	230	110		340
6220	2"	"	1.290	230	120		350
6225	Type TA						
6230	1/2"	EA	0.727	69.00	70.00		140
6240	3/4"	"	0.800	73.00	77.00		150
6245	Type X						
6250	1/2"	EA	0.727	63.00	70.00		130
6260	3/4"	"	0.851	67.00	82.00		150
6270	1"	"	1.000	74.00	96.00		170
6280	1-1/4"	"	1.159	130	110		240
6290	1-1/2"	"	1.290	250	120		370
6300	2"	"	1.455	270	140		410
6305	With union hubs						

CONDUCTORS, CONDUIT AND RACEWAYS

ID Code	Description / Component Descriptions	Output Unit of Meas.	Manhr / Unit	Material Cost	Labor Cost	Equipment Cost	Total Cost
26 - 05338	**BOXES, Cont'd...**						**26 - 05338**
6310	1/2"	EA	0.727	130	70.00		200
6320	3/4"	"	0.800	130	77.00		210
6325	Box covers						
6330	Surface	EA	0.200	27.75	19.25		47.00
6340	Sealing	"	0.200	30.25	19.25		49.50
6350	Dome	"	0.200	41.75	19.25		61.00
6360	1/2" nipple	"	0.200	53.00	19.25		72.00
6370	3/4" nipple	"	0.200	55.00	19.25		74.00
26 - 05339	**PULL AND JUNCTION BOXES**						**26 - 05339**
1050	4"						
1060	Octagon box	EA	0.114	5.84	11.00		16.75
1070	Box extension	"	0.059	9.85	5.68		15.50
1080	Plaster ring	"	0.059	5.40	5.68		11.00
1100	Cover blank	"	0.059	2.38	5.68		8.06
1120	Square box	"	0.114	8.42	11.00		19.50
1140	Box extension	"	0.059	8.25	5.68		14.00
1160	Plaster ring	"	0.059	4.51	5.68		10.25
1180	Cover blank	"	0.059	2.31	5.68		7.99
1190	4-11/16"						
1200	Square box	EA	0.114	17.00	11.00		28.00
1240	Box extension	"	0.059	18.50	5.68		24.25
1260	Plaster ring	"	0.059	11.25	5.68		17.00
1280	Cover blank	"	0.059	4.17	5.68		9.85
1300	Switch and device boxes						
1320	2 gang	EA	0.114	25.50	11.00		36.50
1340	3 gang	"	0.114	44.75	11.00		56.00
1360	4 gang	"	0.160	60.00	15.25		75.00
2000	Device covers						
2020	2 gang	EA	0.059	20.25	5.68		26.00
2040	3 gang	"	0.059	21.00	5.68		26.75
2060	4 gang	"	0.059	28.50	5.68		34.25
2100	Handy box	"	0.114	6.27	11.00		17.25
2120	Extension	"	0.059	5.91	5.68		11.50
2140	Switch cover	"	0.059	3.13	5.68		8.81
2160	Switch box with knockout	"	0.145	9.42	14.00		23.50
2200	Weatherproof cover, spring type	"	0.080	17.50	7.68		25.25
2220	Cover plate, dryer receptacle 1 gang plastic	"	0.100	2.67	9.60		12.25

CONDUCTORS, CONDUIT AND RACEWAYS

ID Code	Component Descriptions	Unit of Meas.	Manhr / Unit	Material Cost	Labor Cost	Equipment Cost	Total Cost
		Output		**Unit Costs**			
26 - 05339	**PULL AND JUNCTION BOXES, Cont'd...**						**26 - 05339**
2240	For 4" receptacle, 2 gang	EA	0.100	4.77	9.60		14.25
2260	Duplex receptacle cover plate, plastic	"	0.059	1.17	5.68		6.85
3005	4", vertical bracket box, 1-1/2" with						
3010	RMX clamps	EA	0.145	12.25	14.00		26.25
3020	BX clamps	"	0.145	13.00	14.00		27.00
3025	4", octagon device cover						
3030	1 switch	EA	0.059	7.12	5.68		12.75
3040	1 duplex recept	"	0.059	7.12	5.68		12.75
3050	4", octagon swivel hanger box, 1/2" hub	"	0.059	19.00	5.68		24.75
3060	3/4" hub	"	0.059	21.75	5.68		27.50
3065	4" octagon adjustable bar hangers						
3070	18-1/2"	EA	0.050	8.84	4.80		13.75
3080	26-1/2"	"	0.050	9.66	4.80		14.50
3085	With clip						
3090	18-1/2"	EA	0.050	6.54	4.80		11.25
3100	26-1/2"	"	0.050	7.34	4.80		12.25
3105	4", square face bracket boxes, 1-1/2"						
3110	RMX	EA	0.145	14.50	14.00		28.50
3120	BX	"	0.145	15.75	14.00		29.75
3130	4" square to round plaster rings	"	0.059	4.82	5.68		10.50
3140	2 gang device plaster rings	"	0.059	4.97	5.68		10.75
3145	Surface covers						
3150	1 gang switch	EA	0.059	4.34	5.68		10.00
3160	2 gang switch	"	0.059	4.44	5.68		10.00
3170	1 single recept	"	0.059	6.54	5.68		12.25
3180	1 20a twist lock recept	"	0.059	8.18	5.68		13.75
3190	1 30a twist lock recept	"	0.059	10.50	5.68		16.25
3200	1 duplex recept	"	0.059	4.05	5.68		9.73
3210	2 duplex recept	"	0.059	4.05	5.68		9.73
3220	Switch and duplex recept	"	0.059	6.76	5.68		12.50
3230	4-11/16" square to round plaster rings	"	0.059	11.25	5.68		17.00
3240	2 gang device plaster rings	"	0.059	9.27	5.68		15.00
3245	Surface covers						
3250	1 gang switch	EA	0.059	12.50	5.68		18.25
3260	2 gang switch	"	0.059	19.25	5.68		25.00
3270	1 single recept	"	0.059	17.25	5.68		23.00
3280	1 20a twist lock recept	"	0.059	17.00	5.68		22.75
3290	1 30a twist lock recept	"	0.059	21.25	5.68		27.00

CONDUCTORS, CONDUIT AND RACEWAYS

ID Code	Component Descriptions	Unit of Meas.	Manhr / Unit	Material Cost	Labor Cost	Equipment Cost	Total Cost
	Description	**Output**		**Unit Costs**			

26 - 05339   **PULL AND JUNCTION BOXES, Cont'd...**   **26 - 05339**

ID Code	Component Descriptions	Unit of Meas.	Manhr / Unit	Material Cost	Labor Cost	Equipment Cost	Total Cost
3300	1 duplex recept	EA	0.059	18.50	5.68		24.25
3310	2 duplex recept	"	0.059	16.25	5.68		22.00
3320	Switch and duplex recept	"	0.059	28.00	5.68		33.75
3325	4" plastic round boxes, ground straps						
3330	Box only	EA	0.145	3.08	14.00		17.00
3340	Box w/clamps	"	0.200	3.59	19.25		22.75
3350	Box w/16" bar	"	0.229	7.62	22.00		29.50
3360	Box w/24" bar	"	0.250	7.60	24.00		31.50
3370	4" plastic round box covers						
3380	Blank cover	EA	0.059	2.01	5.68		7.69
3390	Plaster ring	"	0.059	3.29	5.68		8.97
3395	4" plastic square boxes						
3400	Box only	EA	0.145	2.38	14.00		16.50
3410	Box w/clamps	"	0.200	2.94	19.25		22.25
3420	Box w/hanger	"	0.250	3.63	24.00		27.75
3430	Box w/nails and clamp	"	0.250	5.19	24.00		29.25
3435	4" plastic square box covers						
3440	Blank cover	EA	0.059	1.96	5.68		7.64
3450	1 gang ring	"	0.059	2.38	5.68		8.06
3460	2 gang ring	"	0.059	3.34	5.68		9.02
3470	Round ring	"	0.059	2.65	5.68		8.33

26 - 05340   **FLOOR BOXES**   **26 - 05340**

ID Code	Component Descriptions	Unit of Meas.	Manhr / Unit	Material Cost	Labor Cost	Equipment Cost	Total Cost
1010	Adjustable floor boxes, steel	EA	0.533	37.75	51.00		89.00
1020	Cast bronze round						
1040	1 gang	EA	0.800	56.00	77.00		130
1050	2 gang	"	0.952	100	91.00		190
1060	3 gang	"	1.000	150	96.00		250
1065	Aluminum round						
1070	1 gang	EA	0.800	64.00	77.00		140
1080	2 gang	"	0.952	78.00	91.00		170
1090	3 gang	"	1.000	93.00	96.00		190
1100	Steel plate single recept	"	0.145	21.25	14.00		35.25
1110	Duplex receptacle	"	0.182	20.25	17.50		37.75
1120	Twist lock receptacle	"	0.182	21.50	17.50		39.00
1130	Plug, 3/4"	"	0.145	27.00	14.00		41.00
1140	1" plug	"	0.145	25.25	14.00		39.25
1150	Carpet flange	"	0.145	33.00	14.00		47.00

CONDUCTORS, CONDUIT AND RACEWAYS

ID Code	Description / Component Descriptions	Output / Unit of Meas.	Manhr / Unit	Material Cost	Labor Cost	Equipment Cost	Total Cost
26 - 05340	**FLOOR BOXES, Cont'd...**						**26 - 05340**
1155	Adjustable bronze plates for round cast boxes						
1160	1/2" plug	EA	0.145	11.00	14.00		25.00
1170	3/4" plug	"	0.145	11.00	14.00		25.00
1180	1" plug	"	0.145	13.75	14.00		27.75
1190	1-1/4" plug	"	0.182	15.50	17.50		33.00
1200	2" plug	"	0.200	20.50	19.25		39.75
1210	Combination plug	"	0.200	24.00	19.25		43.25
1220	Duplex receptacle plug	"	0.200	40.50	19.25		60.00
1225	Adjustable aluminum plates for round cast boxes						
1230	1/2" plug	EA	0.145	32.50	14.00		46.50
1240	3/4" plug	"	0.145	33.00	14.00		47.00
1250	1" plug	"	0.145	33.75	14.00		47.75
1260	1-1/4" plug	"	0.182	36.25	17.50		54.00
1270	2" plug	"	0.200	36.50	19.25		56.00
1280	Combination plug	"	0.200	32.75	19.25		52.00
1290	Duplex receptacle plug	"	0.200	55.00	19.25		74.00
1295	Adjustable bronze plates for gang type boxes						
1300	1/2" plug	EA	0.145	34.00	14.00		48.00
1310	3/4" plug	"	0.145	34.50	14.00		48.50
1320	1" plug	"	0.145	35.00	14.00		49.00
1330	1-1/4" plug	"	0.182	36.50	17.50		54.00
1340	2" plug	"	0.200	38.00	19.25		57.00
1350	Carpet plate						
1370	1 gang	EA	0.145	30.00	14.00		44.00
1380	2 gang	"	0.145	45.50	14.00		60.00
1390	3 gang	"	0.200	61.00	19.25		80.00
1395	Adjustable aluminum plates for gang type boxes						
1400	1/2" plug	EA	0.145	30.50	14.00		44.50
1410	3/4" plug	"	0.145	31.25	14.00		45.25
1420	1" plug	"	0.145	31.50	14.00		45.50
1430	1-1/4" plug	"	0.182	36.25	17.50		54.00
1440	2" plug	"	0.200	36.25	19.25		56.00
1460	Duplex recept	"	0.200	30.75	19.25		50.00
1465	Carpet plate						
1470	1 gang	EA	0.145	67.00	14.00		81.00
1480	2 gang	"	0.145	94.00	14.00		110
1490	3 gang	"	0.200	150	19.25		170
1500	4 gang carpet plate	"	0.571	49.50	55.00		100

CONDUCTORS, CONDUIT AND RACEWAYS

ID Code	Description / Component Descriptions	Output / Unit of Meas.	Output / Manhr / Unit	Unit Costs / Material Cost	Unit Costs / Labor Cost	Unit Costs / Equipment Cost	Unit Costs / Total Cost
26 - 05340	**FLOOR BOXES, Cont'd...**						**26 - 05340**
1520	Telephone	EA	0.500	45.25	48.00		93.00
1525	Floor box nozzles, horizontal						
1530	Duplex recept	EA	0.533	77.00	51.00		130
1540	Single recept	"	0.533	100	51.00		150
1550	Double duplex recept	"	0.727	84.00	70.00		150
1560	Vertical with duplex recept	"	0.615	68.00	59.00		130
1570	Double duplex recept	"	0.727	72.00	70.00		140
1580	Floor box bell nozzles split bell	"	0.250	24.75	24.00		48.75
1590	One piece bell	"	0.250	78.00	24.00		100
1595	Floor box standpipe						
1600	1/2" x 3"	EA	0.145	21.50	14.00		35.50
1610	1/2" x 1"	"	0.145	21.25	14.00		35.25
1630	Poke thru floor outlets						
1640	2" floor	EA	1.000	66.00	96.00		160
1650	3" floor	"	1.194	72.00	110		180
1660	4" floor	"	1.290	74.00	120		190
1670	7" floor	"	1.509	78.00	140		220
1680	9" floor	"	1.600	110	150		260
1690	11" floor	"	1.818	110	170		280
1700	13" floor	"	2.000	120	190		310
26 - 05341	**ALUMINUM CONDUIT**						**26 - 05341**
1010	Aluminum conduit						
1020	1/2"	LF	0.030	3.16	2.89		6.05
1040	3/4"	"	0.040	4.06	3.84		7.90
1060	1"	"	0.050	5.69	4.80		10.50
1080	1-1/4"	"	0.059	7.61	5.68		13.25
1100	1-1/2"	"	0.080	9.45	7.68		17.25
1120	2"	"	0.089	12.50	8.53		21.00
1140	2-1/2"	"	0.100	20.00	9.60		29.50
1160	3"	"	0.107	26.25	10.25		36.50
1180	3-1/2"	"	0.123	31.50	11.75		43.25
1200	4"	"	0.145	37.25	14.00		51.00
1220	5"	"	0.182	53.00	17.50		71.00
1240	6"	"	0.200	70.00	19.25		89.00
1490	90 deg. elbow						
1500	1/2"	EA	0.190	24.50	18.25		42.75
1510	3/4"	"	0.250	33.25	24.00		57.00

CONDUCTORS, CONDUIT AND RACEWAYS

	Description		Output		Unit Costs			
ID Code	Component Descriptions		Unit of Meas.	Manhr / Unit	Material Cost	Labor Cost	Equipment Cost	Total Cost
26 - 05341		**ALUMINUM CONDUIT, Cont'd...**					**26 - 05341**	
1520	1"		EA	0.308	46.25	29.50		76.00
1530	1-1/4"		"	0.381	73.00	36.50		110
1540	1-1/2"		"	0.400	98.00	38.50		140
1550	2"		"	0.444	140	42.75		180
1560	2-1/2"		"	0.571	240	55.00		300
1570	3"		"	0.667	380	64.00		440
1580	3-1/2"		"	0.800	580	77.00		660
1590	4"		"	0.889	750	85.00		830
1600	5"		"	1.143	1,900	110		2,010
1610	6"		"	2.222	2,590	210		2,800
1980	Coupling							
2000	1/2"		EA	0.050	4.97	4.80		9.77
2020	3/4"		"	0.059	7.53	5.68		13.25
2040	1"		"	0.080	9.95	7.68		17.75
2060	1-1/4"		"	0.089	12.00	8.53		20.50
2080	1-1/2"		"	0.100	14.00	9.60		23.50
2100	2"		"	0.107	19.75	10.25		30.00
2120	2-1/2"		"	0.123	45.00	11.75		57.00
2140	3"		"	0.123	59.00	11.75		71.00
2160	3-1/2"		"	0.145	80.00	14.00		94.00
2180	4"		"	0.160	97.00	15.25		110
2200	5"		"	0.160	290	15.25		310
2220	6"		"	0.190	460	18.25		480
26 - 05342		**EMT CONDUIT**					**26 - 05342**	
0080	EMT conduit							
0100	1/2"		LF	0.030	0.73	2.89		3.62
1020	3/4"		"	0.040	1.34	3.84		5.18
1030	1"		"	0.050	2.24	4.80		7.04
1040	1-1/4"		"	0.059	3.58	5.68		9.26
1060	1-1/2"		"	0.080	4.54	7.68		12.25
1080	2"		"	0.089	5.67	8.53		14.25
1100	2-1/2"		"	0.100	11.25	9.60		20.75
1120	3"		"	0.123	12.50	11.75		24.25
1140	3-1/2"		"	0.145	17.75	14.00		31.75
1160	4"		"	0.182	17.00	17.50		34.50
2980	90 deg. elbow							
3000	1/2"		EA	0.089	6.92	8.53		15.50

CONDUCTORS, CONDUIT AND RACEWAYS

ID Code	Description / Component Descriptions	Output Unit of Meas.	Manhr / Unit	Material Cost	Labor Cost	Equipment Cost	Total Cost
26 - 05342	**EMT CONDUIT, Cont'd...**						**26 - 05342**
3040	3/4"	EA	0.100	7.60	9.60		17.25
3060	1"	"	0.107	11.75	10.25		22.00
3080	1-1/4"	"	0.123	14.50	11.75		26.25
3100	1-1/2"	"	0.145	17.00	14.00		31.00
3120	2"	"	0.190	24.75	18.25		43.00
3140	2-1/2"	"	0.211	61.00	20.25		81.00
3160	3"	"	0.242	90.00	23.25		110
3170	3-1/2"	"	0.258	120	26.75		150
3180	4"	"	0.286	140	27.50		170
3980	Connector, steel compression						
4000	1/2"	EA	0.089	2.12	8.53		10.75
4040	3/4"	"	0.089	4.05	8.53		12.50
4060	1"	"	0.089	6.10	8.53		14.75
4080	1-1/4"	"	0.107	13.75	10.25		24.00
4100	1-1/2"	"	0.145	20.00	14.00		34.00
4200	2"	"	0.190	29.00	18.25		47.25
4220	2-1/2"	"	0.250	81.00	24.00		110
4240	3"	"	0.286	110	27.50		140
4260	3-1/2"	"	0.308	160	29.50		190
4280	4"	"	0.333	170	32.00		200
4480	Coupling, steel, compression						
4500	1/2"	EA	0.059	3.60	5.68		9.28
4540	3/4"	"	0.059	4.90	5.68		10.50
4560	1"	"	0.059	7.41	5.68		13.00
4580	1-1/4"	"	0.089	13.50	8.53		22.00
4590	1-1/2"	"	0.107	21.00	10.25		31.25
4600	2"	"	0.145	26.25	14.00		40.25
4620	2-1/2"	"	0.222	95.00	21.25		120
4640	3"	"	0.250	120	24.00		140
4660	3-1/2"	"	0.286	180	27.50		210
4680	4"	"	0.308	180	29.50		210
4980	1 hole strap, steel						
5000	1/2"	EA	0.040	0.24	3.84		4.08
5040	3/4"	"	0.040	0.29	3.84		4.13
5060	1"	"	0.040	0.45	3.84		4.29
5080	1-1/4"	"	0.050	0.71	4.80		5.51
5100	1-1/2"	"	0.050	1.09	4.80		5.89
5120	2"	"	0.050	1.72	4.80		6.52

CONDUCTORS, CONDUIT AND RACEWAYS

ID Code	Description Component Descriptions	Output Unit of Meas.	Output Manhr / Unit	Unit Costs Material Cost	Unit Costs Labor Cost	Unit Costs Equipment Cost	Unit Costs Total Cost
26 - 05342	**EMT CONDUIT, Cont'd...**						**26 - 05342**
5140	2-1/2"	EA	0.059	3.59	5.68		9.27
5160	3"	"	0.059	4.01	5.68		9.69
5180	3-1/2"	"	0.059	6.06	5.68		11.75
5200	4"	"	0.059	7.56	5.68		13.25
6000	Connector, steel set screw						
6010	1/2"	EA	0.070	1.62	6.67		8.29
6020	3/4"	"	0.070	2.59	6.67		9.26
6030	1"	"	0.070	4.47	6.67		11.25
6040	1-1/4"	"	0.107	9.52	10.25		19.75
6050	1-1/2"	"	0.145	13.75	14.00		27.75
6060	2"	"	0.182	19.75	17.50		37.25
6070	2-1/2"	"	0.242	64.00	23.25		87.00
6080	3"	"	0.267	76.00	25.50		100
6090	3-1/2"	"	0.296	100	28.50		130
6100	4"	"	0.348	110	33.50		140
6105	Insulated throat						
6110	1/2"	EA	0.070	2.14	6.67		8.81
6120	3/4"	"	0.070	3.48	6.67		10.25
6130	1"	"	0.070	5.75	6.67		12.50
6140	1-1/4"	"	0.107	11.50	10.25		21.75
6150	1-1/2"	"	0.145	16.75	14.00		30.75
6160	2"	"	0.182	24.25	17.50		41.75
6170	2-1/2"	"	0.242	110	23.25		130
6180	3"	"	0.267	140	25.50		170
6190	3-1/2"	"	0.296	180	28.50		210
6200	4"	"	0.348	200	33.50		230
6205	Connector, die cast set screw						
6210	1/2"	EA	0.059	1.10	5.68		6.78
6220	3/4"	"	0.059	1.89	5.68		7.57
6230	1"	"	0.059	3.55	5.68		9.23
6240	1-1/4"	"	0.089	6.22	8.53		14.75
6250	1-1/2"	"	0.107	8.44	10.25		18.75
6260	2"	"	0.145	9.90	14.00		24.00
6270	2-1/2"	"	0.200	22.50	19.25		41.75
6280	3"	"	0.222	27.00	21.25		48.25
6290	3-1/2"	"	0.250	32.25	24.00		56.00
6300	4"	"	0.286	40.75	27.50		68.00
6305	Insulated throat						

CONDUCTORS, CONDUIT AND RACEWAYS

ID Code	Component Descriptions	Unit of Meas.	Manhr / Unit	Material Cost	Labor Cost	Equipment Cost	Total Cost
	Description	**Output**		**Unit Costs**			

26 - 05342   EMT CONDUIT, Cont'd...   **26 - 05342**

ID Code	Component Descriptions	Unit of Meas.	Manhr / Unit	Material Cost	Labor Cost	Equipment Cost	Total Cost
6310	1/2"	EA	0.059	2.36	5.68		8.04
6320	3/4"	"	0.059	3.81	5.68		9.49
6330	1"	"	0.059	6.63	5.68		12.25
6340	1-1/4"	"	0.089	12.75	8.53		21.25
6350	1-1/2"	"	0.107	18.25	10.25		28.50
6360	2"	"	0.145	26.50	14.00		40.50
6370	2-1/2"	"	0.200	120	19.25		140
6380	3"	"	0.222	150	21.25		170
6390	3-1/2"	"	0.250	200	25.75		230
6400	4"	"	0.286	220	27.50		250
6405	Coupling, steel set screw						
6410	1/2"	EA	0.040	2.91	3.84		6.75
6420	3/4"	"	0.040	4.39	3.84		8.23
6430	1"	"	0.040	7.12	3.84		11.00
6440	1-1/4"	"	0.050	15.25	4.80		20.00
6450	1-1/2"	"	0.080	21.75	7.68		29.50
6460	2"	"	0.107	29.25	10.25		39.50
6470	2-1/2"	"	0.160	82.00	15.25		97.00
6480	3"	"	0.190	90.00	18.25		110
6490	3-1/2"	"	0.222	100	21.25		120
6500	4"	"	0.250	110	24.00		130
6505	Diecast set screw						
6510	1/2"	EA	0.040	1.02	3.84		4.86
6520	3/4"	"	0.040	1.64	3.84		5.48
6530	1"	"	0.040	2.72	3.84		6.56
6540	1-1/4"	"	0.050	4.74	4.80		9.54
6550	1-1/2"	"	0.080	6.70	7.68		14.50
6560	2"	"	0.107	9.00	10.25		19.25
6570	2-1/2"	"	0.160	21.00	15.25		36.25
6580	3"	"	0.186	24.50	17.75		42.25
6590	3-1/2"	"	0.222	31.00	21.25		52.00
6600	4"	"	0.250	36.75	24.00		61.00
6605	1 hole malleable straps						
6610	1/2"	EA	0.040	0.47	3.84		4.31
6620	3/4"	"	0.040	0.64	3.84		4.48
6630	1"	"	0.040	1.06	3.84		4.90
6640	1-1/4"	"	0.050	1.86	4.80		6.66
6650	1-1/2"	"	0.050	2.27	4.80		7.07

CONDUCTORS, CONDUIT AND RACEWAYS

ID Code	Description Component Descriptions	Output Unit of Meas.	Manhr / Unit	Unit Costs Material Cost	Labor Cost	Equipment Cost	Total Cost
26 - 05342	**EMT CONDUIT, Cont'd...**						**26 - 05342**
6660	2"	EA	0.050	4.12	4.80		8.92
6670	2-1/2"	"	0.059	9.25	5.68		15.00
6680	3"	"	0.059	13.50	5.68		19.25
6690	3-1/2"	"	0.059	19.75	5.68		25.50
6700	4"	"	0.059	43.50	5.68		49.25
6705	EMT to rigid compression coupling						
6710	1/2"	EA	0.100	5.50	9.60		15.00
6720	3/4"	"	0.100	7.85	9.60		17.50
6730	1"	"	0.150	12.00	14.50		26.50
6735	Set screw couplings						
6740	1/2"	EA	0.100	1.44	9.60		11.00
6750	3/4"	"	0.100	2.17	9.60		11.75
6760	1"	"	0.145	3.63	14.00		17.75
6765	Set screw offset connectors						
6770	1/2"	EA	0.100	3.22	9.60		12.75
6780	3/4"	"	0.100	4.33	9.60		14.00
6790	1"	"	0.145	7.86	14.00		21.75
6795	Compression offset connectors						
6800	1/2"	EA	0.100	5.32	9.60		15.00
6810	3/4"	"	0.100	6.74	9.60		16.25
6820	1"	"	0.145	9.75	14.00		23.75
6825	Type LB set screw condulets						
6830	1/2"	EA	0.229	15.25	22.00		37.25
6840	3/4"	"	0.296	18.50	28.50		47.00
6850	1"	"	0.381	28.25	36.50		65.00
6860	1-1/4"	"	0.444	42.50	42.75		85.00
6870	1-1/2"	"	0.533	56.00	51.00		110
6880	2"	"	0.615	92.00	59.00		150
6890	2-1/2"	"	0.727	150	70.00		220
6900	3"	"	1.000	200	96.00		300
6910	3-1/2"	"	1.333	300	130		430
6920	4"	"	1.600	370	150		520
6925	Type T set screw condulets						
6930	1/2"	EA	0.296	19.00	28.50		47.50
6940	3/4"	"	0.400	23.75	38.50		62.00
6950	1"	"	0.444	34.25	42.75		77.00
6960	1-1/4"	"	0.533	49.25	51.00		100
6970	1-1/2"	"	0.615	61.00	59.00		120

CONDUCTORS, CONDUIT AND RACEWAYS

ID Code	Component Descriptions	Unit of Meas.	Manhr / Unit	Material Cost	Labor Cost	Equipment Cost	Total Cost
26 - 05342	**EMT CONDUIT, Cont'd...**						**26 - 05342**
6980	2"	EA	0.667	80.00	64.00		140
6985	Type C set screw condulets						
6990	1/2"	EA	0.250	15.75	24.00		39.75
7000	3/4"	"	0.296	20.00	28.50		48.50
7010	1"	"	0.381	29.50	36.50		66.00
7020	1-1/4"	"	0.444	51.00	42.75		94.00
7030	1-1/2"	"	0.533	62.00	51.00		110
7040	2"	"	0.381	110	36.50		150
7045	Type LL set screw condulets						
7050	1/2"	EA	0.250	15.75	24.00		39.75
7060	3/4"	"	0.296	19.50	28.50		48.00
7070	1"	"	0.381	29.50	36.50		66.00
7080	1-1/4"	"	0.444	37.25	42.75		80.00
7090	1-1/2"	"	0.533	47.25	51.00		98.00
7100	2"	"	0.615	81.00	59.00		140
7105	Type LR set screw condulets						
7110	1/2"	EA	0.250	15.75	24.00		39.75
7120	3/4"	"	0.296	19.50	28.50		48.00
7130	1"	"	0.381	29.50	36.50		66.00
7140	1-1/4"	"	0.444	36.25	42.75		79.00
7150	1-1/2"	"	0.533	59.00	51.00		110
7160	2"	"	0.615	79.00	59.00		140
7165	Type LB compression condulets						
7170	1/2"	EA	0.296	35.75	28.50		64.00
7180	3/4"	"	0.500	53.00	48.00		100
7190	1"	"	0.500	68.00	48.00		120
7195	Type T compression condulets						
7200	1/2"	EA	0.400	48.00	38.50		87.00
7210	3/4"	"	0.444	63.00	42.75		110
7220	1"	"	0.615	98.00	59.00		160
7225	Condulet covers						
7230	1/2"	EA	0.123	2.20	11.75		14.00
7240	3/4"	"	0.123	2.67	11.75		14.50
7250	1"	"	0.123	3.64	11.75		15.50
7260	1-1/4"	"	0.123	4.21	11.75		16.00
7270	1-1/2"	"	0.145	4.41	14.00		18.50
7280	2"	"	0.145	7.77	14.00		21.75
7290	2-1/2"	"	0.145	9.51	14.00		23.50

CONDUCTORS, CONDUIT AND RACEWAYS

ID Code	Description — Component Descriptions	Unit of Meas.	Manhr / Unit	Material Cost	Labor Cost	Equipment Cost	Total Cost
26 - 05342	**EMT CONDUIT, Cont'd...**						**26 - 05342**
7300	3"	EA	0.182	12.25	17.50		29.75
7310	3-1/2"	"	0.182	14.00	17.50		31.50
7320	4"	"	0.182	14.50	17.50		32.00
7325	Clamp type entrance caps						
7330	1/2"	EA	0.250	11.75	24.00		35.75
7340	3/4"	"	0.296	13.75	28.50		42.25
7350	1"	"	0.400	16.25	38.50		55.00
7360	1-1/4"	"	0.533	18.50	51.00		70.00
7370	1-1/2"	"	0.615	31.25	59.00		90.00
7380	2"	"	0.899	42.75	86.00		130
7390	2-1/2"	"	1.000	72.00	96.00		170
7400	3"	"	1.509	110	140		250
7410	3-1/2"	"	1.739	160	170		330
7420	4"	"	2.222	240	210		450
7425	Slip fitter type entrance caps						
7430	1/2"	EA	0.250	8.62	24.00		32.50
7440	3/4"	"	0.296	10.25	28.50		38.75
7450	1"	"	0.400	12.50	38.50		51.00
7460	1-1/4"	"	0.533	14.00	51.00		65.00
7470	1-1/2"	"	0.615	24.00	59.00		83.00
7480	2"	"	0.899	36.00	86.00		120
7490	2-1/2"	"	1.000	59.00	96.00		160
7500	3"	"	1.509	91.00	140		230
7510	3-1/2"	"	1.739	140	180		320
7520	4"	"	2.222	210	210		420
26 - 05343	**FLEXIBLE CONDUIT**						**26 - 05343**
0080	Flexible conduit, steel						
0100	3/8"	LF	0.030	0.93	2.89		3.82
1020	1/2	"	0.030	1.05	2.89		3.94
1040	3/4"	"	0.040	1.44	3.84		5.28
1060	1"	"	0.040	2.74	3.84		6.58
1080	1-1/4"	"	0.050	3.43	4.80		8.23
1100	1-1/2"	"	0.059	5.68	5.68		11.25
1120	2"	"	0.080	7.00	7.68		14.75
1140	2-1/2"	"	0.089	8.51	8.53		17.00
1160	3"	"	0.107	14.75	10.25		25.00
1200	Flexible conduit, liquid tight						

CONDUCTORS, CONDUIT AND RACEWAYS

ID Code	Component Descriptions	Unit of Meas.	Manhr / Unit	Material Cost	Labor Cost	Equipment Cost	Total Cost
	Description	**Output**		**Unit Costs**			

ID Code	Component Descriptions	Unit of Meas.	Manhr / Unit	Material Cost	Labor Cost	Equipment Cost	Total Cost
26 - 05343	**FLEXIBLE CONDUIT, Cont'd...**					**26 - 05343**	
1210	3/8"	LF	0.030	2.58	2.89		5.47
1220	1/2"	"	0.030	2.91	2.89		5.80
1230	3/4"	"	0.040	3.97	3.84		7.81
1240	1"	"	0.040	5.98	3.84		9.82
1250	1-1/4"	"	0.050	8.22	4.80		13.00
1260	1-1/2"	"	0.059	10.25	5.68		16.00
1270	2"	EA	0.080	13.25	7.68		21.00
1280	2-1/2"	"	0.089	23.50	8.53		32.00
1290	3"	"	0.107	32.50	10.25		42.75
1300	4"	"	0.145	49.25	14.00		63.00
2000	Connector, straight						
2020	3/8"	EA	0.080	4.35	7.68		12.00
2040	1/2"	"	0.080	4.66	7.68		12.25
2060	3/4"	"	0.089	5.93	8.53		14.50
2080	1"	"	0.100	10.50	9.60		20.00
2100	1-1/4"	"	0.107	15.25	10.25		25.50
2120	1-1/2"	"	0.123	22.00	11.75		33.75
2140	2"	"	0.145	36.50	14.00		51.00
2160	2-1/2"	"	0.182	54.00	17.50		72.00
2180	3"	"	0.190	79.00	18.25		97.00
4380	Straight insulated throat connectors						
4400	3/8"	EA	0.123	5.28	11.75		17.00
4440	1/2"	"	0.123	5.28	11.75		17.00
4450	3/4"	"	0.145	7.72	14.00		21.75
4460	1"	"	0.145	12.00	14.00		26.00
4470	1-1/4"	"	0.182	18.25	17.50		35.75
4480	1-1/2"	"	0.211	26.00	20.25		46.25
4490	2"	"	0.229	49.00	22.00		71.00
4500	2-1/2"	"	0.267	270	25.50		300
4510	3"	"	0.333	300	32.00		330
4520	4"	"	0.421	360	40.50		400
4525	90 deg connectors						
4530	3/8"	EA	0.148	7.37	14.25		21.50
4540	1/2"	"	0.148	7.37	14.25		21.50
4550	3/4"	"	0.170	11.75	16.25		28.00
4560	1"	"	0.182	22.75	17.50		40.25
4570	1-1/4"	"	0.229	35.25	22.00		57.00
4580	1-1/2"	"	0.250	42.25	24.00		66.00

CONDUCTORS, CONDUIT AND RACEWAYS

ID Code	Description — Component Descriptions	Output — Unit of Meas.	Manhr / Unit	Unit Costs — Material Cost	Labor Cost	Equipment Cost	Total Cost
26 - 05343	**FLEXIBLE CONDUIT, Cont'd...**						**26 - 05343**
4590	2"	EA	0.267	64.00	25.50		90.00
4600	2-1/2"	"	0.333	280	32.00		310
4610	3"	"	0.381	340	36.50		380
4620	4"	"	0.444	440	42.75		480
4625	90 degree insulated throat connectors						
4630	3/8"	EA	0.145	9.12	14.00		23.00
4640	1/2"	"	0.145	9.12	14.00		23.00
4650	3/4"	"	0.170	13.75	16.25		30.00
4660	1"	"	0.178	26.00	17.00		43.00
4670	1-1/4"	"	0.229	39.75	22.00		62.00
4680	1-1/2"	"	0.250	48.75	24.00		73.00
4690	2"	"	0.267	73.00	25.50		99.00
4700	2-1/2"	"	0.333	370	32.00		400
4710	3"	"	0.381	440	36.50		480
4720	4"	"	0.444	580	42.75		620
4800	Flexible aluminum conduit						
4810	3/8"	LF	0.030	0.54	2.89		3.43
4820	1/2"	"	0.030	0.64	2.89		3.53
4830	3/4"	"	0.040	0.89	3.84		4.73
4840	1"	"	0.040	1.67	3.84		5.51
4850	1-1/4"	"	0.050	2.27	4.80		7.07
4860	1-1/2"	"	0.059	3.74	5.68		9.42
4870	2"	"	0.080	4.54	7.68		12.25
4880	2-1/2"	"	0.089	6.00	8.53		14.50
4890	3"	"	0.107	10.25	10.25		20.50
4900	3-1/2"	"	0.123	11.75	11.75		23.50
4910	4"	"	0.145	13.00	14.00		27.00
5000	Connector, straight						
5020	3/8"	EA	0.100	1.66	9.60		11.25
5040	1/2"	"	0.100	2.31	9.60		12.00
5060	3/4"	"	0.107	2.52	10.25		12.75
5080	1"	"	0.123	9.24	11.75		21.00
5100	1-1/4"	"	0.145	12.75	14.00		26.75
5120	1-1/2"	"	0.182	24.75	17.50		42.25
5140	2"	"	0.190	34.75	18.25		53.00
5160	2-1/2"	"	0.222	49.25	21.25		71.00
5180	3"	"	0.276	82.00	26.50		110
5220	4"	"	0.348	270	33.50		300

CONDUCTORS, CONDUIT AND RACEWAYS

ID Code	Description Component Descriptions	Output Unit of Meas.	Output Manhr / Unit	Unit Costs Material Cost	Unit Costs Labor Cost	Unit Costs Equipment Cost	Unit Costs Total Cost
26 - 05343	**FLEXIBLE CONDUIT, Cont'd...**						**26 - 05343**
6115	Straight insulated throat connectors						
6120	3/8"	EA	0.089	1.62	8.53		10.25
6130	1/2"	"	0.089	3.25	8.53		11.75
6140	3/4"	"	0.089	3.45	8.53		12.00
6150	1"	"	0.100	8.40	9.60		18.00
6160	1-1/4"	"	0.107	13.50	10.25		23.75
6170	1-1/2"	"	0.123	18.75	11.75		30.50
6180	2"	"	0.145	30.25	14.00		44.25
6190	2-1/2"	"	0.182	56.00	17.50		74.00
6200	3"	"	0.190	73.00	18.25		91.00
6210	3-1/2"	"	0.222	260	21.25		280
6220	4"	"	0.276	310	26.50		340
6225	90 deg connectors						
6230	3/8"	EA	0.145	2.66	14.00		16.75
6240	1/2"	"	0.145	4.48	14.00		18.50
6250	3/4"	"	0.145	7.04	14.00		21.00
6260	1"	"	0.170	12.25	16.25		28.50
6270	1-1/4"	"	0.182	24.75	17.50		42.25
6280	1-1/2"	"	0.200	40.25	19.25		60.00
6290	2"	"	0.211	50.00	20.25		70.00
6300	2-1/2"	"	0.229	160	22.00		180
6310	3"	"	0.267	220	25.50		250
6315	90 deg insulated throat connectors						
6320	3/8"	EA	0.145	3.29	14.00		17.25
6330	1/2"	"	0.145	4.98	14.00		19.00
6340	3/4"	"	0.145	8.31	14.00		22.25
6350	1"	"	0.170	13.75	16.25		30.00
6360	1-1/4"	"	0.182	26.25	17.50		43.75
6370	1-1/2"	"	0.200	46.75	19.25		66.00
6380	2"	"	0.211	60.00	20.25		80.00
6390	2-1/2"	"	0.229	160	22.00		180
6400	3"	"	0.267	230	25.50		260
6410	3-1/2"	"	0.333	650	32.00		680
6420	4"	"	0.421	990	40.50		1,030

CONDUCTORS, CONDUIT AND RACEWAYS

ID Code	Description Component Descriptions	Output Unit of Meas.	Output Manhr / Unit	Unit Costs Material Cost	Unit Costs Labor Cost	Unit Costs Equipment Cost	Unit Costs Total Cost
26 - 05344	**GALVANIZED CONDUIT**						**26 - 05344**
1980	Galvanized rigid steel conduit						
2000	1/2"	LF	0.040	3.74	3.84		7.58
2040	3/4"	"	0.050	4.14	4.80		8.94
2060	1"	"	0.059	5.97	5.68		11.75
2080	1-1/4"	"	0.080	8.26	7.68		16.00
2100	1-1/2"	"	0.089	9.71	8.53		18.25
2120	2"	"	0.100	12.25	9.60		21.75
2140	2-1/2"	"	0.145	22.75	14.00		36.75
2160	3"	"	0.182	23.50	17.50		41.00
2180	3-1/2"	"	0.190	34.00	18.25		52.00
2200	4"	"	0.211	38.75	20.25		59.00
2220	5"	"	0.286	72.00	27.50		100
2240	6"	"	0.381	100	36.50		140
2480	90 degree ell						
2500	1/2"	EA	0.250	9.93	24.00		34.00
2540	3/4"	"	0.308	10.50	29.50		40.00
2560	1"	"	0.381	16.00	36.50		53.00
2580	1-1/4"	"	0.444	22.00	42.75		65.00
2590	1-1/2"	"	0.500	27.00	48.00		75.00
2600	2"	"	0.533	39.25	51.00		90.00
2620	2-1/2"	"	0.667	84.00	64.00		150
2640	3"	"	0.889	120	85.00		200
2660	3-1/2"	"	1.000	180	96.00		280
2680	4"	"	1.333	210	130		340
2700	5"	"	2.222	580	210		790
3100	6"	"	3.333	870	320		1,190
3200	Couplings, with set screws						
3220	1/2"	EA	0.050	4.97	4.80		9.77
3260	3/4"	"	0.059	6.56	5.68		12.25
3280	1"	"	0.080	10.50	7.68		18.25
3300	1-1/4"	"	0.100	17.75	9.60		27.25
3320	1-1/2"	"	0.123	23.00	11.75		34.75
3340	2"	"	0.145	52.00	14.00		66.00
3360	2-1/2"	"	0.190	130	18.25		150
3380	3"	"	0.250	150	24.00		170
3400	3-1/2"	"	0.286	210	27.50		240
3420	4"	"	0.308	270	29.50		300
3422	5"	"	0.444	400	42.75		440

CONDUCTORS, CONDUIT AND RACEWAYS

ID Code	Description / Component Descriptions	Output / Unit of Meas.	Output / Manhr / Unit	Unit Costs / Material Cost	Unit Costs / Labor Cost	Unit Costs / Equipment Cost	Unit Costs / Total Cost
26 - 05344	**GALVANIZED CONDUIT, Cont'd...**						**26 - 05344**
3426	6"	EA	0.500	530	48.00		580
3440	Split couplings						
3460	1/2"	EA	0.190	4.23	18.25		22.50
3480	3/4"	"	0.250	5.50	24.00		29.50
3500	1"	"	0.276	7.72	26.50		34.25
3520	1-1/4"	"	0.308	15.25	29.50		44.75
3540	1-1/2"	"	0.381	19.75	36.50		56.00
3560	2"	"	0.571	45.50	55.00		100
3580	2-1/2"	"	0.571	92.00	55.00		150
3600	3"	"	0.727	140	70.00		210
3620	3-1/2"	"	1.000	210	96.00		310
3640	4"	"	1.333	250	130		380
3660	5"	"	1.633	440	160		600
3680	6"	"	2.051	590	200		790
3780	Erickson couplings						
3800	1/2"	EA	0.444	6.04	42.75		48.75
3840	3/4"	"	0.500	7.39	48.00		55.00
3850	1"	"	0.615	13.00	59.00		72.00
3860	1-1/4"	"	0.889	23.25	85.00		110
3870	1-1/2"	"	1.000	30.25	96.00		130
3880	2"	"	1.333	58.00	130		190
3890	2-1/2"	"	1.860	120	180		300
3900	3"	"	2.105	200	200		400
3910	3-1/2"	"	2.500	310	240		550
3920	4"	"	2.667	380	260		640
3940	5"	"	2.963	750	280		1,030
3950	6"	"	3.200	1,090	310		1,400
4000	Seal fittings						
4020	1/2"	EA	0.667	20.00	64.00		84.00
4040	3/4"	"	0.800	22.00	77.00		99.00
4060	1"	"	1.000	28.00	96.00		120
4080	1-1/4"	"	1.143	29.00	110		140
4100	1-1/2"	"	1.333	51.00	130		180
4120	2"	"	1.600	56.00	150		210
4140	2-1/2"	"	1.905	89.00	180		270
4160	3"	"	2.105	110	200		310
4180	3-1/2"	"	2.500	290	240		530
4200	4"	"	2.963	430	280		710

CONDUCTORS, CONDUIT AND RACEWAYS

	Description	Output		Unit Costs			
ID Code	Component Descriptions	Unit of Meas.	Manhr / Unit	Material Cost	Labor Cost	Equipment Cost	Total Cost
26 - 05344	**GALVANIZED CONDUIT, Cont'd...**						**26 - 05344**
4220	5"	EA	4.444	670	430		1,100
4240	6"	"	5.000	740	480		1,220
4980	Entrance fitting (weatherhead), threaded						
5000	1/2"	EA	0.444	9.26	42.75		52.00
5040	3/4"	"	0.500	11.25	48.00		59.00
5060	1"	"	0.571	14.50	55.00		70.00
5080	1-1/4"	"	0.727	19.00	70.00		89.00
5100	1-1/2"	"	0.800	33.25	77.00		110
5120	2"	"	0.889	51.00	85.00		140
5140	2-1/2"	"	1.000	180	96.00		280
5160	3"	"	1.333	250	130		380
5180	3-1/2"	"	1.739	320	170		490
5200	4"	"	2.500	420	240		660
5220	5"	"	3.478	440	330		770
5240	6"	"	4.444	550	430		980
5980	Locknuts						
6000	1/2"	EA	0.050	0.23	4.80		5.03
6040	3/4"	"	0.050	0.30	4.80		5.10
6060	1"	"	0.050	0.47	4.80		5.27
6080	1-1/4"	"	0.050	0.65	4.80		5.45
6100	1-1/2"	"	0.059	1.08	5.68		6.76
6120	2"	"	0.059	1.56	5.68		7.24
6140	2-1/2"	"	0.080	4.36	7.68		12.00
6160	3"	"	0.080	5.59	7.68		13.25
6180	3-1/2"	"	0.080	9.45	7.68		17.25
6200	4"	"	0.089	11.75	8.53		20.25
6220	5"	"	0.089	25.25	8.53		33.75
6240	6"	"	0.089	43.00	8.53		52.00
6250	Plastic conduit bushings						
6260	1/2"	EA	0.123	0.47	11.75		12.25
6300	3/4"	"	0.145	0.72	14.00		14.75
6320	1"	"	0.190	1.01	18.25		19.25
6340	1-1/4"	"	0.222	1.33	21.25		22.50
6360	1-1/2"	"	0.250	1.80	24.00		25.75
6380	2"	"	0.308	4.08	29.50		33.50
6400	2-1/2"	"	0.500	7.83	48.00		56.00
6420	3"	"	0.667	9.09	64.00		73.00
6460	3-1/2"	"	0.800	10.25	77.00		87.00

CONDUCTORS, CONDUIT AND RACEWAYS

ID Code	Description / Component Descriptions	Output / Unit of Meas.	Output / Manhr / Unit	Unit Costs / Material Cost	Unit Costs / Labor Cost	Unit Costs / Equipment Cost	Unit Costs / Total Cost
26 - 05344	**GALVANIZED CONDUIT, Cont'd...**						**26 - 05344**
6480	4"	EA	0.889	14.00	85.00		99.00
6500	5"	"	1.143	26.75	110		140
6600	6"	"	1.600	51.00	150		200
6660	Conduit bushings, steel						
6680	1/2"	EA	0.123	0.65	11.75		12.50
6720	3/4"	"	0.145	0.82	14.00		14.75
6740	1"	"	0.190	1.26	18.25		19.50
6760	1-1/4"	"	0.222	1.80	21.25		23.00
6780	1-1/2"	"	0.250	2.57	24.00		26.50
6800	2"	"	0.308	5.22	29.50		34.75
6820	2-1/2"	"	0.500	9.15	48.00		57.00
6840	3"	"	0.667	11.25	64.00		75.00
6860	3-1/2"	"	0.800	23.50	77.00		100
6880	4"	"	0.889	28.50	85.00		110
6890	5"	"	1.143	59.00	110		170
6900	6"	"	1.600	110	150		260
7500	Pipe cap						
7520	1/2"	EA	0.050	0.60	4.80		5.40
7540	3/4"	"	0.050	0.66	4.80		5.46
7560	1"	"	0.050	1.06	4.80		5.86
7580	1-1/4"	"	0.080	1.82	7.68		9.50
7600	1-1/2"	"	0.080	2.84	7.68		10.50
7620	2"	"	0.080	3.20	7.68		11.00
7640	2-1/2"	"	0.089	5.44	8.53		14.00
7660	3"	"	0.089	6.74	8.53		15.25
7680	3-1/2"	"	0.089	9.24	8.53		17.75
7700	4"	"	0.107	11.75	10.25		22.00
7720	5"	"	0.145	15.75	14.00		29.75
7740	6"	"	0.200	19.75	19.25		39.00
8000	GRS elbows, 36" radius						
8010	2"	EA	0.667	190	64.00		250
8020	2-1/2"	"	0.808	260	78.00		340
8030	3"	"	1.053	350	100		450
8040	3-1/2"	"	1.250	470	120		590
8050	4"	"	1.509	520	140		660
8060	5"	"	2.500	850	240		1,090
8070	6"	"	3.810	900	370		1,270
8075	42" radius						

CONDUCTORS, CONDUIT AND RACEWAYS

ID Code	Description — Component Descriptions	Unit of Meas.	Manhr / Unit	Material Cost	Labor Cost	Equipment Cost	Total Cost
26 - 05344	**GALVANIZED CONDUIT, Cont'd...**						**26 - 05344**
8080	2"	EA	0.808	210	78.00		290
8090	2-1/2"	"	1.000	300	96.00		400
8100	3"	"	1.250	380	120		500
8110	3-1/2"	"	1.509	540	140		680
8120	4"	"	1.739	640	170		810
8130	5"	"	2.857	960	270		1,230
8140	6"	"	4.000	1,010	380		1,390
8145	48" radius						
8150	2"	EA	0.930	240	89.00		330
8160	2-1/2"	"	1.127	330	110		440
8170	3"	"	1.429	430	140		570
8180	3-1/2"	"	1.739	610	170		780
8190	4"	"	2.162	730	210		940
8200	5"	"	3.077	1,080	300		1,380
8210	6"	"	4.444	1,130	430		1,560
8215	Threaded couplings						
8220	1/2"	EA	0.050	2.28	4.80		7.08
8230	3/4"	"	0.059	2.79	5.68		8.47
8240	1"	"	0.080	4.15	7.68		11.75
8250	1-1/4"	"	0.089	5.19	8.53		13.75
8260	1-1/2"	"	0.100	6.37	9.60		16.00
8270	2"	"	0.107	8.67	10.25		19.00
8280	2-1/2"	"	0.123	21.50	11.75		33.25
8290	3"	"	0.145	27.75	14.00		41.75
8300	3-1/2"	"	0.145	37.25	14.00		51.00
8310	4"	"	0.160	37.50	15.25		53.00
8320	5"	"	0.182	86.00	17.50		100
8330	6"	"	0.190	120	18.25		140
8335	Threadless couplings						
8340	1/2"	EA	0.100	9.28	9.60		19.00
8350	3/4"	"	0.123	9.67	11.75		21.50
8360	1"	"	0.145	12.50	14.00		26.50
8370	1-1/4"	"	0.190	14.50	18.25		32.75
8380	1-1/2"	"	0.250	17.50	24.00		41.50
8390	2"	"	0.308	25.50	29.50		55.00
8400	2-1/2"	"	0.500	63.00	48.00		110
8410	3"	"	0.615	76.00	59.00		140
8420	3-1/2"	"	0.808	98.00	78.00		180

CONDUCTORS, CONDUIT AND RACEWAYS

ID Code	Description — Component Descriptions	Output — Unit of Meas.	Manhr / Unit	Unit Costs — Material Cost	Labor Cost	Equipment Cost	Total Cost

26 - 05344 GALVANIZED CONDUIT, Cont'd... **26 - 05344**

ID Code	Description	Unit of Meas.	Manhr / Unit	Material Cost	Labor Cost	Equipment Cost	Total Cost
8430	4"	EA	1.000	110	96.00		210
8440	5"	"	1.250	370	120		490
8450	6"	"	5.333	420	510		930
8455	Threadless connectors						
8460	1/2"	EA	0.100	4.42	9.60		14.00
8470	3/4"	"	0.123	7.06	11.75		18.75
8480	1"	"	0.145	11.25	14.00		25.25
8490	1-1/4"	"	0.190	19.00	18.25		37.25
8500	1-1/2"	"	0.250	29.00	24.00		53.00
8510	2"	"	0.308	55.00	29.50		85.00
8520	2-1/2"	"	0.500	130	48.00		180
8530	3"	"	0.615	170	59.00		230
8540	3-1/2"	"	0.808	220	78.00		300
8550	4"	"	1.000	270	96.00		370
8560	5"	"	1.250	730	120		850
8570	6"	"	1.509	960	140		1,100
8575	Setscrew connectors						
8580	1/2"	EA	0.080	3.22	7.68		11.00
8590	3/4"	"	0.089	4.47	8.53		13.00
8600	1"	"	0.100	6.97	9.60		16.50
8610	1-1/4"	"	0.123	12.25	11.75		24.00
8620	1-1/2"	"	0.145	17.75	14.00		31.75
8630	2"	"	0.190	35.50	18.25		54.00
8640	2-1/2"	"	0.250	110	24.00		130
8650	3"	"	0.308	150	29.50		180
8660	3-1/2"	"	0.381	200	36.50		240
8670	4"	"	0.500	250	48.00		300
8680	5"	"	0.615	420	59.00		480
8690	6"	"	0.808	540	78.00		620
8695	Clamp type entrance caps						
8700	1/2"	EA	0.308	13.00	29.50		42.50
8710	3/4"	"	0.381	15.50	36.50		52.00
8720	1"	"	0.444	21.50	42.75		64.00
8730	1-1/4"	"	0.500	25.00	48.00		73.00
8740	1-1/2"	"	0.615	45.25	59.00		100
8750	2"	"	0.727	54.00	70.00		120
8760	2-1/2"	"	0.941	210	90.00		300
8770	3"	"	1.127	320	110		430

CONDUCTORS, CONDUIT AND RACEWAYS

	Description	Output		Unit Costs			
ID Code	Component Descriptions	Unit of Meas.	Manhr / Unit	Material Cost	Labor Cost	Equipment Cost	Total Cost
26 - 05344	**GALVANIZED CONDUIT, Cont'd...**						**26 - 05344**
8780	3-1/2"	EA	1.379	390	130		520
8790	4"	"	2.424	590	230		820
8795	LB condulets						
8800	1/2"	EA	0.308	15.50	29.50		45.00
8810	3/4"	"	0.381	18.75	36.50		55.00
8820	1"	"	0.444	28.00	42.75		71.00
8830	1-1/4"	"	0.500	48.50	48.00		97.00
8840	1-1/2"	"	0.615	63.00	59.00		120
8850	2"	"	0.727	100	70.00		170
8860	2-1/2"	"	1.000	220	96.00		320
8870	3"	"	1.379	290	130		420
8880	3-1/2"	"	1.739	520	170		690
8890	4"	"	2.105	580	200		780
8895	T condulets						
8900	1/2"	EA	0.381	19.50	36.50		56.00
8910	3/4"	"	0.444	23.25	42.75		66.00
8920	1"	"	0.500	35.00	48.00		83.00
8930	1-1/4"	"	0.571	51.00	55.00		110
8940	1-1/2"	"	0.615	69.00	59.00		130
8950	2"	"	0.727	110	70.00		180
8960	2-1/2"	"	1.127	230	110		340
8970	3"	"	1.509	300	140		440
8980	3-1/2"	"	1.860	550	180		730
8990	4"	"	2.222	610	210		820
8995	X condulets						
9000	1/2"	EA	0.444	28.75	42.75		72.00
9010	3/4"	"	0.500	31.00	48.00		79.00
9020	1"	"	0.571	51.00	55.00		110
9030	1-1/4"	"	0.615	67.00	59.00		130
9040	1-1/2"	"	0.667	86.00	64.00		150
9050	2"	"	0.879	180	84.00		260
9055	Blank steel condulet covers						
9060	1/2"	EA	0.100	4.41	9.60		14.00
9070	3/4"	"	0.100	5.46	9.60		15.00
9080	1"	"	0.100	7.43	9.60		17.00
9090	1-1/4"	"	0.123	9.10	11.75		20.75
9100	1-1/2"	"	0.123	9.56	11.75		21.25
9110	2"	"	0.123	16.00	11.75		27.75

CONDUCTORS, CONDUIT AND RACEWAYS

ID Code	Description	Output		Unit Costs			
	Component Descriptions	Unit of Meas.	Manhr / Unit	Material Cost	Labor Cost	Equipment Cost	Total Cost
26 - 05344	**GALVANIZED CONDUIT, Cont'd...**						**26 - 05344**
9120	2-1/2"	EA	0.145	25.25	14.00		39.25
9130	3"	"	0.145	27.25	14.00		41.25
9140	3-1/2"	"	0.145	29.75	14.00		43.75
9150	4"	"	0.200	32.50	19.25		52.00
9155	Solid condulet gaskets						
9160	1/2"	EA	0.050	3.64	4.80		8.44
9170	3/4"	"	0.050	3.93	4.80		8.73
9180	1"	"	0.050	4.55	4.80		9.35
9190	1-1/4"	"	0.080	5.66	7.68		13.25
9200	1-1/2"	"	0.080	5.96	7.68		13.75
9210	2"	"	0.080	6.68	7.68		14.25
9220	2-1/2"	"	0.100	11.00	9.60		20.50
9230	3"	"	0.100	11.25	9.60		20.75
9240	3-1/2"	"	0.100	14.00	9.60		23.50
9250	4"	"	0.145	14.25	14.00		28.25
9255	One-hole malleable straps						
9260	1/2"	EA	0.040	0.58	3.84		4.42
9270	3/4"	"	0.040	0.80	3.84		4.64
9280	1"	"	0.040	1.15	3.84		4.99
9290	1-1/4"	"	0.050	2.31	4.80		7.11
9300	1-1/2"	"	0.050	2.66	4.80		7.46
9310	2"	"	0.050	5.20	4.80		10.00
9320	2-1/2"	"	0.059	10.75	5.68		16.50
9330	3"	"	0.059	15.50	5.68		21.25
9340	3-1/2"	"	0.059	22.25	5.68		28.00
9350	4"	"	0.080	49.25	7.68		57.00
9360	5"	"	0.080	170	7.68		180
9370	6"	"	0.080	190	7.68		200
9375	One-hole steel straps						
9380	1/2"	EA	0.040	0.15	3.84		3.99
9390	3/4"	"	0.040	0.21	3.84		4.05
9400	1"	"	0.040	0.36	3.84		4.20
9410	1-1/4"	"	0.050	0.51	4.80		5.31
9412	1-1/2"	"	0.050	0.70	4.80		5.50
9413	2"	"	0.050	0.92	4.80		5.72
9414	2-1/2"	"	0.059	3.18	5.68		8.86
9415	3"	"	0.059	4.28	5.68		9.96
9416	3-1/2"	"	0.059	6.36	5.68		12.00

CONDUCTORS, CONDUIT AND RACEWAYS

ID Code	Description / Component Descriptions	Output / Unit of Meas.	Manhr / Unit	Unit Costs / Material Cost	Labor Cost	Equipment Cost	Total Cost
26 - 05344	**GALVANIZED CONDUIT, Cont'd...**						**26 - 05344**
9417	4"	EA	0.080	6.58	7.68		14.25
9450	Bushed chase nipples						
9460	1/2"	EA	0.059	1.23	5.68		6.91
9465	3/4"	"	0.070	1.72	6.67		8.39
9470	1"	"	0.089	3.50	8.53		12.00
9475	1-1/4"	"	0.100	5.24	9.60		14.75
9480	1-1/2"	"	0.123	6.69	11.75		18.50
9485	2"	"	0.145	8.97	14.00		23.00
9490	2-1/2"	"	0.145	22.75	14.00		36.75
9495	3"	"	0.182	29.50	17.50		47.00
9500	3-1/2"	"	0.250	70.00	24.00		94.00
9505	4"	"	0.296	99.00	28.50		130
9508	Offset nipples						
9510	1/2"	EA	0.059	6.36	5.68		12.00
9515	3/4"	"	0.070	6.61	6.67		13.25
9520	1"	"	0.089	8.07	8.53		16.50
9525	1-1/4"	"	0.107	19.50	10.25		29.75
9530	1-1/2"	"	0.123	24.25	11.75		36.00
9535	2"	"	0.145	38.00	14.00		52.00
9540	3"	"	0.182	120	17.50		140
9543	Short elbows						
9545	1/2"	EA	0.145	5.11	14.00		19.00
9550	3/4"	"	0.200	6.99	19.25		26.25
9555	1"	"	0.250	11.50	24.00		35.50
9560	1-1/4"	"	0.296	34.50	28.50		63.00
9565	1-1/2"	"	0.348	47.00	33.50		81.00
9570	2"	"	0.400	85.00	38.50		120
9573	Pulling elbows, female to female						
9575	1/2"	EA	0.250	12.25	24.00		36.25
9580	3/4"	"	0.296	14.50	28.50		43.00
9585	1"	"	0.400	24.00	38.50		63.00
9590	1-1/4"	"	0.533	35.50	51.00		87.00
9600	1-1/2"	"	0.727	66.00	70.00		140
9605	2"	"	0.851	54.00	82.00		140
9650	Grounding locknuts						
9670	1/2"	EA	0.080	3.50	7.68		11.25
9680	3/4"	"	0.080	4.42	7.68		12.00
9690	1"	"	0.080	6.39	7.68		14.00

CONDUCTORS, CONDUIT AND RACEWAYS

ID Code	Description / Component Descriptions	Unit of Meas.	Manhr / Unit	Material Cost	Labor Cost	Equipment Cost	Total Cost
26 - 05344	**GALVANIZED CONDUIT, Cont'd...**						**26 - 05344**
9700	1-1/4"	EA	0.089	6.85	8.53		15.50
9710	1-1/2"	"	0.089	7.14	8.53		15.75
9720	2"	"	0.089	10.50	8.53		19.00
9730	2-1/2"	"	0.100	19.50	9.60		29.00
9740	3"	"	0.100	24.50	9.60		34.00
9750	3-1/2"	"	0.100	40.00	9.60		49.50
9755	4"	"	0.145	54.00	14.00		68.00
9758	Insulated grounding metal bushings						
9760	1/2"	EA	0.190	2.48	18.25		20.75
9765	3/4"	"	0.222	3.66	21.25		25.00
9770	1"	"	0.250	5.23	24.00		29.25
9775	1-1/4"	"	0.308	8.34	29.50		37.75
9780	1-1/2"	"	0.381	10.50	36.50		47.00
9785	2"	"	0.444	15.00	42.75		58.00
9790	2-1/2"	"	0.667	35.75	64.00		100
9795	3"	"	0.808	41.25	78.00		120
9800	3-1/2"	"	0.941	60.00	90.00		150
9805	4"	"	1.053	75.00	100		180
9810	5"	"	1.569	150	150		300
9815	6"	"	1.739	230	170		400
9818	Nipples						
9819	1/2" x						
9820	4"	EA	0.145	6.71	14.00		20.75
9825	6"	"	0.145	8.93	14.00		23.00
9830	8"	"	0.145	15.50	14.00		29.50
9835	10"	"	0.145	17.75	14.00		31.75
9840	12"	"	0.145	20.50	14.00		34.50
9843	3/4" x						
9845	4"	EA	0.145	7.77	14.00		21.75
9850	6"	"	0.145	10.50	14.00		24.50
9855	8"	"	0.145	17.25	14.00		31.25
9860	10"	"	0.145	20.75	14.00		34.75
9865	12"	"	0.145	23.25	14.00		37.25
9868	1" x						
9870	4"	EA	0.145	11.00	14.00		25.00
9875	6"	"	0.145	13.75	14.00		27.75
9880	8"	"	0.145	21.75	14.00		35.75
9885	10"	"	0.145	28.00	14.00		42.00

CONDUCTORS, CONDUIT AND RACEWAYS

ID Code	Description		Output		Unit Costs			
	Component Descriptions		Unit of Meas.	Manhr / Unit	Material Cost	Labor Cost	Equipment Cost	Total Cost
26 - 05344	**GALVANIZED CONDUIT, Cont'd...**							**26 - 05344**
9890	12"		EA	0.145	32.00	14.00		46.00
9893	1-1/4" x							
9895	4"		EA	0.250	16.25	24.00		40.25
9900	6"		"	0.250	21.00	24.00		45.00
9905	8"		"	0.250	34.50	24.00		59.00
9910	10"		"	0.250	43.50	24.00		68.00
9915	12"		"	0.250	51.00	24.00		75.00
9918	1-1/2" x							
9920	4"		EA	0.250	20.50	24.00		44.50
9925	6"		"	0.250	28.25	24.00		52.00
9930	8"		"	0.250	43.25	24.00		67.00
9935	10"		"	0.250	52.00	24.00		76.00
9940	12"		"	0.250	56.00	24.00		80.00
9941	2" x							
9942	4"		EA	0.250	26.75	24.00		51.00
9944	6"		"	0.250	35.50	24.00		60.00
9946	8"		"	0.250	51.00	24.00		75.00
9948	10"		"	0.250	62.00	24.00		86.00
9950	12"		"	0.250	70.00	24.00		94.00
9951	2-1/2" x							
9952	6"		EA	0.300	73.00	28.75		100
9954	8"		"	0.300	96.00	28.75		120
9956	10"		"	0.300	110	28.75		140
9958	12"		"	0.300	130	28.75		160
9959	3" x							
9960	6"		EA	0.300	88.00	28.75		120
9962	8"		"	0.300	110	28.75		140
9964	10"		"	0.300	130	28.75		160
9966	12"		"	0.300	160	28.75		190
9967	3-1/2" x							
9968	6"		EA	0.300	100	28.75		130
9970	8"		"	0.300	130	28.75		160
9972	10"		"	0.300	160	28.75		190
9974	12"		"	0.300	180	28.75		210
9975	4" x							
9976	8"		EA	0.400	150	38.50		190
9978	10"		"	0.400	180	38.50		220
9982	12"		"	0.400	220	38.50		260

CONDUCTORS, CONDUIT AND RACEWAYS

ID Code	Component Descriptions	Unit of Meas.	Manhr / Unit	Material Cost	Labor Cost	Equipment Cost	Total Cost
	Description	**Output**		**Unit Costs**			

26 - 05344	**GALVANIZED CONDUIT, Cont'd...**						**26 - 05344**
9983	5" x						
9984	8"	EA	0.400	270	38.50		310
9985	10"	"	0.400	310	38.50		350
9986	12"	"	0.400	380	38.50		420
9987	6" x						
9988	8"	EA	0.400	330	38.50		370
9990	10"	"	0.400	410	38.50		450
9995	12"	"	0.400	460	38.50		500

26 - 05345	**PLASTIC COATED CONDUIT**						**26 - 05345**
0980	Rigid steel conduit, plastic coated						
1000	1/2"	LF	0.050	10.00	4.80		14.75
1040	3/4"	"	0.059	11.75	5.68		17.50
1060	1"	"	0.080	15.00	7.68		22.75
1080	1-1/4"	"	0.100	19.00	9.60		28.50
1100	1-1/2"	"	0.123	23.25	11.75		35.00
1120	2"	"	0.145	30.25	14.00		44.25
1140	2-1/2"	"	0.190	45.75	18.25		64.00
1160	3"	"	0.222	58.00	21.25		79.00
1180	3-1/2"	"	0.250	71.00	24.00		95.00
1200	4"	"	0.308	86.00	29.50		120
1220	5"	"	0.381	150	36.50		190
1490	90 degree elbows						
1500	1/2"	EA	0.308	39.50	29.50		69.00
1540	3/4"	"	0.381	41.00	36.50		78.00
1560	1"	"	0.444	47.00	42.75		90.00
1580	1-1/4"	"	0.500	58.00	48.00		110
1600	1-1/2"	"	0.615	71.00	59.00		130
1620	2"	"	0.800	99.00	77.00		180
1640	2-1/2"	"	1.143	190	110		300
1660	3"	"	1.333	300	130		430
1680	3-1/2"	"	1.633	390	160		550
1700	4"	"	2.000	430	190		620
1720	5"	"	2.500	1,020	240		1,260
1980	Couplings						
2000	1/2"	EA	0.059	11.25	5.68		17.00
2040	3/4"	"	0.080	11.75	7.68		19.50
2060	1"	"	0.089	15.50	8.53		24.00

CONDUCTORS, CONDUIT AND RACEWAYS

ID Code	Description — Component Descriptions	Output — Unit of Meas.	Manhr / Unit	Unit Costs — Material Cost	Labor Cost	Equipment Cost	Total Cost
26 - 05345	**PLASTIC COATED CONDUIT, Cont'd...**						**26 - 05345**
2080	1-1/4"	EA	0.107	18.25	10.25		28.50
2100	1-1/2"	"	0.123	25.50	11.75		37.25
2120	2"	"	0.145	31.75	14.00		45.75
2140	2-1/2"	"	0.182	80.00	17.50		98.00
2160	3"	"	0.190	93.00	18.25		110
2180	3-1/2"	"	0.200	130	19.25		150
2200	4"	"	0.222	160	21.25		180
2220	5"	"	0.250	450	24.00		470
2980	1 hole conduit straps						
3000	3/4"	EA	0.050	17.50	4.80		22.25
3010	1"	"	0.050	18.00	4.80		22.75
3020	1-1/4"	"	0.059	26.25	5.68		32.00
3030	1-1/2"	"	0.059	27.75	5.68		33.50
3040	2"	"	0.059	40.25	5.68		46.00
3050	3"	"	0.080	73.00	7.68		81.00
3060	3-1/2"	"	0.080	130	7.68		140
3070	4"	"	0.100	140	9.60		150
3075	"L.B." condulets with covers						
3080	1/2"	EA	0.500	85.00	48.00		130
3090	3/4"	"	0.500	94.00	48.00		140
3100	1"	"	0.615	130	59.00		190
3110	1-1/4"	"	0.727	180	70.00		250
3120	1-1/2"	"	0.879	220	84.00		300
3130	2"	"	1.000	330	96.00		430
3140	2-1/2"	"	1.379	610	130		740
3150	3"	"	1.739	760	170		930
3160	3-1/2"	"	2.162	1,120	210		1,330
3170	4"	"	2.500	1,260	240		1,500
3175	"T" condulets with covers						
3180	1/2"	EA	0.571	98.00	55.00		150
3190	3/4"	"	0.615	120	59.00		180
3200	1"	"	0.667	150	64.00		210
3210	1-1/4"	"	0.808	210	78.00		290
3220	1-1/2"	"	0.941	260	90.00		350
3230	2"	"	1.053	380	100		480
3240	2-1/2"	"	1.509	630	140		770
3260	3-1/2"	"	2.222	1,160	210		1,370
3270	4"	"	2.667	1,280	260		1,540

CONDUCTORS, CONDUIT AND RACEWAYS

ID Code	Description / Component Descriptions	Output Unit of Meas.	Manhr / Unit	Material Cost	Labor Cost	Equipment Cost	Total Cost
26 - 05345	**PLASTIC COATED CONDUIT, Cont'd...**						**26 - 05345**
3275	5"	EA	3.333	1,370	320		1,690
26 - 05346	**PLASTIC CONDUIT**						**26 - 05346**
3010	PVC conduit, schedule 40						
3020	1/2"	LF	0.030	1.04	2.89		3.93
3040	3/4"	"	0.030	1.29	2.89		4.18
3060	1"	"	0.040	1.87	3.84		5.71
3080	1-1/4"	"	0.040	2.58	3.84		6.42
3100	1-1/2"	"	0.050	3.08	4.80		7.88
3120	2"	"	0.050	3.68	4.80		8.48
3140	2-1/2"	"	0.059	6.36	5.68		12.00
3160	3"	"	0.059	7.79	5.68		13.50
3180	3-1/2"	"	0.080	10.25	7.68		18.00
3200	4"	"	0.080	11.25	7.68		19.00
3220	5"	"	0.089	17.25	8.53		25.75
3240	6"	"	0.100	23.25	9.60		32.75
3480	Couplings						
3500	1/2"	EA	0.050	0.63	4.80		5.43
3520	3/4"	"	0.050	0.76	4.80		5.56
3540	1"	"	0.050	1.20	4.80		6.00
3560	1-1/4"	"	0.059	1.57	5.68		7.25
3580	1-1/2"	"	0.059	2.18	5.68		7.86
3600	2"	"	0.059	2.86	5.68		8.54
3620	2-1/2"	"	0.059	4.98	5.68		10.75
3640	3"	"	0.080	8.26	7.68		16.00
3660	3-1/2"	"	0.080	9.23	7.68		17.00
3680	4"	"	0.100	13.50	9.60		23.00
3690	5"	"	0.100	32.50	9.60		42.00
3700	6"	"	0.100	41.50	9.60		51.00
3705	90 degree elbows						
3710	1/2"	EA	0.100	2.47	9.60		12.00
3740	3/4"	"	0.123	2.69	11.75		14.50
3760	1"	"	0.123	4.27	11.75		16.00
3780	1-1/4"	"	0.145	5.95	14.00		20.00
3800	1-1/2"	"	0.190	8.07	18.25		26.25
3810	2"	"	0.222	11.25	21.25		32.50
3820	2-1/2"	"	0.250	19.75	24.00		43.75
3830	3"	"	0.308	36.00	29.50		66.00

CONDUCTORS, CONDUIT AND RACEWAYS

	Description	Output		Unit Costs			
ID Code	Component Descriptions	Unit of Meas.	Manhr / Unit	Material Cost	Labor Cost	Equipment Cost	Total Cost
26 - 05346		**PLASTIC CONDUIT, Cont'd...**				**26 - 05346**	
3840	3-1/2"	EA	0.381	47.00	36.50		84.00
3850	4"	"	0.500	59.00	48.00		110
3860	5"	"	0.615	110	59.00		170
3870	6"	"	0.727	180	70.00		250
3900	Terminal adapters						
4000	1/2"	EA	0.100	0.93	9.60		10.50
4040	3/4"	"	0.100	1.51	9.60		11.00
4060	1"	"	0.100	1.88	9.60		11.50
4080	1-1/4"	"	0.160	2.37	15.25		17.50
4100	1-1/2"	"	0.160	3.02	15.25		18.25
4120	2"	"	0.160	4.17	15.25		19.50
4140	2-1/2"	"	0.222	7.12	21.25		28.25
4160	3"	"	0.222	10.00	21.25		31.25
4180	3-1/2"	"	0.222	13.00	21.25		34.25
4200	4"	"	0.381	16.75	36.50		53.00
4220	5"	"	0.381	33.50	36.50		70.00
4240	6"	"	0.381	40.25	36.50		77.00
4250	End bells						
4260	1"	EA	0.100	6.26	9.60		15.75
4340	1-1/4"	"	0.160	7.42	15.25		22.75
4360	1-1/2"	"	0.160	7.71	15.25		23.00
4380	2"	"	0.160	11.50	15.25		26.75
4400	2-1/2"	"	0.222	12.75	21.25		34.00
4420	3"	"	0.222	13.50	21.25		34.75
4440	3-1/2"	"	0.222	14.75	21.25		36.00
4460	4"	"	0.381	16.00	36.50		53.00
4480	5"	"	0.381	25.00	36.50		62.00
4500	6"	"	0.381	27.50	36.50		64.00
5010	LB conduit body						
5020	1/2"	EA	0.190	8.09	18.25		26.25
5040	3/4"	"	0.190	10.50	18.25		28.75
5060	1	"	0.190	11.50	18.25		29.75
5080	1-1/4"	"	0.308	17.50	29.50		47.00
5100	1-1/2"	"	0.308	21.00	29.50		51.00
5120	2"	"	0.308	37.00	29.50		67.00
5140	2-1/2"	"	0.444	130	42.75		170
5150	3"	"	0.533	140	51.00		190
5160	3-1/2"	"	0.615	150	59.00		210

CONDUCTORS, CONDUIT AND RACEWAYS

ID Code	Description / Component Descriptions	Output		Unit Costs			
		Unit of Meas.	Manhr / Unit	Material Cost	Labor Cost	Equipment Cost	Total Cost
26 - 05346	PLASTIC CONDUIT, Cont'd...					26 - 05346	
5170	4"	EA	0.727	150	70.00		220
6010	Direct burial, conduit						
6020	2"	LF	0.050	2.47	4.80		7.27
6040	3"	"	0.059	4.64	5.68		10.25
6080	4"	"	0.080	7.37	7.68		15.00
6100	5"	"	0.089	10.50	8.53		19.00
6120	6"	"	0.100	14.50	9.60		24.00
6130	Encased burial conduit						
6140	2"	LF	0.050	1.57	4.80		6.37
6160	3"	"	0.059	2.61	5.68		8.29
6200	4"	"	0.080	4.00	7.68		11.75
6220	5"	"	0.089	5.73	8.53		14.25
6240	6"	"	0.100	7.45	9.60		17.00
6280	EB and DB duct, 90 degree elbows						
6300	1-1/2"	EA	0.145	16.00	14.00		30.00
6310	2"	"	0.229	17.75	22.00		39.75
6320	3"	"	0.381	25.00	36.50		62.00
6330	3-1/2"	"	0.444	32.75	42.75		76.00
6340	4"	"	0.533	35.50	51.00		87.00
6350	5"	"	0.667	89.00	64.00		150
6360	6"	"	0.899	170	86.00		260
6365	45 degree elbows						
6370	1-1/2"	EA	0.229	18.75	22.00		40.75
6380	2"	"	0.229	19.25	22.00		41.25
6390	3"	"	0.381	25.50	36.50		62.00
6400	3-1/2"	"	0.444	34.50	42.75		77.00
6420	4"	"	0.533	33.25	51.00		84.00
6430	5"	"	0.667	63.00	64.00		130
6450	6"	"	0.899	150	86.00		240
6710	Couplings						
6730	1-1/2"	EA	0.059	1.55	5.68		7.23
6740	2"	"	0.059	1.74	5.68		7.42
6750	3"	"	0.080	6.33	7.68		14.00
6760	3-1/2"	"	0.080	7.53	7.68		15.25
6770	4"	"	0.100	9.91	9.60		19.50
6780	5"	"	0.100	18.00	9.60		27.50
6790	6"	"	0.160	55.00	15.25		70.00
6795	Bell ends						

CONDUCTORS, CONDUIT AND RACEWAYS

ID Code	Description / Component Descriptions	Unit of Meas.	Manhr / Unit	Material Cost	Labor Cost	Equipment Cost	Total Cost
	Description	**Output**		**Unit Costs**			
26 - 05346	**PLASTIC CONDUIT, Cont'd...**						**26 - 05346**
6800	1-1/2"	EA	0.160	9.66	15.25		25.00
6810	2"	"	0.160	12.25	15.25		27.50
6820	3"	"	0.222	15.50	21.25		36.75
6830	3-1/2"	"	0.222	15.75	21.25		37.00
6840	4"	"	0.381	18.25	36.50		55.00
6850	5"	"	0.381	27.50	36.50		64.00
6860	6"	"	0.381	53.00	36.50		90.00
6870	Female adapters, 1-1/2"	"	0.200	2.94	19.25		22.25
7000	45 degree couplings						
7010	1-1/2"	EA	0.070	12.50	6.67		19.25
7020	2"	"	0.070	14.00	6.67		20.75
7030	3"	"	0.100	18.25	9.60		27.75
7040	4"	"	0.145	19.00	14.00		33.00
7050	5"	"	0.145	23.25	14.00		37.25
7060	6"	"	0.145	24.50	14.00		38.50
7065	45 degree elbows						
7070	1/2"	EA	0.123	1.88	11.75		13.75
7080	3/4"	"	0.145	2.37	14.00		16.25
7090	1"	"	0.145	3.35	14.00		17.25
7100	1-1/4"	"	0.182	4.83	17.50		22.25
7110	1-1/2"	"	0.229	6.62	22.00		28.50
7120	2"	"	0.267	9.82	25.50		35.25
7130	2-1/2"	"	0.296	18.25	28.50		46.75
7140	3"	"	0.381	31.50	36.50		68.00
7150	3-1/2"	"	0.444	35.50	42.75		78.00
7160	4"	"	0.615	50.00	59.00		110
7170	5"	"	0.727	80.00	70.00		150
7180	6"	"	0.899	120	86.00		210
7185	Female adapters						
7190	1/2"	EA	0.123	0.98	11.75		12.75
7200	3/4"	"	0.123	1.57	11.75		13.25
7210	1"	"	0.123	1.96	11.75		13.75
7220	1-1/4"	"	0.200	2.54	19.25		21.75
7230	1-1/2"	"	0.200	2.79	19.25		22.00
7240	2"	"	0.200	4.00	19.25		23.25
7250	2-1/2"	"	0.267	6.87	25.50		32.25
7260	3"	"	0.267	11.50	25.50		37.00
7270	3-1/2"	"	0.267	14.25	25.50		39.75

CONDUCTORS, CONDUIT AND RACEWAYS

	Description	Output		Unit Costs			
ID Code	Component Descriptions	Unit of Meas.	Manhr / Unit	Material Cost	Labor Cost	Equipment Cost	Total Cost
26 - 05346	**PLASTIC CONDUIT, Cont'd...**						**26 - 05346**
7280	4"	EA	0.444	18.00	42.75		61.00
7290	5"	"	0.444	40.50	42.75		83.00
7300	6"	"	0.444	46.00	42.75		89.00
7305	Expansion couplings						
7310	1/2"	EA	0.123	39.00	11.75		51.00
7320	3/4"	"	0.123	41.75	11.75		54.00
7330	1"	"	0.145	43.50	14.00		58.00
7340	1-1/4"	"	0.200	43.75	19.25		63.00
7350	1-1/2"	"	0.200	44.00	19.25		63.00
7360	2"	"	0.200	48.50	19.25		68.00
7370	2-1/2"	"	0.296	67.00	28.50		96.00
7380	3"	"	0.296	86.00	28.50		110
7390	3-1/2"	"	0.296	110	28.50		140
7400	4"	"	0.444	120	42.75		160
7410	5"	"	0.444	190	42.75		230
7420	6"	"	0.444	260	42.75		300
7425	Plugs						
7430	2"	EA	0.200	3.14	19.25		22.50
7440	3"	"	0.296	4.63	28.50		33.25
7450	3-1/2"	"	0.296	4.98	28.50		33.50
7460	4"	"	0.444	5.22	42.75		48.00
7470	5"	"	0.444	7.07	42.75		49.75
7480	6"	"	0.500	8.82	48.00		57.00
7485	PVC cement						
7490	1 pint	EA					21.25
7500	1 quart	"					32.50
7510	1 gallon	"					110
7515	Type "T" condulets						
7520	1/2"	EA	0.296	11.00	28.50		39.50
7530	3/4"	"	0.296	12.50	28.50		41.00
7540	1"	"	0.296	14.25	28.50		42.75
7550	1-1/4"	"	0.500	21.00	48.00		69.00
7560	1-1/2"	"	0.500	27.75	48.00		76.00
7570	2"	"	0.500	39.25	48.00		87.00
7875	EB & DB female adapters						
7880	2"	EA	0.250	1.71	24.00		25.75
7890	3"	"	0.381	4.75	36.50		41.25
7900	3-1/2"	"	0.615	6.22	59.00		65.00

CONDUCTORS, CONDUIT AND RACEWAYS

ID Code	Description / Component Descriptions	Unit of Meas.	Manhr / Unit	Material Cost	Labor Cost	Equipment Cost	Total Cost
26 - 05346	**PLASTIC CONDUIT, Cont'd...**						**26 - 05346**
7910	4"	EA	0.727	6.39	70.00		76.00
7920	5"	"	1.000	15.75	96.00		110
7930	6"	"	1.600	20.75	150		170
26 - 05347	**STEEL CONDUIT**						**26 - 05347**
7980	Intermediate metal conduit (IMC)						
8000	1/2"	LF	0.030	3.25	2.89		6.14
8040	3/4"	"	0.040	3.99	3.84		7.83
8060	1"	"	0.050	6.04	4.80		10.75
8080	1-1/4"	"	0.059	7.74	5.68		13.50
8100	1-1/2"	"	0.080	9.68	7.68		17.25
8120	2"	"	0.089	12.75	8.53		21.25
8140	2-1/2"	"	0.119	25.00	11.50		36.50
8160	3"	"	0.145	32.25	14.00		46.25
8180	3-1/2"	"	0.182	37.75	17.50		55.00
8200	4"	"	0.190	41.75	18.25		60.00
8490	90 degree ell						
8500	1/2"	EA	0.250	25.00	24.00		49.00
8540	3/4"	"	0.308	26.00	29.50		56.00
8560	1"	"	0.381	40.00	36.50		77.00
8580	1-1/4"	"	0.444	55.00	42.75		98.00
8600	1-1/2"	"	0.500	68.00	48.00		120
8620	2"	"	0.571	98.00	55.00		150
8640	2-1/2"	"	0.667	180	64.00		240
8660	3"	"	0.889	250	85.00		340
8680	3-1/2"	"	1.143	400	110		510
8690	4"	"	1.333	450	130		580
9260	Couplings						
9280	1/2"	EA	0.050	6.09	4.80		11.00
9290	3/4"	"	0.059	7.49	5.68		13.25
9300	1"	"	0.080	11.00	7.68		18.75
9310	1-1/4"	"	0.089	14.00	8.53		22.50
9320	1-1/2"	"	0.100	17.50	9.60		27.00
9330	2"	"	0.107	23.25	10.25		33.50
9340	2-1/2"	"	0.123	57.00	11.75		69.00
9350	3"	"	0.145	74.00	14.00		88.00
9360	3-1/2"	"	0.145	100	14.00		110
9370	4"	"	0.160	100	15.25		120

CONDUCTORS, CONDUIT AND RACEWAYS

ID Code	Description — Component Descriptions	Output — Unit of Meas.	Output — Manhr / Unit	Unit Costs — Material Cost	Unit Costs — Labor Cost	Unit Costs — Equipment Cost	Unit Costs — Total Cost
26 - 05361	**CABLE TRAY**						**26 - 05361**
1010	Cable tray, 6"	LF	0.059	24.00	5.68		29.75
1020	Ventilated cover	"	0.030	9.68	2.89		12.50
1030	Solid cover	"	0.030	7.54	2.89		10.50
1040	Flat 90	EA	0.500	110	48.00		160
1050	Outside 90	"	0.500	91.00	48.00		140
1060	Inside 90	"	0.500	99.00	48.00		150
1070	Flat 45	"	0.500	82.00	48.00		130
1080	Outside 45	"	0.500	68.00	48.00		120
1090	Inside 45	"	0.500	71.00	48.00		120
1100	Adjustable elbow	"	0.500	100	48.00		150
1110	Support riser	"	0.500	140	48.00		190
1120	Adjustable riser	"	0.500	15.00	48.00		63.00
1130	Tee	"	1.739	140	170		310
1140	Cross	"	1.818	180	170		350
1150	Blind end	"	0.296	11.00	28.50		39.50
1160	Expansion joint	"	0.500	51.00	48.00		99.00
1170	Box connector	"	2.500	34.00	240		270
1180	Standard dropout	"	0.500	5.94	48.00		54.00
1190	2"	"	0.615	29.00	59.00		88.00
1200	3"	"	0.800	51.00	77.00		130
1210	4"	"	1.000	82.00	96.00		180
1220	Cable tray, 9"	LF	0.070	25.50	6.67		32.25
1230	Ventilated cover	"	0.040	10.75	3.84		14.50
1240	Solid cover	"	0.040	8.79	3.84		12.75
1250	Flat 90	EA	0.533	120	51.00		170
1260	Outside 90	"	0.533	98.00	51.00		150
1270	Inside 90	"	0.533	100	51.00		150
1280	Flat 45	"	0.533	85.00	51.00		140
1290	Outside 45	"	0.533	70.00	51.00		120
1300	Inside 45	"	0.533	70.00	51.00		120
1310	Adjustable elbow	"	0.533	110	51.00		160
1320	Support riser	"	0.533	150	51.00		200
1330	Adjustable riser	"	0.533	15.00	51.00		66.00
1340	Tee	"	1.818	150	170		320
1350	Cross	"	1.818	190	170		360
1360	Blind end	"	0.320	12.00	30.75		42.75
1370	Expansion joint	"	0.533	72.00	51.00		120
1380	Box connector	"	2.581	34.50	250		280

CONDUCTORS, CONDUIT AND RACEWAYS

	Description	Output		Unit Costs			
ID Code	Component Descriptions	Unit of Meas.	Manhr / Unit	Material Cost	Labor Cost	Equipment Cost	Total Cost
26 - 05361		**CABLE TRAY, Cont'd...**					**26 - 05361**
1390	Standard dropout	EA	0.533	6.50	51.00		58.00
1400	2"	"	0.667	29.25	64.00		93.00
1410	3"	"	0.667	48.75	64.00		110
1420	4"	"	1.096	82.00	110		190
1430	Cable tray, 12"	LF	0.080	26.00	7.68		33.75
1440	Ventilated cover	"	0.050	10.00	4.80		14.75
1450	Solid cover	"	0.050	9.20	4.80		14.00
1460	Flat 90	EA	0.533	120	51.00		170
1470	Outside 90	"	0.533	98.00	51.00		150
1480	Inside 90	"	0.533	110	51.00		160
1490	Flat 45	"	0.533	90.00	51.00		140
1500	Outside 45	"	0.533	140	51.00		190
1510	Inside 45	"	0.533	120	51.00		170
1520	Adjustable elbow	"	0.533	110	51.00		160
1530	Support riser	"	0.533	150	51.00		200
1540	Adjustable riser	"	0.533	17.00	51.00		68.00
1550	Tee	"	2.000	170	190		360
1560	Cross	"	2.000	210	190		400
1570	Blind end	"	0.348	13.50	33.50		47.00
1580	Expansion joint	"	0.615	77.00	59.00		140
1590	Box connector	"	2.759	36.00	260		300
1600	Standard dropout	"	0.615	7.05	59.00		66.00
1610	2"	"	0.727	30.75	70.00		100
1620	3"	"	0.899	51.00	86.00		140
1630	4"	"	1.159	85.00	110		200
1640	Cable tray, 18"	LF	0.100	27.25	9.60		36.75
1650	Ventilated cover	"	0.059	14.00	5.68		19.75
1660	Solid cover	"	0.059	11.75	5.68		17.50
1670	Flat 90	EA	0.727	140	70.00		210
1680	Outside 90	"	0.727	150	70.00		220
1690	Inside 90	"	0.727	120	70.00		190
1700	Flat 45	"	0.727	100	70.00		170
1710	Outside 45	"	0.727	140	70.00		210
1720	Inside 45	"	0.727	120	70.00		190
1730	Adjustable elbow	"	0.727	150	70.00		220
1740	Support riser	"	0.727	180	70.00		250
1750	Adjustable riser	"	0.727	15.00	70.00		85.00
1760	Tee	"	2.105	190	200		390

CONDUCTORS, CONDUIT AND RACEWAYS

	Description		Output		Unit Costs			
ID Code	Component Descriptions	Unit of Meas.	Manhr / Unit	Material Cost	Labor Cost	Equipment Cost	Total Cost
26 - 05361		**CABLE TRAY, Cont'd...**					**26 - 05361**
1770	Cross	EA	2.105	290	200		490
1780	Blind end	"	0.400	19.75	38.50		58.00
1790	Expansion joint	"	0.727	100	70.00		170
1800	Box connector	"	2.963	42.25	280		320
1810	Standard dropout	"	0.667	10.75	64.00		75.00
1820	2"	"	0.727	34.00	70.00		100
1830	3"	"	0.952	54.00	91.00		140
1840	4"	"	1.194	87.00	110		200
1850	Cable tray, 24"	LF	0.123	26.50	11.75		38.25
1860	Ventilated cover	"	0.070	19.25	6.67		26.00
1870	Solid cover	"	0.070	13.25	6.67		20.00
1880	Flat 90	EA	0.727	180	70.00		250
1890	Outside 90	"	0.727	150	70.00		220
1900	Inside 90	"	0.727	120	70.00		190
1910	Flat 45	"	0.727	120	70.00		190
1920	Outside 45	"	0.727	150	70.00		220
1930	Inside 45	"	0.727	120	70.00		190
1940	Adjustable elbow	"	0.727	150	70.00		220
1950	Support riser	"	0.727	190	70.00		260
1960	Adjustable riser	"	0.727	16.75	70.00		87.00
1970	Tee	"	2.222	220	210		430
1980	Cross	"	2.222	330	210		540
1990	Blind end	"	0.444	22.00	42.75		65.00
2000	Expansion joint	"	0.727	100	70.00		170
2010	Box connector	"	3.478	45.25	330		380
2020	Standard dropout	"	0.667	11.25	64.00		75.00
2030	2"	"	0.800	35.25	77.00		110
2040	3"	"	0.976	56.00	94.00		150
2050	4"	"	1.250	90.00	120		210
2060	Cable tray, 36"	LF	0.145	36.00	14.00		50.00
2070	Ventilated cover	"	0.080	22.25	7.68		30.00
2080	Solid cover	"	0.080	15.50	7.68		23.25
2090	Flat 90	EA	0.851	210	82.00		290
2100	Outside 90	"	0.851	180	82.00		260
2110	Inside 90	"	0.851	150	82.00		230
2120	Flat 45	"	0.851	140	82.00		220
2130	Outside 45	"	0.851	170	82.00		250
2140	Inside 45	"	0.851	140	82.00		220

CONDUCTORS, CONDUIT AND RACEWAYS

	Description		Output		Unit Costs			
ID Code	Component Descriptions	Unit of Meas.	Manhr / Unit	Material Cost	Labor Cost	Equipment Cost	Total Cost
26 - 05361		**CABLE TRAY, Cont'd...**					**26 - 05361**	
2150	Adjustable elbow	EA	0.851	170	82.00		250
2160	Support riser	"	0.851	190	82.00		270
2170	Adjustable riser	"	0.851	17.75	82.00		100
2180	Tee	"	2.286	250	220		470
2190	Cross	"	2.963	370	280		650
2200	Blind end	"	0.471	22.50	45.25		68.00
2210	Expansion joint	"	0.727	110	70.00		180
2220	Box connector	"	3.810	53.00	370		420
2230	Standard dropout	"	0.727	12.75	70.00		83.00
2240	2"	"	0.800	37.50	77.00		110
2250	3"	"	1.000	58.00	96.00		150
2260	4"	"	1.290	90.00	120		210
2265	Reducers						
2270	9" - 6"	EA	0.500	82.00	48.00		130
2280	12" - 9"	"	0.500	85.00	48.00		130
2290	18" - 12"	"	0.615	90.00	59.00		150
2300	24" - 18"	"	0.727	100	70.00		170
2310	36" - 18"	"	0.800	120	77.00		200
2320	36" - 24"	"	0.899	120	86.00		210
2325	Conduit dropouts						
2330	3/4"	EA	0.348	10.75	33.50		44.25
2340	1"	"	0.348	11.25	33.50		44.75
2350	1-1/4"	"	0.400	13.00	38.50		52.00
2360	1-1/2"	"	0.500	15.00	48.00		63.00
2370	2"	"	0.533	17.25	51.00		68.00
2380	2-1/2"	"	0.727	19.50	70.00		90.00
2390	3"	"	0.800	22.75	77.00		100
2395	Wall brackets						
2400	6"	EA	0.145	42.25	14.00		56.00
2410	9"	"	0.145	48.75	14.00		63.00
2420	12"	"	0.200	51.00	19.25		70.00
2430	18"	"	0.250	62.00	24.00		86.00
2440	24"	"	0.296	72.00	28.50		100
2450	36"	"	0.400	89.00	38.50		130

CONDUCTORS, CONDUIT AND RACEWAYS

ID Code	Description — Component Descriptions	Unit of Meas.	Manhr / Unit	Material Cost	Labor Cost	Equipment Cost	Total Cost
26 - 05362	**FIBERGLASS CABLE TRAY**						**26 - 05362**
1000	Fiberglass cable tray, 6"	LF	0.040	38.00	3.84		41.75
1010	Tray cover	"	0.030	10.50	2.89		13.50
1015	Horizontal						
1020	90	EA	0.276	410	26.50		440
1030	45	"	0.276	170	26.50		200
1040	30	"	0.276	180	26.50		210
1045	Inside						
1050	90	EA	0.276	160	26.50		190
1060	45	"	0.276	130	26.50		160
1070	30	"	0.276	170	26.50		200
1080	Horizontal tee	"	0.727	270	70.00		340
1090	Horizontal cross	"	1.096	330	110		440
1100	Splice plate	"	0.145	15.50	14.00		29.50
1110	Floor flange	"	0.348	17.00	33.50		51.00
1120	Panel flange	"	1.250	17.50	120		140
1130	End plate	"	0.145	11.75	14.00		25.75
1140	Nylon rivet	"	0.050	0.42	4.80		5.22
1150	Barrier strip	"	0.050	5.81	4.80		10.50
1160	Hold down clamp	"	0.050	2.55	4.80		7.35
1170	Drop out	"	0.533	18.50	51.00		70.00
1180	Cover stand off	"	0.050	4.89	4.80		9.69
1190	Wall bracket	"	0.200	69.00	19.25		88.00
1200	Sealer	"					25.00
1205	Outside						
1210	90	EA	0.276	160	26.50		190
1220	45	"	0.276	130	26.50		160
1230	30	"	0.276	170	26.50		200
1240	Fiberglass cable tray, 9"	LF	0.050	40.00	4.80		44.75
1250	Tray cover	"	0.040	11.00	3.84		14.75
1255	Horizontal						
1260	90	EA	0.276	430	26.50		460
1270	45	"	0.276	180	26.50		210
1280	30	"	0.276	180	26.50		210
1285	Inside						
1290	90	EA	0.276	170	26.50		200
1300	45	"	0.276	130	26.50		160
1310	30	"	0.276	180	26.50		210
1320	Horizontal tee	"	0.727	280	70.00		350

CONDUCTORS, CONDUIT AND RACEWAYS

ID Code	Description — Component Descriptions	Output — Unit of Meas.	Manhr / Unit	Unit Costs — Material Cost	Labor Cost	Equipment Cost	Total Cost
26 - 05362	**FIBERGLASS CABLE TRAY, Cont'd...**					**26 - 05362**	
1330	Horizontal cross	EA	1.096	340	110		450
1340	Splice plate	"	0.145	16.25	14.00		30.25
1350	Floor flange	"	0.348	18.00	33.50		52.00
1360	Panel flange	"	1.250	18.50	120		140
1370	End plate	"	0.145	12.25	14.00		26.25
1380	Nylon rivet	"	0.050	0.44	4.80		5.24
1390	Barrier strap	"	0.050	6.10	4.80		11.00
1400	Hold down clamp	"	0.050	2.69	4.80		7.49
1410	Drop out	"	0.533	19.25	51.00		70.00
1420	Cover stand off	"	0.050	5.10	4.80		9.90
1430	Wall bracket	"	0.200	72.00	19.25		91.00
1440	Sealer	"					25.00
1445	Outside						
1450	90	EA	0.276	170	26.50		200
1460	45	"	0.276	140	26.50		170
1470	30	"	0.276	180	26.50		210
1480	Fiberglass cable tray, 12"	LF	0.059	41.75	5.68		47.50
1490	Tray cover	"	0.050	11.75	4.80		16.50
1495	Horizontal						
1500	90	EA	0.348	450	33.50		480
1510	45	"	0.348	200	33.50		230
1520	30	"	0.348	200	33.50		230
1525	Inside						
1530	90	EA	0.348	170	33.50		200
1540	45	"	0.348	140	33.50		170
1550	Inside 30	"	0.348	180	33.50		210
1560	Horizontal tee	"	0.952	300	91.00		390
1570	Horizontal cross	"	1.290	350	120		470
1580	Splice plate	"	0.160	17.00	15.25		32.25
1590	Floor flange	"	0.364	18.75	35.00		54.00
1600	Panel flange	"	1.600	19.50	150		170
1610	End plate	"	0.160	12.75	15.25		28.00
1620	Nylon rivet	"	0.050	0.45	4.80		5.25
1630	Barrier strip	"	0.050	6.38	4.80		11.25
1670	Hold down clamp	"	0.050	2.83	4.80		7.63
1680	Drop out	"	0.533	28.75	51.00		80.00
1690	Cover stand off	"	0.050	5.32	4.80		10.00
1700	Wall bracket	"	0.200	76.00	19.25		95.00

CONDUCTORS, CONDUIT AND RACEWAYS

ID Code	Description Component Descriptions	Output Unit of Meas.	Output Manhr / Unit	Unit Costs Material Cost	Unit Costs Labor Cost	Unit Costs Equipment Cost	Unit Costs Total Cost
26 - 05362	**FIBERGLASS CABLE TRAY, Cont'd...**					**26 - 05362**	
1710	Sealer	EA					25.00
1715	Outside						
1720	90	EA	0.348	170	33.50		200
1730	45	"	0.348	140	33.50		170
1740	30	"	0.348	200	33.50		230
1750	Fiberglass cable tray, 18"	LF	0.070	44.00	6.73		51.00
1760	Tray cover	"	0.059	21.75	5.68		27.50
1765	Horizontal						
1770	90	EA	0.400	480	38.50		520
1780	45	"	0.400	200	38.50		240
1790	30	"	0.400	210	38.50		250
1795	Inside						
1800	90	EA	0.400	180	38.50		220
1810	45	"	0.400	140	38.50		180
1820	30	"	0.400	200	38.50		240
1830	Horizontal tee	"	1.538	310	150		460
1840	Horizontal cross	"	1.356	370	130		500
1850	Splice plate	"	0.160	18.00	15.25		33.25
1860	Floor flange	"	0.381	19.75	36.50		56.00
1870	Panel flange	"	1.702	20.50	160		180
1880	End plate	"	0.170	12.75	16.25		29.00
1890	Nylon rivet	"	0.050	0.48	4.80		5.28
1900	Barrier strip	"	0.050	6.74	4.80		11.50
1910	Hold down clamp	"	0.050	2.98	4.80		7.78
1920	Drop out	"	0.533	41.75	51.00		93.00
1930	Cover stand off	"	0.050	5.67	4.80		10.50
1940	Wall bracket	"	0.200	79.00	19.25		98.00
1950	Sealer	"					25.00
1955	Outside						
1960	90	EA	0.400	180	38.50		220
1970	45	"	0.400	140	38.50		180
1980	30	"	0.400	200	38.50		240
1990	Fiberglass cable tray, 24"	LF	0.080	46.00	7.68		54.00
2000	Tray cover	"	0.070	25.75	6.67		32.50
2005	Horizontal						
2010	90	EA	0.500	480	48.00		530
2020	45	"	0.500	210	48.00		260
2030	30	"	0.500	210	48.00		260

CONDUCTORS, CONDUIT AND RACEWAYS

	Description	Output		Unit Costs			
ID Code	Component Descriptions	Unit of Meas.	Manhr / Unit	Material Cost	Labor Cost	Equipment Cost	Total Cost

26 - 05362 — FIBERGLASS CABLE TRAY, Cont'd... — 26 - 05362

ID Code	Component Descriptions	Unit of Meas.	Manhr / Unit	Material Cost	Labor Cost	Equipment Cost	Total Cost
2035	Inside						
2040	90	EA	0.500	180	48.00		230
2050	45	"	0.500	170	48.00		220
2060	30	"	0.500	200	48.00		250
2070	Horizontal tee	"	1.000	330	96.00		430
2080	Horizontal cross	"	1.509	380	140		520
2090	Splice plate	"	0.160	18.75	15.25		34.00
2100	Floor flange	"	0.400	20.75	38.50		59.00
2110	Panel flange	"	1.818	21.50	170		190
2120	End plate	"	0.170	14.00	16.25		30.25
2130	Nylon Rivet	"	0.050	0.49	4.80		5.29
2140	Barrier strip	"	0.050	7.02	4.80		11.75
2150	Hold down clamp	"	0.050	3.12	4.80		7.92
2160	Drop out	"	0.533	55.00	51.00		110
2170	Cover stand off	"	0.050	5.89	4.80		10.75
2180	Wall bracket	"	0.200	87.00	19.25		110
2190	Sealer	"					25.00
2195	Outside						
2200	90	EA	0.500	180	48.00		230
2210	45	"	0.500	160	48.00		210
2220	30	"	0.500	200	48.00		250
2230	Fiberglass cable tray, 30"	LF	0.089	53.00	8.53		62.00
2240	Tray cover	"	0.080	36.50	7.68		44.25
2245	Horizontal						
2250	90	EA	0.615	500	59.00		560
2260	45	"	0.615	210	59.00		270
2270	30	"	0.615	210	59.00		270
2275	Inside						
2280	90	EA	0.615	200	59.00		260
2290	45	"	0.615	160	59.00		220
2300	30	"	0.615	210	59.00		270
2310	Horizontal tee	"	1.096	370	110		480
2320	Horizontal cross	"	1.600	440	150		590
2330	Splice plate	"	0.160	19.75	15.25		35.00
2340	Floor flange	"	0.400	21.75	38.50		60.00
2350	Panel flange	"	1.905	22.50	180		200
2360	End plate	"	0.200	14.75	19.25		34.00
2370	Nylon rivet	"	0.050	0.53	4.80		5.33

CONDUCTORS, CONDUIT AND RACEWAYS

ID Code	Description	Output		Unit Costs			
	Component Descriptions	Unit of Meas.	Manhr / Unit	Material Cost	Labor Cost	Equipment Cost	Total Cost
26 - 05362	**FIBERGLASS CABLE TRAY, Cont'd...**						**26 - 05362**
2380	Barrier strip	EA	0.050	7.38	4.80		12.25
2390	Hold down clamp	"	0.050	3.26	4.80		8.06
2400	Dropout	"	0.615	63.00	59.00		120
2410	Cover stand off	"	0.050	6.24	4.80		11.00
2420	Wall bracket	"	0.200	98.00	19.25		120
2430	Sealer	"					68.00
2435	Outside						
2440	90	EA	0.615	200	59.00		260
2450	45	"	0.615	160	59.00		220
2460	30	"	0.615	210	59.00		270
2470	Fiberglass cable tray, 36"	LF	0.100	61.00	9.60		71.00
2480	Tray cover	"	0.089	48.50	8.53		57.00
2485	Horizontal						
2490	90	EA	0.667	500	64.00		560
2500	45	"	0.667	210	64.00		270
2510	30	"	0.667	210	64.00		270
2515	Inside						
2520	90	EA	0.667	200	64.00		260
2530	45	"	0.667	160	64.00		220
2540	30	"	0.667	210	64.00		270
2550	Horizontal tee	"	1.194	410	110		520
2560	Horizontal cross	"	1.702	480	160		640
2570	Splice plate	"	0.160	20.75	15.25		36.00
2580	Floor flange	"	0.444	22.75	42.75		66.00
2590	Panel flange	"	2.105	23.75	200		220
2600	End plate	"	0.200	15.50	19.25		34.75
2610	Nylon rivet	"	0.050	0.59	4.80		5.39
2620	Barrier strip	"	0.050	7.80	4.80		12.50
2630	Hold down clamp	"	0.050	3.40	4.80		8.20
2640	Drop out	"	0.667	73.00	64.00		140
2650	Cover stand off	"	0.050	6.52	4.80		11.25
2660	Wall bracket	"	0.200	110	19.25		130
2670	Sealer	"					25.00
2675	Outside						
2680	90	EA	0.667	200	64.00		260
2690	45	"	0.667	170	64.00		230
2700	30	"	0.667	210	64.00		270
2705	Reducers						

CONDUCTORS, CONDUIT AND RACEWAYS

	Description	Output		Unit Costs			
ID Code	Component Descriptions	Unit of Meas.	Manhr / Unit	Material Cost	Labor Cost	Equipment Cost	Total Cost
26 - 05362	**FIBERGLASS CABLE TRAY, Cont'd...**						**26 - 05362**
2710	12" - 6"	EA	0.296	260	28.50		290
2720	12" - 9"	"	0.296	260	28.50		290
2730	18" - 6"	"	0.348	230	33.50		260
2740	18" - 9"	"	0.348	230	33.50		260
2750	18" - 12"	"	0.400	240	38.50		280
2760	24" - 6"	"	0.400	270	38.50		310
2770	24" - 9"	"	0.400	280	38.50		320
2780	24" - 12"	"	0.400	280	38.50		320
2790	24" - 18"	"	0.444	280	42.75		320
2800	30" - 9"	"	0.400	280	38.50		320
2810	30" - 12"	"	0.400	280	38.50		320
2820	30" - 18"	"	0.444	280	42.75		320
2830	30" - 24"	"	0.500	310	48.00		360
2840	36" - 9"	"	0.500	300	48.00		350
2850	36" - 12"	"	0.500	310	48.00		360
2860	36" - 18"	"	0.533	310	51.00		360
2870	36" - 24"	"	0.615	330	59.00		390
2880	36" - 30"	"	0.727	330	70.00		400
26 - 05394	**UNDERFLOOR DUCT**						**26 - 05394**
1010	Underfloor blank duct, insert duct						
1020	7/8"	LF	0.050	26.75	4.80		31.50
1030	1-3/8"	"	0.050	28.25	4.80		33.00
1040	1-7/8"	"	0.050	28.75	4.80		33.50
1050	Box opening plugs	EA	0.145	14.50	14.00		28.50
1060	Duct end plugs	"	0.145	16.75	14.00		30.75
1070	Sleeve couplings	"	0.348	38.50	33.50		72.00
1080	Expansion couplings	"	0.348	220	33.50		250
1090	Vertical elbow	"	0.348	180	33.50		210
1100	Offset elbow	"	0.348	150	33.50		180
1110	Horizontal elbow	"	0.348	360	33.50		390
1120	Adjustable elbow	"	0.348	66.00	33.50		100
1130	Cabinet connector	"	1.194	53.00	110		160
1140	Y take-off	"	0.615	160	59.00		220
1150	Underfloor duct leveling legs	"	0.145	16.50	14.00		30.50
1155	Conduit adapters						
1160	1/2"	EA	0.250	66.00	24.00		90.00
1170	3/4"	"	0.250	67.00	24.00		91.00

CONDUCTORS, CONDUIT AND RACEWAYS

ID Code	Description — Component Descriptions	Output — Unit of Meas.	Output — Manhr / Unit	Unit Costs — Material Cost	Unit Costs — Labor Cost	Unit Costs — Equipment Cost	Unit Costs — Total Cost
26 - 05394	**UNDERFLOOR DUCT, Cont'd...**						**26 - 05394**
1180	1"	EA	0.250	57.00	24.00		81.00
1190	1-1/4"	"	0.296	75.00	28.50		100
1200	2"	"	0.348	100	33.50		130
1205	Reducer bushings						
1210	1-1/4" x 3/4"	EA	0.200	33.25	19.25		53.00
1220	1-1/4" x 1"	"	0.200	38.50	19.25		58.00
1230	2" x 1-1/2"	"	0.250	45.25	24.00		69.00
1235	Support couplers						
1240	1 standard	EA	0.250	93.00	24.00		120
1250	2 standard	"	0.296	93.00	28.50		120
1260	3 standard	"	0.348	93.00	33.50		130
1265	Supports						
1270	1 duct	EA	0.145	93.00	14.00		110
1280	2 duct	"	0.170	93.00	16.25		110
1290	3 duct	"	0.190	93.00	18.25		110
1300	4 duct	"	0.250	130	24.00		150
1310	5 duct	"	0.296	140	28.50		170
1315	Single level junction box						
1320	1 standard	EA	0.800	910	77.00		990
1330	2 standard	"	1.509	1,350	140		1,490
1340	3 standard	"	2.963	2,610	280		2,890
1350	4 standard	"	4.000	3,340	380		3,720
1355	Two level junction boxes						
1360	1 standard	EA	1.000	990	96.00		1,090
1370	2 standard	"	2.000	1,550	190		1,740
1380	Sealing compound	"					20.25
1390	Insert adapters	"	0.145	49.50	14.00		64.00
1400	Ellipsoids	"	0.348	55.00	33.50		89.00
1410	Insert closing cap	"	0.145	5.78	14.00		19.75
1420	Marker Screws	"	0.145	15.25	14.00		29.25
1430	Access boxes	"	1.000	1,090	96.00		1,190
1440	Closing caps	"	0.145	5.78	14.00		19.75
1450	Afterset markers	"	0.145	15.00	14.00		29.00
1460	Cell markers	"	0.145	33.00	14.00		47.00
1470	Tie down straps	"	0.145	30.50	14.00		44.50
1480	Plastic grommets	"	0.400	15.25	38.50		54.00
1490	Metal grommets	"	0.400	110	38.50		150
1495	Receptacle						

CONDUCTORS, CONDUIT AND RACEWAYS

ID Code	Description	Output		Unit Costs			
	Component Descriptions	Unit of Meas.	Manhr / Unit	Material Cost	Labor Cost	Equipment Cost	Total Cost
26 - 05394	**UNDERFLOOR DUCT, Cont'd...**					**26 - 05394**	
1500	Duplex, 20a	EA	0.500	160	48.00		210
1505	Single						
1510	30a	EA	0.500	210	48.00		260
1520	50a	"	0.615	240	59.00		300
1530	Double duplex	"	0.615	170	59.00		230
1540	Single, 20a	"	0.444	150	42.75		190
1550	Double single	"	0.500	180	48.00		230
1560	Twist lock	"	0.444	160	42.75		200
1570	1 conduit opening	"	0.444	160	42.75		200
1580	2 conduit openings	"	0.500	180	48.00		230
1590	1 bushed opening	"	0.444	160	42.75		200
1600	2 bushed openings	"	0.500	180	48.00		230
1605	Amphenol connector						
1610	1"	EA	0.500	260	48.00		310
1620	2"	"	0.533	290	51.00		340
1630	5"	"	0.615	340	59.00		400
1635	Standpipes						
1640	Aluminum	EA	0.250	190	24.00		210
1650	Brass	"	0.250	200	24.00		220
1655	Abandonment plates						
1660	Aluminum	EA	0.250	55.00	24.00		79.00
1670	Brass	"	0.250	91.00	24.00		110
1675	Split bell caps						
1680	Aluminum	EA	0.250	310	24.00		330
1690	Brass	"	0.250	400	24.00		420
1695	Flush floor receptacles						
1700	Aluminum	EA	0.727	260	70.00		330
1710	Brass	"	0.727	290	70.00		360
1715	Flush floor telephone						
1720	Aluminum	EA	0.500	190	48.00		240
1730	Brass	"	0.500	230	48.00		280
1735	Super underfloor duct, blank duct						
1740	1/2"	LF	0.059	52.00	5.68		58.00
1750	7/8"	"	0.059	54.00	5.68		60.00
1760	1-3/8"	"	0.059	55.00	5.68		61.00
1770	1-7/8"	"	0.059	62.00	5.68		68.00
1780	Box opening plugs	EA	0.145	12.00	14.00		26.00
1790	End plugs	"	0.145	12.75	14.00		26.75

CONDUCTORS, CONDUIT AND RACEWAYS

ID Code	Description / Component Descriptions	Output / Unit of Meas.	Manhr / Unit	Unit Costs / Material Cost	Labor Cost	Equipment Cost	Total Cost
26 - 05394	**UNDERFLOOR DUCT, Cont'd...**						**26 - 05394**
1800	Conduit adapters	EA	0.727	50.00	70.00		120
1810	Sleeve coupling	"	0.444	40.75	42.75		84.00
1820	Expansion coupling	"	0.444	140	42.75		180
1830	Reducing coupling	"	0.444	170	42.75		210
1840	Vertical elbow	"	0.444	110	42.75		150
1850	Offset elbow	"	0.444	110	42.75		150
1860	Horizontal elbow	"	0.444	270	42.75		310
1870	Adjustable elbow	"	0.444	110	42.75		150
1880	Cabinet connector	"	1.509	81.00	140		220
1890	Super underfloor duct Y take-off	"	0.727	160	70.00		230
1900	Leveling legs	"	0.145	14.75	14.00		28.75
1905	Support couplers						
1910	1 super	EA	0.250	72.00	24.00		96.00
1920	2 super	"	0.296	97.00	28.50		130
1930	1 super, 1 standard	"	0.296	89.00	28.50		120
1940	2 super, 2 standard	"	0.348	110	33.50		140
1945	Single level junction boxes						
1950	1 super	EA	1.509	940	140		1,080
1960	2 super	"	2.000	2,040	190		2,230
1970	4 super	"	3.478	2,610	330		2,940
1975	Double level junction boxes						
1980	1 super	EA	1.194	1,170	110		1,280
1990	2 super	"	1.509	1,540	140		1,680
26 - 05436	**TRENCH DUCT**						**26 - 05436**
1000	Trench duct, with cover						
1010	9"	LF	0.170	240	16.25		260
1020	12"	"	0.200	270	19.25		290
1030	18"	"	0.267	360	25.50		390
1040	24"	"	0.348	470	33.50		500
1050	30"	"	0.400	530	38.50		570
1060	36"	"	0.571	620	55.00		680
1065	Tees						
1070	9"	EA	1.739	880	170		1,050
1080	12"	"	2.000	1,030	190		1,220
1090	18"	"	2.222	1,320	210		1,530
1100	24"	"	2.500	1,890	240		2,130
1110	30"	"	2.963	2,470	280		2,750

CONDUCTORS, CONDUIT AND RACEWAYS

ID Code	Description — Component Descriptions	Output — Unit of Meas.	Output — Manhr / Unit	Unit Costs — Material Cost	Unit Costs — Labor Cost	Unit Costs — Equipment Cost	Unit Costs — Total Cost
26 - 05436	**TRENCH DUCT, Cont'd...**						**26 - 05436**
1120	36"	EA	3.376	3,220	320		3,540
1125	Vertical elbows						
1130	9"	EA	0.800	310	77.00		390
1140	12"	"	1.096	340	110		450
1150	18"	"	1.356	390	130		520
1160	24"	"	1.667	480	160		640
1170	30"	"	2.000	530	190		720
1180	36"	"	2.500	580	240		820
1185	Cabinet connectors						
1190	9"	EA	2.000	490	190		680
1200	12"	"	2.105	550	200		750
1210	18"	"	2.424	640	230		870
1220	24"	"	2.500	780	240		1,020
1230	30"	"	2.759	930	260		1,190
1240	36"	"	2.963	1,020	280		1,300
1245	End closers						
1250	9"	EA	0.615	91.00	59.00		150
1260	12"	"	0.667	100	64.00		160
1270	18"	"	0.800	160	77.00		240
1280	24"	"	1.096	210	110		320
1290	30"	"	1.290	260	120		380
1300	36"	"	1.455	310	140		450
1305	Horizontal elbows						
1310	9"	EA	1.509	830	140		970
1320	12"	"	1.739	960	170		1,130
1330	18"	"	2.105	1,170	200		1,370
1340	24"	"	2.500	1,840	240		2,080
1350	30"	"	2.857	2,470	270		2,740
1360	36"	"	3.200	3,220	310		3,530
1365	Crosses						
1370	9"	EA	2.000	1,450	190		1,640
1380	12"	"	2.222	1,540	210		1,750
1390	18"	"	2.500	1,840	240		2,080
1400	24"	"	2.759	2,350	260		2,610
1410	30"	"	3.200	3,030	310		3,340
1420	36"	"	3.478	3,810	330		4,140

TRANSFORMERS

ID Code	Description — Component Descriptions	Output — Unit of Meas.	Output — Manhr / Unit	Unit Costs — Material Cost	Unit Costs — Labor Cost	Unit Costs — Equipment Cost	Unit Costs — Total Cost
26 - 12001	**MEDIUM-VOLTAGE TRANSFORMERS**						**26 - 12001**
0080	Floor mtd, one phase, int. dry, 480v-120/240v						
0100	3 kva	EA	1.818	890	170		1,060
1080	5 kva	"	3.077	1,180	300		1,480
1100	7.5 kva	"	3.478	1,590	330		1,920
1120	10 kva	"	3.810	1,980	370		2,350
1140	15 kva	"	4.301	2,650	410		3,060
1160	25 kva	"	7.547	4,860	720		5,580
1180	37.5 kva	"	9.412	5,190	900		6,090
1200	50 kva	"	10.256	6,170	980		7,150
1220	75 kva	"	10.667	8,130	1,020		9,150
1240	100 kva	"	11.594	10,670	1,110		11,780
1980	Three phase, 480v-120/208v						
2000	15 kva	EA	6.015	2,950	580		3,530
2040	30 kva	"	9.412	3,550	900		4,450
2060	45 kva	"	10.811	4,720	1,040		5,760
2080	75 kva	"	10.959	7,100	1,050		8,150
2100	112.5 kva	"	12.698	9,310	1,220		10,530
2120	150 kva	"	13.559	11,080	1,300		12,380
2140	225 kva	"	15.385	16,960	1,480		18,440
3000	Single phase, dry type, 2400v						
3020	167 kva	EA	22.472	30,230	2,160		32,390
3030	250 kva	"	29.963	39,420	2,880		42,300
3040	333 kva	"	37.559	48,420	3,610		52,030
3045	5000v						
3050	167 kva	EA	22.472	32,890	2,160		35,050
3060	250 kva	"	29.963	40,720	2,880		43,600
3070	333 kva	"	37.559	50,920	3,610		54,530
3075	8660v						
3080	167 kva	EA	27.491	34,030	2,640		36,670
3090	250 kva	"	34.934	46,010	3,350		49,360
3100	333 kva	"	67.797	54,630	6,510		61,140
3105	1500v						
3110	167 kva	EA	27.491	38,370	2,640		41,010
3120	250 kva	"	34.934	49,680	3,350		53,030
3130	333 kva	"	42.553	59,350	4,090		63,440
3135	Three phase, dry type transformer, 2400v						
3140	225 kva	EA	25.000	39,820	2,400		42,220
3150	300 kva	"	27.491	48,050	2,640		50,690

TRANSFORMERS

ID Code	Description — Component Descriptions	Output — Unit of Meas.	Manhr / Unit	Unit Costs — Material Cost	Labor Cost	Equipment Cost	Total Cost
26 - 12001	**MEDIUM-VOLTAGE TRANSFORMERS, Cont'd...**					**26 - 12001**	
3160	500 kva	EA	42.553	56,110	4,090		60,200
3170	750 kva	"	52.632	70,260	5,050		75,310
3175	5000v						
3180	225 kva	EA	25.000	43,080	2,400		45,480
3190	300 kva	"	27.491	53,410	2,640		56,050
3200	500 kva	"	42.553	65,430	4,090		69,520
3210	750 kva	"	52.632	90,120	5,050		95,170
3215	8660v						
3220	225 kva	EA	29.963	51,060	2,880		53,940
3230	300 kva	"	32.520	61,610	3,120		64,730
3240	500 kva	"	47.619	71,950	4,570		76,520
3250	750 kva	"	57.554	90,100	5,530		95,630
3255	1500v						
3260	225 kva	EA	29.963	57,330	2,880		60,210
3270	300 kva	"	32.520	67,270	3,120		70,390
3280	500 kva	"	47.619	83,620	4,570		88,190
3290	750 kva	"	57.554	102,350	5,530		107,880
3295	Buck boost transformers						
3300	.25 kva	EA	1.000	260	96.00		360
3310	.50 kva	"	1.250	330	120		450
3320	.75 kva	"	1.509	430	140		570
3330	1.00 kva	"	1.739	530	170		700
3340	1.50 kva	"	2.000	660	190		850
3350	2.00 kva	"	2.500	840	240		1,080
3360	3.00 kva	"	2.963	1,120	280		1,400
26 - 22001	**HIGH-VOLTAGE TRANSFORMERS**					**26 - 22001**	
0080	Pad mounted, single phase, dry type, 480v-120/240v						
0100	15 kva	EA	8.000	2,610	770		3,380
1020	25 kva	"	8.989	3,520	860		4,380
1040	37.5 kva	"	10.000	5,120	960		6,080
1060	50 kva	"	10.959	6,060	1,050		7,110
2080	3 phase						
2100	225 kva	EA	25.000	15,180	2,400		17,580
2120	300 kva	"	30.769	19,520	2,950		22,470
2140	500 kva	"	38.095	30,720	3,660		34,380
2160	750 kva	"	47.059	49,520	4,520		54,040
2180	1000 kva	"	50.000	60,000	4,800		64,800

TRANSFORMERS

ID Code	Description — Component Descriptions	Output — Unit of Meas.	Output — Manhr / Unit	Unit Costs — Material Cost	Unit Costs — Labor Cost	Unit Costs — Equipment Cost	Unit Costs — Total Cost
26 - 22001	**HIGH-VOLTAGE TRANSFORMERS, Cont'd...**						**26 - 22001**
2200	1500 kva	EA	57.143	70,760	5,490		76,250
2240	Substation transformers, outdoor, 5 kv - 208v						
2260	112.5 kva	EA	21.978	23,380	2,110		25,490
2270	150 kva	"	24.024	25,410	2,310		27,720
2280	225 kva	"	27.972	28,810	2,690		31,500
2290	300 kva	"	29.963	33,200	2,880		36,080
2300	500 kva	"	44.944	44,050	4,310		48,360
2310	750 kva	"	55.172	61,000	5,300		66,300
2320	1000 kva	"	65.041	75,230	6,240		81,470
2325	15 kv, 208v						
2330	112 kva	EA	27.972	29,270	2,690		31,960
2340	150 kva	"	29.963	29,930	2,880		32,810
2350	225 kva	"	34.934	30,930	3,350		34,280
2360	300 kva	"	40.000	34,870	3,840		38,710
2370	500 kva	"	50.000	49,020	4,800		53,820
2380	750 kva	"	60.150	49,020	5,770		54,790
2390	1000 kva	"	70.175	73,680	6,740		80,420
2395	5kv, 480v						
2400	112kva	EA	21.978	22,360	2,110		24,470
2410	150 kva	"	24.024	23,690	2,310		26,000
2420	225 kva	"	27.972	26,650	2,690		29,340
2430	300 kva	"	29.963	29,930	2,880		32,810
2440	500 kva	"	44.944	40,130	4,310		44,440
2450	750 kva	"	55.172	54,610	5,300		59,910
2460	1000 kva	"	65.041	64,150	6,240		70,390
2470	1500 kva	"	74.766	83,880	7,180		91,060
2480	2000 kva	"	89.888	103,300	8,630		111,930
2490	2500 kva	"	109.589	122,700	10,520		133,220
2495	15 kv, 480v						
2500	112.5 kva	EA	27.972	28,620	2,690		31,310
2510	150 kva	"	29.963	29,270	2,880		32,150
2520	225 kva	"	34.934	29,600	3,350		32,950
2530	300 kva	"	40.000	32,240	3,840		36,080
2540	500 kva	"	50.000	45,720	4,800		50,520
2550	750 kva	"	60.150	55,590	5,770		61,360
2560	1000 kva	"	70.175	64,150	6,740		70,890
2570	1500 kva	"	80.000	84,220	7,680		91,900
2580	2000 kva	"	89.888	103,630	8,630		112,260

TRANSFORMERS

ID Code	Description — Component Descriptions	Output — Unit of Meas.	Output — Manhr / Unit	Unit Costs — Material Cost	Unit Costs — Labor Cost	Unit Costs — Equipment Cost	Unit Costs — Total Cost
26 - 22001	**HIGH-VOLTAGE TRANSFORMERS, Cont'd...**						**26 - 22001**
2590	2500 kva	EA	119.403	123,040	11,460		134,500
2980	Pad mounted 3 phase, 15 kv outdoor						
3000	50 kva	EA	10.256	7,250	980		8,230
3020	75 kva	"	11.765	10,190	1,130		11,320
3040	112 kva	"	12.903	13,130	1,240		14,370
3060	150 kva	"	14.545	18,640	1,400		20,040
3080	225 kva	"	15.385	22,450	1,480		23,930
3100	300 kva	"	17.021	27,630	1,630		29,260
3120	500 kva	"	27.586	33,860	2,650		36,510
3140	750 kva	"	36.364	43,180	3,490		46,670
3160	1000 kva	"	44.444	51,120	4,270		55,390
3180	1500 kva	"	53.333	65,290	5,120		70,410
3480	Dry type, for power gear, 5 kv indoor						
3500	75 kva	EA	16.000	20,040	1,540		21,580
3540	112.5 kva	"	18.605	22,450	1,790		24,240
3560	150 kva	"	21.053	25,900	2,020		27,920
3600	225 kva	"	23.529	32,120	2,260		34,380
3620	300 kva	"	25.000	37,980	2,400		40,380
3640	500 kva	"	27.586	50,190	2,650		52,840
3660	750 kva	"	36.364	65,960	3,490		69,450

SERVICE AND DISTRIBUTION

ID Code	Description — Component Descriptions	Output — Unit of Meas.	Output — Manhr / Unit	Unit Costs — Material Cost	Unit Costs — Labor Cost	Unit Costs — Equipment Cost	Unit Costs — Total Cost
26 - 24130	**SWITCHBOARDS**						**26 - 24130**
0580	Switchboard, 90" high, no main disconnect, 208/120v						
0581	400a	EA	7.921	3,770	760		4,530
1000	600a	"	8.000	5,860	770		6,630
1020	1000a	"	8.000	7,370	770		8,140
1040	1200a	"	10.000	7,800	960		8,760
1060	1600a	"	11.940	8,570	1,150		9,720
1080	2000a	"	14.035	9,200	1,350		10,550
1100	2500a	"	16.000	9,310	1,540		10,850
1520	277/480v						
1540	600a	EA	8.163	6,720	780		7,500
1560	800a	"	8.163	7,370	780		8,150
1580	1600a	"	11.940	9,280	1,150		10,430
1600	2000a	"	14.035	9,930	1,350		11,280
1620	2500a	"	16.000	10,570	1,540		12,110

SERVICE AND DISTRIBUTION

ID Code	Description (Component Descriptions)	Output (Unit of Meas.)	Output (Manhr / Unit)	Unit Costs (Material Cost)	Unit Costs (Labor Cost)	Unit Costs (Equipment Cost)	Unit Costs (Total Cost)
26 - 24130	**SWITCHBOARDS, Cont'd...**						**26 - 24130**
1640	3000a	EA	27.586	12,170	2,650		14,820
1660	4000a	"	29.630	14,740	2,840		17,580
1680	Main breaker sections, 600v						
1700	1200a, GFI	EA	16.985	27,920	1,630		29,550
1705	1600a, GFI	"	19.512	32,980	1,870		34,850
1710	2000a, GFI	"	20.000	35,000	1,920		36,920
1720	2500a, GFI	"	25.000	38,700	2,400		41,100
1730	3000a, GFI	"	29.963	46,500	2,880		49,380
1740	4000a, GFI	"	34.934	51,140	3,350		54,490
1745	Switchboard meter sections, 600v						
1750	400a	EA	8.000	5,040	770		5,810
1760	600a	"	10.000	7,740	960		8,700
1780	800a	"	11.004	9,410	1,060		10,470
1790	1000a	"	13.491	11,440	1,300		12,740
1800	2000a	"	16.000	14,140	1,540		15,680
1810	2500a	"	20.000	15,810	1,920		17,730
1820	3000a	"	25.000	16,490	2,400		18,890
1830	4000a	"	29.963	18,510	2,880		21,390
2000	Insulated case, draw out compartment, 208/120v						
2040	800a	EA	2.500	3,050	240		3,290
2060	1600a	"	2.963	3,650	280		3,930
2080	2000a	"	3.478	4,570	330		4,900
2100	2500a	"	3.478	5,490	330		5,820
2120	3000a	"	4.000	9,150	380		9,530
2140	4000a	"	4.790	12,820	460		13,280
2240	Accessories for power trip breakers						
2260	Shunt trip	EA	0.500	1,460	48.00		1,510
2280	Key interlock	"	2.222	710	210		920
2300	Lifting and transport truck	"	4.494	4,570	430		5,000
2320	Lifting device	"	1.333	470	130		600
2500	Bus duct connection, 3 phase, 4 wire						
2520	225a	EA	2.963	660	280		940
2540	400a	"	2.963	820	280		1,100
2560	600a	"	3.333	1,070	320		1,390
2580	800a	"	4.000	1,170	380		1,550
2680	2500a	"	6.015	2,180	580		2,760
2700	3000a	"	7.477	3,280	720		4,000
2720	4000a	"	8.791	5,890	840		6,730

SERVICE AND DISTRIBUTION

ID Code	Component Descriptions	Unit of Meas.	Manhr / Unit	Material Cost	Labor Cost	Equipment Cost	Total Cost
	Description	**Output**		**Unit Costs**			
26 - 24130	**SWITCHBOARDS, Cont'd...**						**26 - 24130**
3020	Provision for mounting current transformers						
3040	800a & below primary	EA	2.963	2,460	280		2,740
3060	1000 to 1500a primary	"	2.963	3,060	280		3,340
3080	2000 to 6000a primary	"	2.963	3,690	280		3,970
3100	Provision for mounting potential transformers						
3120	2000a max	EA	3.810	8,810	370		9,180
3140	Switchboard instruments						
3160	Voltmeter	EA	1.000	3,160	96.00		3,260
3180	Ammeter, incoming line	"	1.000	3,000	96.00		3,100
3220	Wattmeter	"	1.000	5,020	96.00		5,120
3240	Varmeter	"	1.000	5,190	96.00		5,290
3260	Power factor meter	"	1.000	6,000	96.00		6,100
3280	Frequency meter	"	1.000	6,720	96.00		6,820
3300	Recording voltmeter	"	2.000	13,420	190		13,610
3340	Wattmeter	"	2.000	14,280	190		14,470
3360	Power factor meter	"	2.000	16,670	190		16,860
3400	Frequency meter	"	2.000	16,670	190		16,860
3420	Instrument phase select switch	"	0.500	600	48.00		650
5460	Enclosure, 90" high, 3 phase, 4 wire						
5480	1000a	EA	6.838	5,450	660		6,110
5540	1200a	"	7.018	5,760	670		6,430
5560	1600a	"	8.602	6,410	830		7,240
5580	2000a	"	13.333	6,720	1,280		8,000
5590	5500a	"	15.686	8,330	1,510		9,840
5600	3000a	"	18.182	9,280	1,750		11,030
5620	4000a	"	23.529	12,170	2,260		14,430
6120	Circuit breakers, 600v, 100a, frame						
6150	15-30a, 1 pole	EA	0.296	210	28.50		240
6160	15-60a, 2 pole	"	0.348	550	33.50		580
6170	70-100a, 2 pole	"	0.400	680	38.50		720
6180	15-60a, 3 pole	"	0.444	700	42.75		740
6190	70-100a, 3 pole	"	0.500	810	48.00		860
6195	Bolt on breakers, 600v, 225a frame, 110-225a						
6200	2 pole	EA	0.615	1,590	59.00		1,650
6210	3 pole	"	1.096	1,990	110		2,100
6220	400a frame, 250-400a, 2 pole	"	1.250	2,920	120		3,040
6230	800a frame						
6240	450-600a, 2 pole	EA	1.905	5,130	180		5,310

SERVICE AND DISTRIBUTION

ID Code	Description / Component Descriptions	Unit of Meas.	Output / Manhr / Unit	Material Cost	Labor Cost	Equipment Cost	Total Cost
26 - 24130	**SWITCHBOARDS, Cont'd...**					**26 - 24130**	
6250	700-800a, 2 pole	EA	2.500	5,990	240		6,230
6260	450-600a, 3 pole	"	4.211	5,870	400		6,270
6270	700-800a, 3 pole	"	4.444	7,590	430		8,020
6290	Bolt on branch breakers, 600v						
6300	1000-2000a, 2 pole	EA	5.333	10,650	510		11,160
6310	2500a, 2 pole	"	10.753	19,670	1,030		20,700
6320	1000-2000a, 3 pole	"	8.000	13,590	770		14,360
6330	2500a, 3 pole	"	11.004	23,990	1,060		25,050
6340	3000a, 3 pole	"	20.000	45,660	1,920		47,580
6345	Metal clad substation switch board, selector switch						
6350	600a, 5kv	EA	42.105	35,240	4,040		39,280
6360	15kv	"	47.904	39,100	4,600		43,700
6365	Fused switch, 600a						
6370	5kv	EA	34.934	25,650	3,350		29,000
6380	15kv	"	34.934	35,560	3,350		38,910
6385	1200a						
6390	5kv	EA	40.000	28,520	3,840		32,360
6400	15kv	"	40.000	38,210	3,840		42,050
6405	Oil cutout switch						
6410	5 kv	EA	15.009	9,930	1,440		11,370
6420	15 kv	"	18.018	13,780	1,730		15,510
6430	Liquid air terminal section	"	8.000	1,280	770		2,050
6440	Dry air terminal section	"	8.502	2,560	820		3,380
6450	Auxiliary compartment	"	29.963	18,260	2,880		21,140
26 - 24160	**PANELBOARDS**					**26 - 24160**	
1000	Indoor load center, 1 phase 240v main lug only						
1020	30a - 2 spaces	EA	2.000	40.00	190		230
1030	100a - 8 spaces	"	2.424	130	230		360
1040	150a - 16 spaces	"	2.963	330	280		610
1050	200a - 24 spaces	"	3.478	690	330		1,020
1060	200a - 42 spaces	"	4.000	710	380		1,090
1100	Main circuit breaker						
1110	100a - 8 spaces	EA	2.424	410	230		640
1120	100a - 16 spaces	"	2.759	440	260		700
1130	150a - 16 spaces	"	2.963	720	280		1,000
1140	150a - 24 spaces	"	3.200	850	310		1,160
1150	200a - 24 spaces	"	3.478	800	330		1,130

SERVICE AND DISTRIBUTION

ID Code	Description — Component Descriptions	Output — Unit of Meas.	Manhr / Unit	Material Cost	Labor Cost	Equipment Cost	Total Cost
				Unit Costs			

ID Code	Component Descriptions	Unit of Meas.	Manhr / Unit	Material Cost	Labor Cost	Equipment Cost	Total Cost
26 - 24160	**PANELBOARDS, Cont'd...**						**26 - 24160**
1160	200a - 42 spaces	EA	3.636	1,140	350		1,490
1500	3 phase, 480/277v, main lugs only, 120a, 30 circuits	"	3.478	2,070	330		2,400
1510	277/480v, 4 wire, flush surface						
1520	225a, 30 circuits	EA	4.000	3,380	380		3,760
1540	400a, 30 circuits	"	5.000	4,580	480		5,060
1560	600a, 42 circuits	"	6.015	8,840	580		9,420
2000	208/120v, main circuit breaker, 3 phase, 4 wire						
2030	100a						
2035	12 circuits	EA	5.096	1,790	490		2,280
2040	20 circuits	"	6.299	2,230	600		2,830
2060	30 circuits	"	7.018	3,270	670		3,940
2070	225a						
2080	30 circuits	EA	7.767	2,790	750		3,540
2100	42 circuits	"	9.524	5,060	910		5,970
2110	400a						
2120	30 circuits	EA	14.815	6,900	1,420		8,320
2140	42 circuits	"	16.000	8,270	1,540		9,810
2180	600a, 42 circuits	"	18.182	16,090	1,750		17,840
2510	120/208v, flush, 3 ph., 4 wire, main only						
2515	100a						
2520	12 circuits	EA	5.096	1,270	490		1,760
2540	20 circuits	"	6.299	1,750	600		2,350
2560	30 circuits	"	7.018	2,600	670		3,270
2570	225a						
2580	30 circuits	EA	7.767	2,640	750		3,390
2600	42 circuits	"	9.524	3,340	910		4,250
2610	400a						
2620	30 circuits	EA	14.815	5,060	1,420		6,480
2640	42 circuits	"	16.000	7,380	1,540		8,920
2680	600a, 42 circuits	"	18.182	11,500	1,750		13,250
2700	Panelboard accessories						
2720	Grounding bus	EA	0.348	59.00	33.50		93.00
2740	Handle lock device	"	0.145	28.75	14.00		42.75
2750	Factory assembled panel						
2760	1 pole space	EA	0.348	36.75	33.50		70.00
2780	2 pole space	"	0.145	78.00	14.00		92.00
2800	3 pole space	"	0.133	120	12.75		130
3000	Panelboards 1 phase, 240/120v main circuit breaker						

SERVICE AND DISTRIBUTION

ID Code	Component Descriptions	Unit of Meas.	Manhr / Unit	Material Cost	Labor Cost	Equipment Cost	Total Cost
26 - 24160	**PANELBOARDS, Cont'd...**						**26 - 24160**
3020	Single phase, 3 wire, 120/240v flush						
3040	100a, 20 circuits	EA	3.478	1,860	330		2,190
3060	225a, 30 circuits	"	4.000	3,540	380		3,920
3500	240/120v, main lugs only						
3510	100a						
3520	8 circuits	EA	2.963	890	280		1,170
3540	12 circuits	"	2.963	960	280		1,240
3560	20 circuits	"	2.963	1,000	280		1,280
3570	225a						
3580	24 circuits	EA	3.478	1,070	330		1,400
3600	30 circuits	"	3.810	1,220	370		1,590
3620	42 circuits	"	3.810	1,350	370		1,720
4000	Distribution panelboards, 3 ph, main breaker						
4010	225a	EA	16.000	3,380	1,540		4,920
4020	400a	"	18.018	6,200	1,730		7,930
4030	600a	"	21.978	9,670	2,110		11,780
4040	800a	"	24.024	12,420	2,310		14,730
4050	1000a	"	27.972	16,110	2,690		18,800
4060	1200a	"	29.963	19,800	2,880		22,680
4065	Single phase						
4070	225a	EA	14.011	2,990	1,350		4,340
4080	400a	"	16.000	5,520	1,540		7,060
4090	600a	"	20.000	7,810	1,920		9,730
4100	800a	"	24.024	10,590	2,310		12,900
4110	1000a	"	27.972	13,790	2,690		16,480
4120	1200a	"	29.963	17,480	2,880		20,360
4125	Fusible distribution panelboards, 3 phase, 600v						
4130	100a	EA	14.011	3,210	1,350		4,560
4140	200a	"	16.000	3,670	1,540		5,210
4150	400a	"	20.000	6,440	1,920		8,360
4160	600a	"	24.024	8,250	2,310		10,560
4170	800a	"	27.972	12,420	2,690		15,110
4175	Single phase						
4180	100a	EA	11.994	2,750	1,150		3,900
4190	200a	"	14.011	3,210	1,350		4,560
4200	400a	"	18.018	6,900	1,730		8,630
4210	600a	"	21.978	7,360	2,110		9,470
4220	800a	"	25.974	10,130	2,490		12,620

SERVICE AND DISTRIBUTION

ID Code	Component Descriptions	Unit of Meas.	Manhr / Unit	Material Cost	Labor Cost	Equipment Cost	Total Cost
	Description	**Output**		**Unit Costs**			

26 - 24160 — PANELBOARDS, Cont'd... — 26 - 24160

ID Code	Component Descriptions	Unit of Meas.	Manhr / Unit	Material Cost	Labor Cost	Equipment Cost	Total Cost
4225	Hospital panels, operating room						
4230	3kv - 208v	EA	6.154	9,730	590		10,320
4240	3kv - 277v	"	6.154	10,170	590		10,760
4250	5kv - 208v	"	6.154	10,170	590		10,760
4260	5kv - 277v	"	6.154	11,070	590		11,660
4265	Coronary care						
4270	3kv - 208v	EA	7.273	11,500	700		12,200
4280	3kv - 277v	"	7.273	12,400	700		13,100
4290	5kv - 208v	"	7.273	12,400	700		13,100
4300	5kv - 277v	"	7.273	13,270	700		13,970
4305	Intensive care						
4310	3kv - 208v	EA	8.000	12,830	770		13,600
4320	3kv - 277v	"	8.000	13,270	770		14,040
4330	5kv - 208v	"	8.000	13,270	770		14,040
4340	5kv - 277v	"	8.000	14,170	770		14,940
4350	15kv - 208v	"	11.994	25,230	1,150		26,380
4360	15kv - 277v	"	11.994	26,100	1,150		27,250
4370	25kv - 208v	"	16.000	27,000	1,540		28,540
4380	25kv - 277v	"	16.000	27,870	1,540		29,410
4385	Explosion proof, 240v, m.l.b. 20a, single phase						
4390	6 breakers	EA	11.004	7,070	1,060		8,130
4400	8 breakers	"	11.747	7,970	1,130		9,100
4410	10 breakers	"	12.500	9,300	1,200		10,500
4420	12 breakers	"	13.245	10,610	1,270		11,880
4430	14 breakers	"	14.011	12,830	1,350		14,180
4440	16 breakers	"	14.011	15,500	1,350		16,850
4450	18 breakers	"	15.504	16,810	1,490		18,300
4460	20 breakers	"	16.260	17,720	1,560		19,280
4470	22 breakers	"	16.985	18,140	1,630		19,770
4480	24 breakers	"	17.738	19,030	1,700		20,730

26 - 24190 — MOTOR CONTROLS — 26 - 24190

ID Code	Component Descriptions	Unit of Meas.	Manhr / Unit	Material Cost	Labor Cost	Equipment Cost	Total Cost
0080	Motor generator set, 3 phase, 480/277v, w/controls						
0100	10kw	EA	27.586	16,760	2,650		19,410
1020	15kw	"	30.769	21,860	2,950		24,810
1040	20kw	"	32.000	24,260	3,070		27,330
1060	25kw	"	34.783	27,980	3,340		31,320
1080	30kw	"	36.364	31,310	3,490		34,800

SERVICE AND DISTRIBUTION

ID Code	Component Descriptions	Unit of Meas.	Manhr / Unit	Material Cost	Labor Cost	Equipment Cost	Total Cost
	Description	**Output**		**Unit Costs**			
26 - 24190	**MOTOR CONTROLS, Cont'd...**					**26 - 24190**	
1100	40kw	EA	38.095	34,160	3,660		37,820
1120	50kw	"	40.000	38,120	3,840		41,960
1140	60kw	"	44.444	43,180	4,270		47,450
1160	75kw	"	50.000	44,090	4,800		48,890
1180	100kw	"	61.538	50,880	5,910		56,790
1200	125kw	"	66.667	92,070	6,400		98,470
1220	150kw	"	66.667	101,770	6,400		108,170
1240	200kw	"	72.727	116,310	6,980		123,290
1260	250kw	"	72.727	125,990	6,980		132,970
1280	300kw	"	80.000	145,380	7,680		153,060
2010	2 pole, 230 volt starter, w/NEMA-1						
2020	1 hp, 9a, size 00	EA	1.000	190	96.00		290
2040	2 hp, 18a, size 0	"	1.000	210	96.00		310
2060	3 hp, 27a, size 1	"	1.000	300	96.00		400
2080	5 hp, 45a, size 1p	"	1.000	300	96.00		400
2100	7-1/2 hp, 45a, size 2	"	1.000	720	96.00		820
2120	15 hp, 90a, size 3	"	1.000	1,090	96.00		1,190
2130	2 pole, w/NEMA-4 enclosure						
2140	2 hp, 18a, size 1	EA	1.600	620	150		770
2160	5 hp, 45a, size 1p	"	1.600	790	150		940
2180	7-1/2 hp, 45a, size 2	"	1.600	1,230	150		1,380
2500	3 pole, 2 hp, 9a, 200-575v starter						
2520	W/NEMA-1, size 00	EA	1.333	380	130		510
2540	W/NEMA-4 enclosure, size 00	"	1.739	620	170		790
2550	5hp, 18a						
2560	W/NEMA-1 enclosure, size 0	EA	1.333	480	130		610
2580	W/NEMA-4 enclosure, size 0	"	1.739	940	170		1,110
2590	7.5-10hp, 27a						
2600	7.5-10hp 27a, w/NEMA-1 enclosure, size 1	EA	1.333	540	130		670
2620	W/NEMA-4 enclosure size 1	"	1.739	1,010	170		1,180
2630	10-25hp, 45a						
2640	W/NEMA-1 enclosure, size 2	EA	1.333	1,070	130		1,200
2660	W/NEMA-4 enclosure, size 2	"	1.739	2,010	170		2,180
2670	25-50hp, 90a						
2680	W/NEMA-1 enclosure, size 3	EA	1.739	1,680	170		1,850
2700	W/NEMA-4 enclosure, size 3	"	2.500	1,580	240		1,820
2710	40-100hp, 135a						
2720	W/NEMA-1 enclosure, size 4	EA	2.500	2,700	240		2,940

SERVICE AND DISTRIBUTION

ID Code	Description — Component Descriptions	Unit of Meas.	Manhr / Unit	Material Cost	Labor Cost	Equipment Cost	Total Cost
26 - 24190	**MOTOR CONTROLS, Cont'd...**						**26 - 24190**
2740	W/NEMA-4 enclosure, size 4	EA	3.478	4,160	330		4,490
2745	75-200hp, 270a						
2750	W/NEMA-1 enclosure, size 5	EA	5.517	6,360	530		6,890
2760	W/NEMA-4 enclosure, size 5	"	7.018	8,060	670		8,730
2800	Magnetic starter accessories						
2820	On-off-auto selector switch kit	EA	0.320	43.75	30.75		75.00
2840	With pilot light	"	0.348	82.00	33.50		120
3000	Control center main lug only, 208v, 3 phase						
3010	600a	EA	11.994	2,550	1,150		3,700
3020	1200a	"	16.000	5,450	1,540		6,990
3025	Main circuit breakers, 208v, 3 phase						
3030	400a	EA	10.000	5,090	960		6,050
3040	600a	"	14.011	5,670	1,350		7,020
3050	800a	"	16.000	6,510	1,540		8,050
3060	1000a	"	18.018	7,080	1,730		8,810
3070	1200a	"	20.000	13,030	1,920		14,950
3075	Non-reversing starters						
3080	Size 1	EA	0.727	600	70.00		670
3090	Size 2	"	1.250	1,170	120		1,290
3100	Size 3	"	1.509	1,840	140		1,980
3110	Size 4	"	1.739	3,790	170		3,960
3120	Reversing starters						
3140	Size 1	EA	0.727	1,350	70.00		1,420
3150	Size 2	"	1.096	3,130	110		3,240
3155	Fusible switch, non-revolving starters						
3160	Size 1	EA	0.727	1,270	70.00		1,340
3170	Size 2	"	1.250	1,600	120		1,720
3180	Size 3	"	1.509	2,060	140		2,200
3190	Size 4	"	1.739	3,280	170		3,450
3195	Reversing starters						
3200	Size 1	EA	0.727	1,870	70.00		1,940
3210	Size 2	"	1.096	1,970	110		2,080
3215	Two speed, non-reversing starter						
3220	Size 1	EA	0.727	1,880	70.00		1,950
3230	Size 2	"	1.096	2,760	110		2,870
3235	Magnetic starter, 600v, 2 pole, NEMA 3R						
3240	Size 0, 2 hp	EA	1.000	480	96.00		580
3250	Size 1, 5hp	"	1.096	630	110		740

SERVICE AND DISTRIBUTION

ID Code	Description — Component Descriptions	Output — Unit of Meas.	Manhr / Unit	Unit Costs — Material Cost	Labor Cost	Equipment Cost	Total Cost
26 - 24190	**MOTOR CONTROLS, Cont'd...**						**26 - 24190**
3255	NEMA 3R						
3260	Size 2, 7.5 hp	EA	1.143	1,310	110		1,420
3270	Size 3, 15 hp	"	1.194	2,910	110		3,020
3275	NEMA 7						
3280	Size 0, 2 hp	EA	1.739	1,250	170		1,420
3290	Size 1, 5 hp	"	2.000	1,410	190		1,600
3300	Size 2, 7.5 hp	"	2.222	2,450	210		2,660
3305	NEMA 12						
3310	Size 0, 2 hp	EA	1.509	320	140		460
3320	Size 1, 5 hp	"	1.739	420	170		590
3330	Size 2, 7.5 hp	"	2.000	790	190		980
3340	Size 3, 15 hp	"	2.222	1,250	210		1,460
3345	3 pole, NEMA 1						
3420	Size 6	EA	10.000	13,690	960		14,650
3430	Size 7	"	11.994	18,450	1,150		19,600
3440	Size 8	"	16.000	25,600	1,540		27,140
3445	NEMA 4						
3520	Size 6	EA	14.011	17,260	1,350		18,610
3530	Size 7	"	16.000	22,020	1,540		23,560
3540	Size 8	"	20.000	29,170	1,920		31,090
3545	NEMA 3R						
3550	Size 0	EA	1.250	340	120		460
3560	Size 1	"	1.356	390	130		520
3570	Size 2	"	1.818	730	170		900
3580	Size 3	"	1.905	1,140	180		1,320
3590	Size 4	"	2.759	2,850	260		3,110
3595	NEMA 7						
3600	Size 0	EA	2.000	1,400	190		1,590
3610	Size 1	"	2.162	1,480	210		1,690
3620	Size 2	"	2.222	2,360	210		2,570
3630	Size 3	"	2.759	3,550	260		3,810
3640	Size 4	"	4.444	5,930	430		6,360
3650	Size 5	"	9.744	13,750	940		14,690
3660	Size 6	"	16.000	32,360	1,540		33,900
3665	NEMA 12						
3670	Size 00	EA	1.739	240	170		410
3680	Size 0	"	1.860	260	180		440
3690	Size 1	"	1.905	370	180		550

SERVICE AND DISTRIBUTION

ID Code	Description — Component Descriptions	Output — Unit of Meas.	Manhr / Unit	Unit Costs — Material Cost	Labor Cost	Equipment Cost	Total Cost
26 - 24190	**MOTOR CONTROLS, Cont'd...**						**26 - 24190**
3700	Size 2	EA	2.000	660	190		850
3710	Size 3	"	2.500	1,070	240		1,310
3720	Size 4	"	3.478	2,450	330		2,780
3730	Size 5	"	7.273	5,930	700		6,630
3740	Size 6	"	12.500	14,290	1,200		15,490
3750	Size 7	"	14.011	20,490	1,350		21,840
3760	Size 8	"	20.000	30,210	1,920		32,130
3765	Reversing magnetic starters, 600v, 3 pole, NEMA 1						
3770	Size 00	EA	1.250	1,040	120		1,160
3780	Size 0	"	1.290	1,200	120		1,320
3790	Size 1	"	1.356	1,370	130		1,500
3800	Size 2	"	1.509	2,720	140		2,860
3810	Size 3	"	1.739	4,710	170		4,880
3820	Size 4	"	2.000	10,790	190		10,980
3830	Size 5	"	5.333	23,280	510		23,790
3840	Size 6	"	9.501	52,160	910		53,070
3850	Size 7	"	11.004	71,140	1,060		72,200
3860	Size 8	"	18.018	101,320	1,730		103,050
3865	NEMA 4						
3870	Size 0	EA	1.739	1,750	170		1,920
3880	Size 1	"	1.818	2,140	170		2,310
3890	Size 2	"	1.905	3,830	180		4,010
3900	Size 3	"	2.000	5,860	190		6,050
3910	Size 4	"	2.500	13,320	240		13,560
3920	Size 5	"	7.273	24,280	700		24,980
3930	Size 6	"	12.500	54,480	1,200		55,680
3940	Size 7	"	15.009	71,320	1,440		72,760
3950	Size 8	"	20.000	98,780	1,920		100,700
3965	NEMA 7						
3967	Size 0	EA	2.000	3,610	190		3,800
3970	Size 1	"	2.222	3,760	210		3,970
3980	Size 2	"	2.500	6,480	240		6,720
3990	Size 3	"	2.963	10,170	280		10,450
3995	NEMA 12						
4000	Size 0	EA	1.739	1,560	170		1,730
4010	Size 1	"	1.905	1,760	180		1,940
4020	Size 2	"	2.000	3,270	190		3,460
4030	Size 3	"	2.222	5,860	210		6,070

SERVICE AND DISTRIBUTION

ID Code	Description — Component Descriptions	Output — Unit of Meas.	Output — Manhr / Unit	Unit Costs — Material Cost	Unit Costs — Labor Cost	Unit Costs — Equipment Cost	Unit Costs — Total Cost
26 - 24190	**MOTOR CONTROLS, Cont'd...**						**26 - 24190**
4040	Size 4	EA	2.500	12,650	240		12,890
4050	Size 5	"	7.273	27,930	700		28,630
4060	Size 6	"	14.011	59,050	1,350		60,400
4070	Size 7	"	16.000	78,870	1,540		80,410
4080	Size 8	"	20.000	110,420	1,920		112,340
4085	Electrically held lighting contactors, NEMA 1, 20a						
4090	2 pole	EA	1.000	590	96.00		690
4100	3 pole	"	1.250	650	120		770
4110	4 pole	"	1.509	820	140		960
4120	6 pole	"	2.000	830	190		1,020
4130	8 pole	"	2.500	1,080	240		1,320
4140	10 pole	"	2.963	1,240	280		1,520
4150	12 pole	"	3.478	1,440	330		1,770
4155	30a						
4160	2 pole	EA	1.000	640	96.00		740
4165	3 pole	"	1.250	690	120		810
4170	4 pole	"	1.509	850	140		990
4175	5 pole	"	1.739	1,060	170		1,230
4180	60a						
4185	2 pole	EA	1.000	1,300	96.00		1,400
4190	3 pole	"	1.250	1,380	120		1,500
4195	4 pole	"	1.509	1,720	140		1,860
4200	5 pole	"	1.739	2,500	170		2,670
4205	100a						
4210	2 pole	EA	1.250	2,140	120		2,260
4215	3 pole	"	1.739	2,290	170		2,460
4220	4 pole	"	2.222	2,820	210		3,030
4225	5 pole	"	2.759	3,850	260		4,110
4230	200a						
4280	2 pole	EA	2.759	5,030	260		5,290
4290	3 pole	"	2.963	5,370	280		5,650
4300	4 pole	"	3.200	7,160	310		7,470
4305	300a						
4310	2 pole	EA	4.211	9,480	400		9,880
4320	3 pole	"	5.333	9,940	510		10,450
4325	400a						
4330	2 pole	EA	4.211	19,350	400		19,750
4340	3 pole	"	5.333	21,880	510		22,390

SERVICE AND DISTRIBUTION

ID Code	Description		Output		Unit Costs			
	Component Descriptions	Unit of Meas.	Manhr / Unit	Material Cost	Labor Cost	Equipment Cost	Total Cost
26 - 24190		**MOTOR CONTROLS, Cont'd...**					**26 - 24190**	
4345	600a							
4350	2 pole	EA	6.667	23,810	640		24,450	
4360	3 pole	"	9.249	26,500	890		27,390	
4365	800a							
4370	2 pole	EA	8.000	28,420	770		29,190	
4380	3 pole	"	11.004	31,490	1,060		32,550	
4385	Mechanically held lighting contactors, NEMA 1, 20a							
4390	2 pole	EA	1.000	650	96.00		750	
4400	3 pole	"	1.250	690	120		810	
4410	4 pole	"	1.509	730	140		870	
4420	6 pole	"	2.000	1,200	190		1,390	
4430	8 pole	"	2.500	1,300	240		1,540	
4440	10 pole	"	2.963	1,460	280		1,740	
4445	30a							
4450	2 pole	EA	1.000	690	96.00		790	
4460	3 pole	"	1.250	730	120		850	
4470	4 pole	"	1.509	750	140		890	
4480	5 pole	"	1.739	950	170		1,120	
4485	60a							
4490	2 pole	EA	1.000	1,380	96.00		1,480	
4500	3 pole	"	1.250	1,440	120		1,560	
4510	4 pole	"	1.509	1,730	140		1,870	
4520	5 pole	"	1.739	2,230	170		2,400	
4525	100a							
4530	2 pole	EA	1.250	1,920	120		2,040	
4540	3 pole	"	1.739	2,030	170		2,200	
4550	4 pole	"	2.000	2,440	190		2,630	
4560	5 pole	"	2.500	3,270	240		3,510	
4565	200a							
4570	2 pole	EA	1.739	4,950	170		5,120	
4580	3 pole	"	2.500	5,600	240		5,840	
4590	4 pole	"	3.200	6,820	310		7,130	
4595	300a							
4600	2 pole	EA	4.211	8,830	400		9,230	
4610	3 pole	"	5.333	9,610	510		10,120	
4615	400a							
4620	2 pole	EA	4.211	21,110	400		21,510	
4630	3 pole	"	5.333	23,810	510		24,320	

SERVICE AND DISTRIBUTION

ID Code	Description — Component Descriptions	Output — Unit of Meas.	Output — Manhr / Unit	Unit Costs — Material Cost	Unit Costs — Labor Cost	Unit Costs — Equipment Cost	Unit Costs — Total Cost
26 - 24190	**MOTOR CONTROLS, Cont'd...**						**26 - 24190**
4635	600a						
4640	2 pole	EA	6.667	25,360	640		26,000
4650	3 pole	"	8.889	28,420	850		29,270
4655	800a						
4660	2 pole	EA	8.000	29,970	770		30,740
4670	3 pole	"	11.429	33,420	1,100		34,520
4675	AC relays, control type, open, 15a, 600v						
4680	2 pole	EA	1.000	260	96.00		360
4690	3 pole	"	1.250	290	120		410
4700	4 pole	"	1.509	350	140		490
4710	6 pole	"	2.000	440	190		630
4720	8 pole	"	2.500	520	240		760
4730	10 pole	"	2.963	840	280		1,120
4740	12 pole	"	3.478	890	330		1,220

BASIC MATERIALS

ID Code	Description — Component Descriptions	Output — Unit of Meas.	Output — Manhr / Unit	Unit Costs — Material Cost	Unit Costs — Labor Cost	Unit Costs — Equipment Cost	Unit Costs — Total Cost
26 - 25003	**BUS DUCT**						**26 - 25003**
1000	Bus duct, 100a, plug-in						
1010	10', 600v	EA	2.759	560	260		820
1020	With ground	"	4.211	740	400		1,140
1030	10', 277/480v	"	2.759	710	260		970
1040	With ground	"	4.211	890	400		1,290
1050	Cable tap box	"	2.500	320	240		560
1060	End closure	"	0.400	380	38.50		420
1070	Edgewise hanger	"	0.727	47.50	70.00		120
1080	Flatwise hanger	"	0.727	47.50	70.00		120
1090	Outside elbow	"	0.800	440	77.00		520
1100	Inside elbow	"	0.800	440	77.00		520
1110	Outside tee	"	1.100	570	110		680
1120	Inside tee	"	1.100	570	110		680
1130	Outlet cover	"	0.400	33.00	38.50		72.00
1140	Wall flange	"	0.400	92.00	38.50		130
1145	Circuit breakers, with enclosure						
1147	1 pole						
1150	15a-60a	EA	1.000	520	96.00		620
1160	70a-100a	"	1.250	600	120		720
1165	2 pole						

BASIC MATERIALS

ID Code	Component Descriptions	Unit of Meas.	Manhr / Unit	Material Cost	Labor Cost	Equipment Cost	Total Cost
	Description	**Output**		**Unit Costs**			

26 - 25003   **BUS DUCT, Cont'd...**   **26 - 25003**

ID Code	Component Descriptions	Unit of Meas.	Manhr / Unit	Material Cost	Labor Cost	Equipment Cost	Total Cost
1170	15a-60a	EA	1.100	730	110		840
1180	70a-100a	"	1.301	870	120		990
1185	3 pole						
1190	15a-60a	EA	1.159	830	110		940
1200	70a-100a	"	1.509	970	140		1,110
1205	Bus duct, copper feeder duct, 277/480v, 4 wire						
1210	800a	LF	0.400	820	38.50		860
1220	1000a	"	0.500	930	48.00		980
1230	1200a	"	0.533	1,080	51.00		1,130
1240	1350a	"	0.615	1,280	59.00		1,340
1250	1600a	"	0.727	1,550	70.00		1,620
1260	2000a	"	0.800	1,940	77.00		2,020
1270	2500a	"	0.851	2,370	82.00		2,450
1280	3000a	"	0.952	3,010	91.00		3,100
1285	Weatherproof						
1290	800a	LF	0.444	890	42.75		930
1300	1000a	"	0.533	1,030	51.00		1,080
1310	1350a	"	0.667	1,370	64.00		1,430
1320	1600a	"	0.727	1,650	70.00		1,720
1330	2000a	"	0.833	2,040	80.00		2,120
1340	2500a	"	0.899	2,510	86.00		2,600
1350	3000a	"	0.976	3,200	94.00		3,290
1360	4000a	"	1.509	4,090	140		4,230
1370	5000a	"	1.818	4,920	170		5,090
1375	Plug-in feeder duct, 277/480v, 4 wire						
1380	400a	LF	0.400	460	38.50		500
1390	600a	"	0.444	470	42.75		510
1400	800a	"	0.500	730	48.00		780
1410	1000a	"	0.500	820	48.00		870
1420	1200a	"	0.533	980	51.00		1,030
1430	1350a	"	0.615	1,100	59.00		1,160
1440	1600a	"	0.727	1,320	70.00		1,390
1450	2000a	"	0.800	1,600	77.00		1,680
1460	2500a	"	0.851	1,960	82.00		2,040
1470	3000a	"	0.952	2,470	91.00		2,560
1475	Ground bus, material only						
1480	225a	LF					96.00
1490	400a	"					99.00

BASIC MATERIALS

ID Code	Description Component Descriptions	Output Unit of Meas.	Output Manhr / Unit	Unit Costs Material Cost	Unit Costs Labor Cost	Unit Costs Equipment Cost	Unit Costs Total Cost
26 - 25003	**BUS DUCT, Cont'd...**						**26 - 25003**
1500	600a	LF					100
1510	800a	"					110
1520	1000a	"					110
1530	1200a	"					140
1540	1350a	"					150
1550	1600a	"					200
1560	2000a	"					240
1570	2500a	"					320
1580	3000a	"					440
1590	4000a	"					530
1600	5000a	"					670
1605	Copper flanged ends, 277/480v, 4 wire						
1610	225a	EA	2.500	1,780	240		2,020
1620	400a	"	2.759	2,130	260		2,390
1630	600a	"	2.963	2,790	280		3,070
1640	800a	"	3.077	3,240	300		3,540
1650	1000a	"	3.200	3,690	310		4,000
1660	1200a	"	3.265	3,970	310		4,280
1670	1350a	"	3.333	4,110	320		4,430
1680	1600a	"	3.478	4,810	330		5,140
1690	2000a	"	3.478	5,510	330		5,840
1700	2500a	"	3.636	5,580	350		5,930
1710	3000a	"	3.810	7,980	370		8,350
1720	4000a	"	4.444	10,730	430		11,160
1730	5000a	"	4.706	12,410	450		12,860
1735	Bus duct, copper elbows, 277/480v-4w						
1740	225a-1000a	EA	2.105	3,830	200		4,030
1750	1200a-3000a	"	2.500	6,590	240		6,830
1760	4000a-5000a	"	2.963	16,340	280		16,620
1765	Tees, 277/480v-4w						
1770	225a-1000a	EA	2.222	6,070	210		6,280
1780	1200a-3000a	"	2.581	8,550	250		8,800
1790	4000a-5000a	"	2.963	23,390	280		23,670
1795	Crosses, 277/480v-4w						
1800	225a-1000a	EA	2.222	6,440	210		6,650
1810	1200a-3000a	"	2.581	9,970	250		10,220
1820	4000a-5000a	"	2.963	15,720	280		16,000
1825	Copper end closures, 277/480v-4w						

BASIC MATERIALS

ID Code	Component Descriptions	Unit of Meas.	Manhr / Unit	Material Cost	Labor Cost	Equipment Cost	Total Cost
	Description	**Output**		**Unit Costs**			
26 - 25003		**BUS DUCT, Cont'd...**					**26 - 25003**
1830	225a-1000a	EA	0.899	580	86.00		670
1840	1200a-3000a	"	1.194	770	110		880
1850	4000a-5000a	"	1.667	960	160		1,120
1855	Tap boxes, 277/480v-4w						
1860	225a	EA	3.478	2,630	330		2,960
1870	400a	"	4.444	2,640	430		3,070
1880	600a	"	7.273	2,880	700		3,580
1890	800a	"	8.000	3,150	770		3,920
1900	1000a	"	10.000	3,520	960		4,480
1910	1200a	"	11.004	3,590	1,060		4,650
1920	1350a	"	13.008	4,370	1,250		5,620
1930	1600a	"	14.011	4,860	1,350		6,210
1940	2000a	"	16.985	5,730	1,630		7,360
1950	2500a	"	22.989	6,940	2,210		9,150
1960	3000a	"	27.972	7,230	2,690		9,920
1970	4000a	"	37.915	7,560	3,640		11,200
1980	5000a	"	44.944	7,560	4,310		11,870
1985	Circuit breaker, adapter cubicle						
1990	225a	EA	1.509	7,960	140		8,100
2000	400a	"	1.600	9,400	150		9,550
2010	600a	"	1.702	13,930	160		14,090
2020	800a	"	1.818	15,920	170		16,090
2030	1000a	"	1.905	18,440	180		18,620
2040	1200a	"	2.000	22,100	190		22,290
2050	1600a	"	2.105	26,900	200		27,100
2060	2000a	"	2.222	31,480	210		31,690
2065	Transformer taps, 1 phase 277/480v						
2070	600a	EA	7.273	1,370	700		2,070
2080	800a	"	8.000	1,440	770		2,210
2090	1000a	"	10.000	1,750	960		2,710
2100	1200a	"	11.004	1,910	1,060		2,970
2110	1350a	"	13.008	2,010	1,250		3,260
2120	1600a	"	14.011	2,290	1,350		3,640
2130	2000a	"	16.985	2,620	1,630		4,250
2140	2500a	"	22.989	3,190	2,210		5,400
2150	3000a	"	27.972	3,750	2,690		6,440
2160	4000a	"	37.915	4,420	3,640		8,060
2170	5000a	"	45.977	5,550	4,410		9,960

BASIC MATERIALS

ID Code	Description / Component Descriptions	Output / Unit of Meas.	Output / Manhr / Unit	Unit Costs / Material Cost	Unit Costs / Labor Cost	Unit Costs / Equipment Cost	Unit Costs / Total Cost
26 - 25003	**BUS DUCT, Cont'd...**						**26 - 25003**
2175	3 phase, 480v, 3 wire						
2180	600a	EA	7.273	3,090	700		3,790
2190	800a	"	8.000	3,350	770		4,120
2200	1000a	"	10.000	3,830	960		4,790
2210	1200a	"	11.004	4,130	1,060		5,190
2220	1350a	"	13.008	4,370	1,250		5,620
2230	1600a	"	14.011	5,270	1,350		6,620
2240	2000a	"	16.985	5,900	1,630		7,530
2250	2500a	"	22.989	6,850	2,210		9,060
2260	3000a	"	27.972	7,980	2,690		10,670
2270	4000a	"	37.915	9,190	3,640		12,830
2280	5000a	"	45.977	10,600	4,410		15,010
2285	3 phase, 4 wire, 277/480v						
2290	600a	EA	7.273	2,680	700		3,380
2300	800a	"	8.000	2,870	770		3,640
2310	1000a	"	10.000	3,300	960		4,260
2320	1200a	"	11.004	3,620	1,060		4,680
2330	1350a	"	13.008	3,800	1,250		5,050
2340	1600a	"	14.011	4,460	1,350		5,810
2350	2000a	"	16.985	5,080	1,630		6,710
2360	2500a	"	22.989	5,940	2,210		8,150
2370	3000a	"	27.972	6,950	2,690		9,640
2380	4000a	"	40.816	8,060	3,920		11,980
2390	5000a	"	45.977	9,120	4,410		13,530
2395	Transformer connection, 4 wire, 277/480v						
2400	600a	EA	2.759	5,560	260		5,820
2410	800a	"	2.857	5,790	270		6,060
2420	1000a	"	2.963	6,070	280		6,350
2430	1200a	"	3.077	6,220	300		6,520
2440	1350a	"	3.200	6,330	310		6,640
2450	1600a	"	3.333	6,800	320		7,120
2460	2000a	"	3.478	7,080	330		7,410
2470	2500a	"	3.636	8,000	350		8,350
2480	3000a	"	3.810	8,770	370		9,140
2490	4000a	"	4.444	11,620	430		12,050
2500	5000a	"	4.706	15,530	450		15,980
2505	Unfused reducers, 3 wire, 480v, 3 phase						
2510	400a	EA	2.500	1,000	240		1,240

BASIC MATERIALS

ID Code	Description — Component Descriptions	Output Unit of Meas.	Output Manhr / Unit	Unit Costs Material Cost	Unit Costs Labor Cost	Unit Costs Equipment Cost	Unit Costs Total Cost
26 - 25003	**BUS DUCT, Cont'd...**						**26 - 25003**
2520	600a	EA	3.810	1,070	370		1,440
2530	800a	"	4.706	1,340	450		1,790
2540	1000a	"	5.000	1,590	480		2,070
2550	1200a	"	5.333	2,680	510		3,190
2560	1350a	"	5.714	3,410	550		3,960
2570	1600a	"	6.154	3,730	590		4,320
2580	2000a	"	6.400	5,000	610		5,610
2590	2500a	"	6.667	6,190	640		6,830
2600	3000a	"	7.273	7,460	700		8,160
2610	4000a	"	8.753	9,710	840		10,550
2620	5000a	"	10.796	11,710	1,040		12,750
2630	4 wire, 277/480v, 3 phase, 400a	"	2.759	1,530	260		1,790
2640	600a	"	4.000	1,970	380		2,350
2650	800a	"	5.000	2,430	480		2,910
2660	1000a	"	5.333	2,850	510		3,360
2670	1200a	"	5.714	4,890	550		5,440
2680	1350a	"	6.154	6,240	590		6,830
2690	1600a	"	6.667	6,770	640		7,410
2700	2000a	"	6.957	9,120	670		9,790
2710	2500a	"	7.273	11,250	700		11,950
2720	3000a	"	8.247	13,600	790		14,390
2730	4000a	"	9.744	17,680	940		18,620
2740	5000a	"	11.994	21,300	1,150		22,450
2745	Circuit breaker reducers, 4 wire, 277/480v						
2750	400a	EA	2.222	3,360	210		3,570
2760	600a	"	3.478	3,640	330		3,970
2770	800a	"	4.211	4,490	400		4,890
2780	1000a	"	4.706	5,290	450		5,740
2790	1350a	"	5.333	11,580	510		12,090
2800	1600a	"	5.714	12,570	550		13,120
2810	2000a	"	6.154	16,920	590		17,510
2820	2500a	"	6.667	20,940	640		21,580
2830	3000a	"	6.957	25,240	670		25,910
2840	4000a	"	7.273	32,800	700		33,500
2850	5000a	"	10.000	39,510	960		40,470
2855	Expansion fittings, 4 wire, 277/480v						
2860	225a	EA	2.500	2,320	240		2,560
2870	400a	"	3.810	2,440	370		2,810

BASIC MATERIALS

ID Code	Component Descriptions	Unit of Meas.	Manhr / Unit	Material Cost	Labor Cost	Equipment Cost	Total Cost
	Description	**Output**		**Unit Costs**			
26 - 25003	**BUS DUCT, Cont'd...**						**26 - 25003**
2880	600a	EA	4.706	2,720	450		3,170
2890	800a	"	5.000	3,180	480		3,660
2900	1000a	"	5.333	3,600	510		4,110
2910	1200a	"	5.714	4,900	550		5,450
2920	1350a	"	5.926	4,900	570		5,470
2930	1600a	"	6.154	6,310	590		6,900
2940	2000a	"	6.667	7,140	640		7,780
2950	2500a	"	7.273	8,640	700		9,340
2960	3000a	"	8.753	10,090	840		10,930
2970	4000a	"	10.796	13,140	1,040		14,180
2980	5000a	"	11.994	13,800	1,150		14,950
2985	Wall flanges						
2990	225a-2500a	EA	4.000	420	380		800
3000	3000a-5000a	"	6.154	610	590		1,200
3010	Weather seals	"	1.000	620	96.00		720
3020	Roof flanges	"	4.000	1,320	380		1,700
3030	Fire barriers	"	1.509	630	140		770
3040	Spring hangers	"	1.739	180	170		350
3050	Sway brace collars	"	1.250	37.75	120		160
3055	Hook sticks, material only						
3060	8'	EA					280
3070	14'	"					460
3075	Fusible switches, 240v, 3 phase						
3080	30a	EA	1.000	690	96.00		790
3090	60a	"	1.250	840	120		960
3100	100a	"	1.509	1,120	140		1,260
3110	200a	"	2.105	1,960	200		2,160
3120	400a	"	4.000	3,170	380		3,550
3130	600a	"	6.154	6,090	590		6,680
3135	208v, 4 wire						
3140	30a	EA	1.194	840	110		950
3150	60a	"	1.356	870	130		1,000
3160	100a	"	1.818	1,250	170		1,420
3170	200a	"	2.759	2,190	260		2,450
3180	400a	"	5.000	3,750	480		4,230
3190	600a	"	8.000	4,720	770		5,490
3195	600v						
3200	30a	EA	1.000	800	96.00		900

BASIC MATERIALS

ID Code	Component Descriptions	Unit of Meas.	Manhr / Unit	Material Cost	Labor Cost	Equipment Cost	Total Cost
	Description	**Output**		**Unit Costs**			
26 - 25003	**BUS DUCT, Cont'd...**						**26 - 25003**
3210	60a	EA	1.250	860	120		980
3220	100a	"	1.509	1,230	140		1,370
3230	200a	"	2.105	2,170	200		2,370
3240	400a	"	4.000	3,680	380		4,060
3250	600a	"	6.154	4,550	590		5,140
3260	800a	"	6.667	7,850	640		8,490
3270	1000a	"	8.000	9,280	770		10,050
3280	1200a	"	11.004	14,750	1,060		15,810
3290	1600a	"	11.994	14,930	1,150		16,080
3295	480v, 4 wire						
3300	30a	EA	1.194	910	110		1,020
3310	60a	"	1.356	960	130		1,090
3320	100a	"	1.818	1,420	170		1,590
3330	200a	"	2.759	2,430	260		2,690
3340	400a	"	5.000	4,120	480		4,600
3350	600a	"	8.000	5,180	770		5,950
3360	800a	"	8.247	8,500	790		9,290
3370	1000a	"	11.004	10,040	1,060		11,100
3380	1200a	"	11.994	15,650	1,150		16,800
3390	1600a	"	14.011	16,360	1,350		17,710
3395	Fusible combination starters, 600v, 3 phase						
3400	Size 0	EA	1.290	2,030	120		2,150
3410	Size 1	"	1.600	2,280	150		2,430
3420	Size 2	"	1.818	2,900	170		3,070
3430	Size 3	"	2.581	4,730	250		4,980
3435	Circuit breaker combination starters, 600v, 3 phase						
3440	Size 0	EA	1.290	2,210	120		2,330
3450	Size 1	"	1.600	2,300	150		2,450
3460	Size 2	"	1.818	3,340	170		3,510
3470	Size 3	"	2.581	4,350	250		4,600
3475	Fusible combination contactors, 600v, 3 phase						
3480	30a	EA	1.290	1,540	120		1,660
3490	60a	"	1.600	1,890	150		2,040
3500	100a	"	1.818	2,840	170		3,010
3510	200a	"	2.581	4,630	250		4,880
3515	Circuit breaker, combination contactors, 600v, 3 phase						
3520	30a	EA	1.290	1,580	120		1,700
3530	60a	"	1.600	1,640	150		1,790

BASIC MATERIALS

ID Code	Component Descriptions	Unit of Meas.	Manhr / Unit	Material Cost	Labor Cost	Equipment Cost	Total Cost
26 - 25003	**BUS DUCT, Cont'd...**						**26 - 25003**
3540	100a	EA	1.818	2,300	170		2,470
3550	200a	"	2.581	3,030	250		3,280
3555	Fusible contactor electrically held, 480v, 4 wire						
3560	30a	EA	1.290	1,540	120		1,660
3570	60a	"	1.600	1,960	150		2,110
3580	100a	"	1.818	2,810	170		2,980
3590	200a	"	2.759	5,890	260		6,150
3595	Mechanically held						
3600	30a	EA	1.290	1,670	120		1,790
3610	60a	"	1.600	2,390	150		2,540
3620	100a	"	1.860	3,310	180		3,490
3630	200a	"	2.759	6,800	260		7,060
3650	Circuit breakers, 240v, 3 phase						
3660	15a-60a	EA	1.159	670	110		780
3670	70a-100a	"	1.600	750	150		900
3675	600v, 3 phase						
3680	15a-60a	EA	1.159	730	110		840
3700	125a-225a	"	2.286	1,870	220		2,090
3710	250a-400a	"	4.211	3,860	400		4,260
3720	500a-600a	"	5.333	5,500	510		6,010
3730	700a-800a	"	8.000	6,620	770		7,390
3740	900a-1000a	"	10.000	7,880	960		8,840
3750	1200a-1600a	"	11.004	12,490	1,060		13,550
3755	120/208v, 4 wire						
3760	15a-60a	EA	1.290	620	120		740
3770	70a-100a	"	1.860	720	180		900
3775	277/480v, 4 wire						
3780	15a-60a	EA	1.290	760	120		880
3790	70a-100a	"	1.905	820	180		1,000
3800	125a-225a	"	2.759	2,000	260		2,260
3810	250a-400a	"	5.000	4,050	480		4,530
3820	500a-600a	"	8.000	5,760	770		6,530
3830	700a-800a	"	8.000	6,880	770		7,650
3840	900a-1000a	"	11.004	8,130	1,060		9,190
3850	1200a-1600a	"	14.011	12,400	1,350		13,750
3855	600v, 3 phase, 65,000 AIC						
3860	60a	EA	1.159	950	110		1,060
3870	70a-100a	"	1.600	1,020	150		1,170

BASIC MATERIALS

ID Code	Description / Component Descriptions	Output / Unit of Meas.	Manhr / Unit	Unit Costs / Material Cost	Labor Cost	Equipment Cost	Total Cost
26 - 25003	**BUS DUCT, Cont'd...**						**26 - 25003**
3880	125a-225a	EA	2.286	3,350	220		3,570
3890	250a-400a	"	4.211	5,350	400		5,750
3900	500a-600a	"	6.154	6,340	590		6,930
3910	700a-800a	"	8.000	7,510	770		8,280
3920	900a-1000a	"	10.000	8,510	960		9,470
3925	277/480v, 4 wire, 65,000 AIC						
3930	15a-60a	EA	1.290	1,020	120		1,140
3940	70a-100a	"	1.905	1,110	180		1,290
3950	125a-225a	"	2.759	3,450	260		3,710
3960	250a-400a	"	5.000	5,550	480		6,030
3970	500a-600a	"	6.154	6,620	590		7,210
3980	700a-800a	"	8.000	7,760	770		8,530
3990	900a-1000a	"	10.349	12,670	990		13,660
3995	600v, 3 phase, current limiting						
4000	15a-60a	EA	1.159	2,460	110		2,570
4010	70a-100a	"	1.600	3,040	150		3,190
4020	125a-225a	"	2.286	5,680	220		5,900
4030	250a-400a	"	4.211	6,750	400		7,150
4040	500a-600a	"	6.154	10,130	590		10,720
4050	700a-800a	"	8.000	11,430	770		12,200
4060	900a-1000a	"	10.000	12,720	960		13,680
4065	277/480v, 4 wire, current limiting						
4070	15a-60a	EA	1.290	2,530	120		2,650
4080	70a-100a	"	40.000	3,250	3,840		7,090
4090	125a-225a	"	2.759	5,830	260		6,090
4100	250a-400a	"	5.000	6,950	480		7,430
4110	500a-600a	"	6.154	7,740	590		8,330
4120	700a-800a	"	8.000	11,690	770		12,460
4130	900a-1000a	"	11.004	12,980	1,060		14,040
4135	Capacitors, 3 phase, 240v						
4140	5 kvar	EA	5.000	1,190	480		1,670
4150	7.5 kvar	"	6.154	1,490	590		2,080
4160	10 kvar	"	8.097	1,740	780		2,520
4170	15 kvar	"	9.195	2,280	880		3,160
4175	480v						
4180	2.5 kvar	EA	2.759	510	260		770
4190	5 kvar	"	4.706	770	450		1,220
4200	7.5 kvar	"	5.714	940	550		1,490

BASIC MATERIALS

ID Code	Description / Component Descriptions	Output Unit of Meas.	Output Manhr / Unit	Unit Costs Material Cost	Unit Costs Labor Cost	Unit Costs Equipment Cost	Unit Costs Total Cost
26 - 25003	**BUS DUCT, Cont'd...**						**26 - 25003**
4210	10 kvar	EA	8.097	1,040	780		1,820
4220	15 kvar	"	8.999	1,250	860		2,110
4230	20 kvar	"	11.494	1,560	1,100		2,660
4240	25 kvar	"	13.008	1,950	1,250		3,200
4250	30 kvar	"	14.210	2,300	1,360		3,660
4255	Transformers, 3 phase, 480v						
4260	1.0 kva	EA	1.739	680	170		850
4270	1.5 kva	"	2.000	750	190		940
4280	2 kva	"	2.500	810	240		1,050
4290	3 kva	"	2.759	960	260		1,220
4300	5 kva	"	4.000	1,320	380		1,700
4310	7.5 kva	"	5.000	1,590	480		2,070
4320	10 kva	"	5.333	1,860	510		2,370
26 - 27130	**METERING**						**26 - 27130**
0490	Outdoor wp meter sockets, 1 gang, 240v, 1 phase						
0510	Includes sealing ring, 100a	EA	1.509	72.00	140		210
0520	150a	"	1.778	86.00	170		260
0530	200a	"	2.000	110	190		300
0570	Die cast hubs, 1-1/4"	"	0.320	10.00	30.75		40.75
0580	1-1/2"	"	0.320	11.50	30.75		42.25
0590	2"	"	0.320	14.00	30.75		44.75
1000	Indoor meter center, main switch single phase, 240v						
1010	400a	EA	8.000	2,440	770		3,210
1020	600a	"	11.004	4,270	1,060		5,330
1030	800a	"	11.696	6,680	1,120		7,800
1035	Main breaker						
1040	400a	EA	8.000	4,590	770		5,360
1050	600a	"	11.004	6,190	1,060		7,250
1060	800a	"	11.696	7,220	1,120		8,340
1070	1000a	"	16.000	9,950	1,540		11,490
1080	1200a	"	16.495	10,880	1,580		12,460
1090	1600a	"	18.018	19,610	1,730		21,340
1095	Terminal box						
1100	800a	EA	10.000	680	960		1,640
1110	1600a	"	18.018	2,350	1,730		4,080
1115	Main switch, three phase, 208v						
1120	400a	EA	8.502	2,490	820		3,310

BASIC MATERIALS

ID Code	Component Descriptions	Unit of Meas.	Manhr / Unit	Material Cost	Labor Cost	Equipment Cost	Total Cost
	Description	**Output**		**Unit Costs**			

26 - 27130 — METERING, Cont'd... — **26 - 27130**

ID Code	Component Descriptions	Unit of Meas.	Manhr / Unit	Material Cost	Labor Cost	Equipment Cost	Total Cost
1130	600a	EA	11.994	4,490	1,150		5,640
1140	800a	"	13.491	9,020	1,300		10,320
1145	Main breaker						
1150	400a	EA	8.502	4,580	820		5,400
1160	600a	"	11.994	7,220	1,150		8,370
1170	800a	"	13.491	9,590	1,300		10,890
1180	1000a	"	16.985	12,270	1,630		13,900
1190	1200a	"	18.018	17,100	1,730		18,830
1200	1600a	"	20.997	25,190	2,020		27,210
1205	Terminal box						
1210	800a	EA	13.008	780	1,250		2,030
1220	1600a	"	20.997	2,480	2,020		4,500
1225	Indoor meter center						
1230	2 meters	EA	5.000	720	480		1,200
1240	3 meters	"	6.154	910	590		1,500
1250	4 meters	"	7.273	1,170	700		1,870
1260	5 meters	"	8.000	1,480	770		2,250
1270	6 meters	"	8.999	1,890	860		2,750
1275	Plug on breakers, single phase, 208v						
1280	60a	EA	0.250	40.00	24.00		64.00
1290	70a	"	0.250	80.00	24.00		100
1300	80a	"	0.250	110	24.00		130
1310	90a	"	0.250	120	24.00		140
1320	100a	"	0.348	120	33.50		150
1330	Indoor meter center, single phase, 125a breakers						
1340	3 meters	EA	6.154	1,040	590		1,630
1350	4 meters	"	7.273	1,240	700		1,940
1360	5 meters	"	8.000	1,550	770		2,320
1370	6 meters	"	8.502	1,790	820		2,610
1380	7 meters	"	10.000	2,280	960		3,240
1390	8 meters	"	11.004	2,490	1,060		3,550
1400	10 meters	"	11.994	3,100	1,150		4,250
1405	150a breakers						
1410	3 meters	EA	6.154	3,530	590		4,120
1420	4 meters	"	7.273	4,710	700		5,410
1430	6 meters	"	8.000	7,060	770		7,830
1440	7 meters	"	10.000	8,250	960		9,210
1450	8 meters	"	11.004	9,430	1,060		10,490

BASIC MATERIALS

ID Code	Description / Component Descriptions	Output Unit of Meas.	Output Manhr / Unit	Unit Costs Material Cost	Unit Costs Labor Cost	Unit Costs Equipment Cost	Unit Costs Total Cost
26 - 27130	**METERING, Cont'd...**						**26 - 27130**
1455	200a breakers						
1460	3 meters	EA	6.154	2,900	590		3,490
1470	4 meters	"	7.273	3,930	700		4,630
1480	6 meters	"	8.000	5,830	770		6,600
1490	7 meters	"	10.000	6,850	960		7,810
1500	8 meters	"	11.004	7,870	1,060		8,930
1505	Indoor meter center, three phase, 125a breakers						
1510	3 meters	EA	6.154	1,020	590		1,610
1520	4 meters	"	7.273	1,220	700		1,920
1530	5 meters	"	8.000	1,530	770		2,300
1540	6 meters	"	8.999	1,750	860		2,610
1550	7 meters	"	10.000	2,240	960		3,200
1560	8 meters	"	11.004	2,460	1,060		3,520
1570	10 meters	"	11.994	3,060	1,150		4,210
1575	150a breakers						
1580	3 meters	EA	6.154	4,310	590		4,900
1590	4 meters	"	7.273	5,770	700		6,470
1600	6 meters	"	8.502	8,750	820		9,570
1610	7 meters	"	11.004	10,100	1,060		11,160
1620	8 meters	"	11.994	11,530	1,150		12,680
1625	200a breakers						
1630	3 meters	EA	6.667	5,140	640		5,780
1640	4 meters	"	7.273	6,870	700		7,570
1650	6 meters	"	8.999	9,050	860		9,910
1660	7 meters	"	11.004	10,580	1,060		11,640
1670	8 meters	"	11.994	12,070	1,150		13,220
1675	NEMA 3R, meter center, main switch, 1 phase, 240v						
1680	400a	EA	8.000	2,500	770		3,270
1690	600a	"	10.000	4,850	960		5,810
1700	800a	"	11.004	7,460	1,060		8,520
1705	Main breaker						
1710	400a	EA	8.000	6,210	770		6,980
1720	600a	"	10.000	7,940	960		8,900
1730	800a	"	12.308	9,750	1,180		10,930
1740	1000a	"	15.009	11,500	1,440		12,940
1750	1200a	"	16.000	16,230	1,540		17,770
1755	Terminal box						
1760	225a	EA	7.273	600	700		1,300

BASIC MATERIALS

ID Code	Description / Component Descriptions	Unit of Meas.	Manhr / Unit	Material Cost	Labor Cost	Equipment Cost	Total Cost
26 - 27130	**METERING, Cont'd...**						**26 - 27130**
1770	800a	EA	11.494	900	1,100		2,000
1780	1600a	"	18.018	2,960	1,730		4,690
1785	NEMA 3R, three phase, 280v						
1790	400a	EA	8.502	2,900	820		3,720
1800	600a	"	11.994	5,470	1,150		6,620
1810	800a	"	13.008	8,330	1,250		9,580
1815	Main breaker						
1820	400a	EA	8.502	7,190	820		8,010
1830	600a	"	11.994	9,230	1,150		10,380
1840	800a	"	13.008	11,690	1,250		12,940
1850	1000a	"	16.985	13,120	1,630		14,750
1860	1200a	"	18.018	17,980	1,730		19,710
1865	Terminal box						
1870	225a	EA	8.000	690	770		1,460
1880	800a	"	13.008	1,000	1,250		2,250
1890	1600a	"	20.997	3,080	2,020		5,100
1895	NEMA 3R meter center, single phase, 208v, 100a						
1900	2 meters	EA	5.000	750	480		1,230
1910	3 meters	"	6.154	900	590		1,490
1920	4 meters	"	7.273	1,310	700		2,010
1930	5 meters	"	8.000	1,580	770		2,350
1940	6 meters	"	8.999	2,500	860		3,360
2000	125a, 3 meters	"	6.154	1,990	590		2,580
2010	4 meters	"	7.273	2,540	700		3,240
2020	6 meters	"	8.239	3,160	790		3,950
2030	7 meters	"	10.000	3,710	960		4,670
2040	8 meters	"	11.004	4,530	1,060		5,590
2050	150a, 3 meters	"	6.154	3,920	590		4,510
2060	4 meters	"	7.273	5,240	700		5,940
2070	6 meters	"	8.502	9,440	820		10,260
2080	7 meters	"	10.000	11,020	960		11,980
2090	8 meters	"	11.004	12,580	1,060		13,640
2095	NEMA 3R center, 3 phase, 208v, 125a breakers						
2100	3 meters	EA	6.154	1,040	590		1,630
2110	4 meters	"	7.273	1,420	700		2,120
2120	6 meters	"	8.502	2,120	820		2,940
2130	7 meters	"	10.000	2,480	960		3,440
2140	8 meters	"	11.004	2,830	1,060		3,890

BASIC MATERIALS

ID Code	Description — Component Descriptions	Output — Unit of Meas.	Output — Manhr / Unit	Unit Costs — Material Cost	Unit Costs — Labor Cost	Unit Costs — Equipment Cost	Unit Costs — Total Cost
26 - 27130	**METERING, Cont'd...**						**26 - 27130**
2145	150a						
2150	3 meters	EA	6.667	4,690	640		5,330
2160	4 meters	"	7.273	6,140	700		6,840
2170	6 meters	"	8.999	9,210	860		10,070
2180	7 meters	"	10.499	10,750	1,010		11,760
2190	8 meters	"	11.494	12,290	1,100		13,390
2195	200a						
2200	3 meters	EA	6.667	3,630	640		4,270
2210	4 meters	"	7.273	4,840	700		5,540
2220	6 meters	"	8.999	7,250	860		8,110
2230	7 meters	"	11.004	8,490	1,060		9,550
2240	8 meters	"	11.494	9,700	1,100		10,800
2245	NEMA 3R, center plug-on breakers, 208v, 1 phase						
2250	60a	EA	0.250	55.00	24.00		79.00
2260	70a	"	0.250	110	24.00		130
2270	90a	"	0.250	160	24.00		180
2280	100a	"	0.348	170	33.50		200
2290	125a	"	0.400	330	38.50		370
26 - 27268	**RECEPTACLES**						**26 - 27268**
0490	Contractor grade duplex receptacles, 15a 120v						
0510	Duplex	EA	0.200	2.35	19.25		21.50
1000	125 volt, 20a, duplex, standard grade	"	0.200	17.50	19.25		36.75
1040	Ground fault interrupter type	"	0.296	57.00	28.50		86.00
1520	250 volt, 20a, 2 pole, single, grounding type	"	0.200	29.25	19.25		48.50
1540	120/208v, 4 pole, single receptacle, twist lock						
1560	20a	EA	0.348	34.75	33.50		68.00
1580	50a	"	0.348	66.00	33.50		100
1590	125/250v, 3 pole, flush receptacle						
1600	30a	EA	0.296	35.25	28.50		64.00
1620	50a	"	0.296	43.50	28.50		72.00
1640	60a	"	0.348	110	33.50		140
1660	277v, 20a, 2 pole, grounding type, twist lock	"	0.200	19.00	19.25		38.25
2020	Dryer receptacle, 250v, 30a/50a, 3 wire	"	0.296	26.25	28.50		55.00
2040	Clock receptacle, 2 pole, grounding type	"	0.200	17.50	19.25		36.75
3000	125v, 20a single recept. grounding type						
3010	Standard grade	EA	0.200	19.00	19.25		38.25
3020	Specification	"	0.200	22.75	19.25		42.00

BASIC MATERIALS

	Description	Output		Unit Costs			
ID Code	Component Descriptions	Unit of Meas.	Manhr / Unit	Material Cost	Labor Cost	Equipment Cost	Total Cost
26 - 27268		**RECEPTACLES, Cont'd...**					**26 - 27268**
3030	Hospital	EA	0.200	23.75	19.25		43.00
3040	Isolated ground orange	"	0.250	80.00	24.00		100
3045	Duplex						
3050	Specification grade	EA	0.200	19.00	19.25		38.25
3060	Hospital	"	0.200	39.00	19.25		58.00
3070	Isolated ground orange	"	0.250	80.00	24.00		100
3080	250v, 20a, duplex, 2 pole, grounding, spec. grade	"	0.200	31.00	19.25		50.00
3090	Combination recepts, 20a, 125v and 250v, duplex	"	0.200	41.00	19.25		60.00
3100	GFI hospital grade recepts, 20a, 125v, duplex	"	0.296	86.00	28.50		110
3105	125/250v, 3 pole, 3 wire surface recepts						
3110	30a	EA	0.296	29.75	28.50		58.00
3120	50a	"	0.296	33.00	28.50		62.00
3130	60a	"	0.348	73.00	33.50		110
3135	Cord set, 3 wire, 6' cord						
3140	30a	EA	0.296	26.50	28.50		55.00
3150	50a	"	0.296	37.50	28.50		66.00
3155	125/250v, 3 pole, 3 wire cap						
3160	30a	EA	0.400	26.25	38.50		65.00
3170	50a	"	0.400	47.75	38.50		86.00
3180	60a	"	0.444	61.00	42.75		100
26 - 28130		**FUSES**					**26 - 28130**
1000	Fuse, one-time, 250v						
1010	30a	EA	0.050	4.05	4.80		8.85
1020	60a	"	0.050	6.85	4.80		11.75
1040	100a	"	0.050	28.50	4.80		33.25
1060	200a	"	0.050	69.00	4.80		74.00
1080	400a	"	0.050	160	4.80		160
1100	600a	"	0.050	270	4.80		270
1120	600v						
1140	30a	EA	0.050	20.50	4.80		25.25
1160	60a	"	0.050	32.50	4.80		37.25
1180	100a	"	0.050	61.00	4.80		66.00
1200	200a	"	0.050	160	4.80		160
1220	400a	"	0.050	340	4.80		340
1500	Fusetron, 600v						
1526	200a	EA	0.050	140	4.80		140
1528	400a	"	0.050	290	4.80		290

BASIC MATERIALS

ID Code	Description / Component Descriptions	Output / Unit of Meas.	Manhr / Unit	Unit Costs / Material Cost	Labor Cost	Equipment Cost	Total Cost
26 - 28130	**FUSES, Cont'd...**						**26 - 28130**
2000	Fuse, amp-trap, K1, 250v						
2020	30a	EA	0.050	13.25	4.80		18.00
2040	60a	"	0.050	24.25	4.80		29.00
2060	100a	"	0.050	54.00	4.80		59.00
2080	200a	"	0.050	120	4.80		120
2100	400a	"	0.050	270	4.80		270
2120	600a	"	0.050	360	4.80		360
2140	600v						
2160	30a	EA	0.050	39.50	4.80		44.25
2180	60a	"	0.050	71.00	4.80		76.00
2200	100a	"	0.050	140	4.80		140
2220	200a	"	0.050	210	4.80		210
2240	400a	"	0.050	440	4.80		440
2500	K5, 250v						
2520	30a	EA	0.050	9.90	4.80		14.75
2540	60a	"	0.050	18.00	4.80		22.75
2560	100a	"	0.050	40.75	4.80		45.50
2580	200a	"	0.050	89.00	4.80		94.00
2600	400a	"	0.050	160	4.80		160
2620	600a	"	0.050	260	4.80		260
2640	600v						
2660	30a	EA	0.050	22.00	4.80		26.75
2680	60a	"	0.050	37.75	4.80		42.50
2700	100a	"	0.050	78.00	4.80		83.00
2720	200a	"	0.050	160	4.80		160
2740	400a	"	0.050	310	4.80		310
2750	600a	"	0.050	450	4.80		450
3020	J, 600v						
3040	30a	EA	0.050	35.75	4.80		40.50
3060	60a	"	0.050	60.00	4.80		65.00
3080	100a	"	0.050	110	4.80		110
3100	200a	"	0.050	210	4.80		210
3120	400a	"	0.050	430	4.80		430
3520	L, 600v						
3540	1200a	EA	0.400	1,030	38.50		1,070
3560	1600a	"	0.400	1,320	38.50		1,360
3580	2000a	"	0.400	1,770	38.50		1,810
3600	2500a	"	0.400	2,350	38.50		2,390

BASIC MATERIALS

ID Code	Component Descriptions	Unit of Meas.	Manhr / Unit	Material Cost	Labor Cost	Equipment Cost	Total Cost
	Description	**Output**		**Unit Costs**			
26 - 28130	**FUSES, Cont'd...**						**26 - 28130**
3620	3000a	EA	0.400	2,700	38.50		2,740
3640	4000a	"	0.400	3,700	38.50		3,740
3660	5000a	"	0.400	5,820	38.50		5,860
4000	Fuse cl-ay 250v						
4020	600a	EA	0.296	860	28.50		890
4140	1200a	"	0.296	860	28.50		890
4160	1600a	"	0.296	1,060	28.50		1,090
4180	2000a	"	0.296	1,370	28.50		1,400
4200	600v						
4220	1200a	EA	0.296	1,020	28.50		1,050
4240	1600a	"	0.296	1,190	28.50		1,220
4260	2000a	"	0.296	1,430	28.50		1,460
5000	Reducers, 600v						
5010	60a-30a	EA	0.145	26.25	14.00		40.25
5020	100a-30a	"	0.145	92.00	14.00		110
5030	100a-60a	"	0.145	59.00	14.00		73.00
5040	200a-60a	"	0.250	230	24.00		250
5050	200a-100a	"	0.250	87.00	24.00		110
5060	400a-100a	"	0.348	410	33.50		440
5070	400a-200a	"	0.348	350	33.50		380
5080	600a-100a	"	0.400	550	38.50		590
5090	600a-200a	"	0.400	620	38.50		660
5100	600a-400a	"	0.400	550	38.50		590
26 - 28161	**CIRCUIT BREAKERS**						**26 - 28161**
0950	Molded case, 240v, 15-60a, bolt-on						
1000	1 pole	EA	0.250	29.25	24.00		53.00
1060	2 pole	"	0.348	62.00	33.50		96.00
1080	70-100a, 2 pole	"	0.533	180	51.00		230
1100	15-60a, 3 pole	"	0.400	220	38.50		260
1120	70-100a, 3 pole	"	0.615	360	59.00		420
1980	480v, 2 pole						
2000	15-60a	EA	0.296	450	28.50		480
2080	70-100a	"	0.400	580	38.50		620
2090	3 pole						
2100	15-60a	EA	0.400	580	38.50		620
2120	70-100a	"	0.444	680	42.75		720
2140	70-225a	"	0.615	1,410	59.00		1,470

BASIC MATERIALS

ID Code	Description / Component Descriptions	Output Unit of Meas.	Manhr / Unit	Unit Costs Material Cost	Labor Cost	Equipment Cost	Total Cost
26 - 28161	**CIRCUIT BREAKERS, Cont'd...**					**26 - 28161**	
4000	Draw out air circuit breakers						
4040	600a	EA	16.000	19,540	1,540		21,080
4080	800a	"	18.182	25,260	1,750		27,010
4140	1600a	"	24.242	40,560	2,330		42,890
4180	2000a	"	27.586	54,350	2,650		57,000
4220	3000a	"	32.000	94,320	3,070		97,390
4260	4000a	"	38.095	144,870	3,660		148,530
5000	Load center circuit breakers, 240v						
5010	1 pole, 10-60a	EA	0.250	31.25	24.00		55.00
5015	2 pole						
5020	10-60a	EA	0.400	63.00	38.50		100
5030	70-100a	"	0.667	190	64.00		250
5040	110-150a	"	0.727	410	70.00		480
5045	3 pole						
5050	10-60a	EA	0.500	180	48.00		230
5060	70-100a	"	0.727	270	70.00		340
5065	Load center, GFI breakers, 240v						
5070	1 pole, 15-30a	EA	0.296	230	28.50		260
5080	2 pole, 15-30a	"	0.400	410	38.50		450
5090	Key operated breakers, 240v, 1 pole, 10-30a	"	0.296	150	28.50		180
5095	Tandem breakers, 240v						
5100	1 pole, 15-30a	EA	0.400	51.00	38.50		90.00
5110	2 pole, 15-30a	"	0.533	94.00	51.00		140
5120	Bolt-on, GFI breakers, 240v, 1 pole, 15-30a	"	0.348	220	33.50		250
5130	Enclosed breaker, 120v, 1 pole, 15-50a, NEMA 1	"	0.800	250	77.00		330
5135	240v, 2 pole						
5140	15-60a, NEMA 1	EA	1.250	360	120		480
5150	70-100a, NEMA 1	"	1.739	450	170		620
5155	3 pole						
5160	15-60a, NEMA 1	EA	1.509	450	140		590
5170	70-100a, NEMA 1	"	2.222	600	210		810
5175	Enclosed circuit breakers						
5180	120v, 1 pole, NEMA 3R, 15-50a	EA	0.899	490	86.00		580
5185	240v, 2 pole, NEMA 3R						
5190	15-60a	EA	1.250	600	120		720
5200	70-100a	"	1.739	690	170		860
5205	3 pole, NEMA 3R						
5210	15-60a	EA	1.509	670	140		810

BASIC MATERIALS

ID Code	Component Descriptions	Unit of Meas.	Manhr / Unit	Material Cost	Labor Cost	Equipment Cost	Total Cost
	Description	**Output**		**Unit Costs**			

26 - 28161	**CIRCUIT BREAKERS, Cont'd...**						26 - 28161
5220	70-100a	EA	2.222	810	210		1,020
5225	480v, NEMA 1						
5230	1 pole, 15-50a	EA	0.800	310	77.00		390
5240	2 pole, 15-60a	"	1.250	540	120		660
5250	70-100a	"	1.509	670	140		810
5255	3 pole, NEMA 1						
5260	15-60a	EA	1.509	650	140		790
5270	70-100a	"	2.222	760	210		970
5280	480v, 1 pole, 15-50a, NEMA 3R	"	1.000	510	96.00		610
5285	2 pole						
5290	2 pole, 15-60a, NEMA 3R	EA	1.250	760	120		880
5300	70-100a, NEMA 3R	"	1.739	870	170		1,040
5305	3 pole						
5310	15-60a, NEMA 3R	EA	1.509	850	140		990
5320	70-100a, NEMA 3R	"	2.222	960	210		1,170
5340	70-100a, NEMA 1	"	1.739	720	170		890
5345	3 pole						
5350	15-60a, NEMA 1	EA	1.739	720	170		890
5360	70-100a, NEMA 1	"	2.222	870	210		1,080
5365	Enclosed breakers, 600v, 2 phase, NEMA 3R						
5370	15-60a	EA	1.250	780	120		900
5380	70-100a	"	1.739	960	170		1,130
5385	3 phase, NEMA 3R						
5390	15-60a	EA	1.509	900	140		1,040
5400	70-100a	"	2.222	1,030	210		1,240
5405	600v, 3 phase, NEMA 1						
5410	125a	EA	2.222	1,270	210		1,480
5420	150a	"	2.963	1,950	280		2,230
5430	175a	"	2.963	1,950	280		2,230
5440	200a	"	2.963	2,920	280		3,200
5450	225a	"	2.963	2,920	280		3,200
5460	250a	"	6.154	3,580	590		4,170
5470	300a	"	6.154	3,580	590		4,170
5480	350a	"	6.154	4,970	590		5,560
5490	400a	"	6.154	4,970	590		5,560
5500	500a	"	9.744	6,820	940		7,760
5510	600a	"	9.744	6,820	940		7,760
5520	700a	"	10.753	7,480	1,030		8,510

BASIC MATERIALS

ID Code	Description — Component Descriptions	Output — Unit of Meas.	Output — Manhr / Unit	Unit Costs — Material Cost	Unit Costs — Labor Cost	Unit Costs — Equipment Cost	Unit Costs — Total Cost
26 - 28161	**CIRCUIT BREAKERS, Cont'd...**						**26 - 28161**
5530	800a	EA	10.753	9,430	1,030		10,460
5540	900a	"	15.009	9,430	1,440		10,870
5550	1000a	"	15.009	11,880	1,440		13,320
5560	1200a	"	18.519	15,000	1,780		16,780
5570	1400a	"	18.519	20,830	1,780		22,610
5580	1600a	"	24.024	20,830	2,310		23,140
5590	1800a	"	29.963	29,220	2,880		32,100
5600	2000a	"	29.963	31,170	2,880		34,050
5605	600v, 3 phase, NEMA 3R						
5610	125-225a	EA	2.222	2,440	210		2,650
5620	250-400a	"	5.714	4,560	550		5,110
5630	500-600a	"	9.744	6,840	940		7,780
5640	700-800a	"	11.004	9,120	1,060		10,180
5650	900-1000a	"	15.009	10,350	1,440		11,790
5660	1000-1200a	"	19.002	24,430	1,820		26,250
5670	1400-1600a	"	24.024	24,740	2,310		27,050
5680	1800-2000a	"	29.963	25,050	2,880		27,930
26 - 28164	**SWITCHES**						**26 - 28164**
0100	Oil switches, medium voltage, bus components						
1000	Switches, 277/120v, toggle device only	EA	1.600	810	150		960
1020	With oil 35kv, g&w gram 44, 4 way switch	"	8.000	26,960	770		27,730
1030	Weatherproof enclosure						
1040	3 way switch	EA	10.000	33,300	960		34,260
1060	4 way switch	"	10.959	31,710	1,050		32,760
1080	Fused interrupter load, 35kv, 20a						
1100	1 pole	EA	16.000	38,060	1,540		39,600
1120	2 pole	"	17.021	41,230	1,630		42,860
1140	3 way	"	17.021	40,360	1,630		41,990
1160	4 way	"	18.182	43,250	1,750		45,000
1180	30a, 1 pole	"	16.000	34,600	1,540		36,140
1220	3 way	"	17.021	40,360	1,630		41,990
1240	4 way	"	18.182	43,250	1,750		45,000
1250	Weatherproof switch, including box & cover, 20a						
1260	1 pole	EA	16.000	41,210	1,540		42,750
1280	2 pole	"	17.021	44,400	1,630		46,030
1300	3 way	"	18.182	43,250	1,750		45,000
1320	4 way	"	18.182	46,140	1,750		47,890

BASIC MATERIALS

ID Code	Description / Component Descriptions	Output / Unit of Meas.	Manhr / Unit	Unit Costs / Material Cost	Labor Cost	Equipment Cost	Total Cost
26 - 28164	**SWITCHES, Cont'd...**						**26 - 28164**
1340	3 way, oil switch, 15kv enclosure	EA	11.940	34,870	1,150		36,020
1350	Pedestal for 35kv double breaker switch	"	5.000	1,380	480		1,860
1360	Bus terminal connector, 2	"	2.500	1,200	240		1,440
1370	2 to 3	"	2.500	1,470	240		1,710
1380	Support connector, 3	"	1.600	780	150		930
1390	Tee connector, 2 to 3	"	2.000	1,120	190		1,310
1400	Flexible bus stud connector	"	1.739	960	170		1,130
1410	End cap 3	"	1.333	1,030	130		1,160
1420	Weldment connection, 3	"	1.000	430	96.00		530
1640	Plate switch, 1 gang	"	0.050	1.04	4.80		5.84
2000	Start stop stations, manual motor starters	"	0.727	100	70.00		170
2020	Lockout switch	"	0.250	17.25	24.00		41.25
2040	Forward-reverse switch	"	0.727	130	70.00		200
2060	On-off switch	"	0.727	130	70.00		200
2080	Open-close switch	"	0.727	130	70.00		200
2100	Forward-reverse-stop switch	"	1.000	210	96.00		310
2120	Standard 3 button switch any standard legend	"	1.000	140	96.00		240
2140	Standard 3 button with lockout	"	1.000	140	96.00		240
2480	Manual motor starters, tog, 115/230v						
2500	Size 1 gp	EA	1.000	220	96.00		320
2540	Size 2	"	1.000	280	96.00		380
2550	Button						
2560	Size 0	EA	1.000	160	96.00		260
2580	Size 1	"	1.000	240	96.00		340
2600	Size 2	"	1.000	300	96.00		400
2610	3-phase						
2620	Size 0	EA	1.333	360	130		490
2640	Size 1	"	1.333	440	130		570
3000	Time & float switches	"	1.600	540	150		690
3020	Astronomical time switch, 40a, 240v	"	1.000	850	96.00		950
3060	Timer switch 0-5 minute, with box	"	0.500	41.50	48.00		90.00
3080	Single pole/single throw time, 277v, NEMA-1	"	0.727	150	70.00		220
3220	Single toggle switch, 20a, 120v, with pilot	"	0.250	37.25	24.00		61.00
3240	3-way toggle	"	0.296	100	28.50		130
4000	Photoelectric switches						
4010	1000 watt						
4020	105-135v	EA	0.727	54.00	70.00		120
4040	208-277v	"	0.727	73.00	70.00		140

BASIC MATERIALS

ID Code	Component Descriptions	Unit of Meas.	Manhr / Unit	Material Cost	Labor Cost	Equipment Cost	Total Cost
	Description	**Output**		**Unit Costs**			

26 - 28164 — SWITCHES, Cont'd... — **26 - 28164**

ID Code	Component Descriptions	Unit of Meas.	Manhr / Unit	Material Cost	Labor Cost	Equipment Cost	Total Cost
4080	3000 watt, 105-130v						
4100	Double throw	EA	1.000	210	96.00		310
4120	Single throw	"	1.000	190	96.00		290
4140	Double pole/single throw, 210-250v	"	1.333	250	130		380
4970	Dimmer switch and switch plate						
4990	600w	EA	0.308	49.75	29.50		79.00
5005	1000w	"	0.348	82.00	33.50		120
5008	Dimmer switch incandescent						
5010	1500w	EA	0.702	160	67.00		230
5020	2000w	"	0.748	200	72.00		270
5025	Fluorescent						
5030	12 lamps	EA	0.500	110	48.00		160
5040	20 lamps	"	0.552	250	53.00		300
5050	30 lamps	"	0.602	440	58.00		500
5060	40 lamps	"	0.702	430	67.00		500
5065	Time clocks with skip, 40a, 120v						
5070	SPST	EA	0.748	150	72.00		220
5080	SPDT	"	0.748	320	72.00		390
5090	DPST	"	0.748	220	72.00		290
5100	DPDT	"	1.000	250	96.00		350
5115	Astronomic time clocks with skip, 40a, 120v						
5120	DPST	EA	0.748	220	72.00		290
5130	SPST	"	1.000	310	96.00		410
5140	SPDT	"	0.748	230	72.00		300
5150	Raintight time clocks, 40a, 120v						
5160	SPDT	EA	1.000	230	96.00		330
5170	DPST	"	1.000	220	96.00		320
5171	Contractor grade wall switch 15a, 120v						
5172	Single pole	EA	0.160	2.62	15.25		17.75
5173	Three way	"	0.200	4.78	19.25		24.00
5174	Four way	"	0.267	16.00	25.50		41.50
5175	Specification grade toggle switches, 20a, 120-277v						
5180	Single pole	EA	0.200	5.75	19.25		25.00
5190	Double pole	"	0.296	13.75	28.50		42.25
5200	3 way	"	0.250	15.00	24.00		39.00
5210	4 way	"	0.296	45.25	28.50		74.00
5215	30a, 120-277v						
5220	Single pole	EA	0.200	37.75	19.25		57.00

BASIC MATERIALS

ID Code	Description — Component Descriptions	Unit of Meas.	Manhr / Unit	Material Cost	Labor Cost	Equipment Cost	Total Cost
		Output		Unit Costs			

26 - 28164   **SWITCHES, Cont'd...**   **26 - 28164**

ID Code	Component Descriptions	Unit of Meas.	Manhr / Unit	Material Cost	Labor Cost	Equipment Cost	Total Cost
5230	Double pole	EA	0.296	52.00	28.50		81.00
5240	3 way	"	0.250	52.00	24.00		76.00
5245	Specification grade key switches, 20a, 120-277v						
5250	Single pole	EA	0.200	39.75	19.25		59.00
5260	Double pole	"	0.296	50.00	28.50		79.00
5270	3 way	"	0.250	43.00	24.00		67.00
5280	4 way	"	0.296	83.00	28.50		110
5285	Red pilot light handle switches, 20a, 120-277v						
5290	Single pole	EA	0.200	42.25	19.25		62.00
5300	Double pole	"	0.296	49.00	28.50		78.00
5310	3 way	"	0.250	74.00	24.00		98.00
5315	30a, 120-277v						
5320	Single pole	EA	0.200	53.00	19.25		72.00
5330	Double pole	"	0.296	64.00	28.50		93.00
5340	3 way	"	0.250	97.00	24.00		120
5345	Momentary contact switches, 20a						
5350	SPDT, ivory	EA	0.250	49.50	24.00		74.00
5360	SPDT, locking	"	0.296	65.00	28.50		94.00
5365	Maintained contact switches						
5370	SPDT ivory	EA	0.250	100	24.00		120
5380	DPDT ivory	"	0.250	100	24.00		120
5390	SPDT locking	"	0.296	120	28.50		150
5400	DPDT locking	"	0.348	120	33.50		150
5420	Mercury switch, 3 way	"	0.250	24.75	24.00		48.75
5430	Door switches, open on or off	"	0.500	61.00	48.00		110
5440	Combination switch and pilot light, single pole	"	0.296	20.00	28.50		48.50
5450	3 way	"	0.348	24.50	33.50		58.00
5460	Combination switch and receptacle, single pole	"	0.296	28.50	28.50		57.00
5470	3 way	"	0.296	35.00	28.50		64.00
5480	Combination two switches, single pole/single pole	"	0.250	23.50	24.00		47.50
5490	3 way	"	0.400	29.00	38.50		68.00
5495	Switch plates, plastic ivory						
5510	1 gang	EA	0.080	0.62	7.68		8.30
5520	2 gang	"	0.100	1.48	9.60		11.00
5530	3 gang	"	0.119	2.31	11.50		13.75
5540	4 gang	"	0.145	5.92	14.00		20.00
5550	5 gang	"	0.160	6.20	15.25		21.50
5560	6 gang	"	0.182	7.31	17.50		24.75

BASIC MATERIALS

ID Code	Description — Component Descriptions	Unit of Meas.	Manhr / Unit	Material Cost	Labor Cost	Equipment Cost	Total Cost
26 - 28164	**SWITCHES, Cont'd...**						**26 - 28164**
5565	Stainless steel						
5570	1 gang	EA	0.080	5.33	7.68		13.00
5580	2 gang	"	0.100	7.40	9.60		17.00
5590	3 gang	"	0.123	11.25	11.75		23.00
5600	4 gang	"	0.145	19.50	14.00		33.50
5610	5 gang	"	0.160	22.75	15.25		38.00
5620	6 gang	"	0.182	28.50	17.50		46.00
5625	Brass						
5630	1 gang	EA	0.080	9.94	7.68		17.50
5640	2 gang	"	0.100	21.25	9.60		30.75
5650	3 gang	"	0.123	33.00	11.75		44.75
5660	4 gang	"	0.145	38.00	14.00		52.00
5670	5 gang	"	0.160	47.00	15.25		62.00
5680	6 gang	"	0.182	57.00	17.50		75.00
26 - 28166	**SAFETY SWITCHES**						**26 - 28166**
0080	Fused, 3 phase, 30 amp, 600v, heavy duty						
1010	NEMA 1	EA	1.143	350	110		460
1020	NEMA 3r	"	1.143	790	110		900
1040	NEMA 4	"	1.600	2,210	150		2,360
1060	NEMA 12	"	1.739	710	170		880
1070	60a						
1080	NEMA 1	EA	1.143	490	110		600
1100	NEMA 3r	"	1.143	930	110		1,040
1120	NEMA 4	"	1.600	2,440	150		2,590
1140	NEMA 12	"	1.739	840	170		1,010
1150	100a						
1160	NEMA 1	EA	1.739	830	170		1,000
1200	NEMA 3r	"	1.739	1,450	170		1,620
1220	NEMA 4	"	2.000	5,190	190		5,380
1240	NEMA 12	"	2.500	1,260	240		1,500
1250	200a						
1260	NEMA 1	EA	2.500	1,230	240		1,470
1280	NEMA 3r	"	2.500	2,010	240		2,250
1300	NEMA 4	"	2.759	6,820	260		7,080
1320	NEMA 12	"	3.478	1,860	330		2,190
1500	400a						
1520	NEMA 1	EA	5.517	2,990	530		3,520

BASIC MATERIALS

ID Code	Component Descriptions	Unit of Meas.	Manhr / Unit	Material Cost	Labor Cost	Equipment Cost	Total Cost
	Description	**Output**		**Unit Costs**			
26 - 28166	**SAFETY SWITCHES, Cont'd...**					**26 - 28166**	
1540	NEMA 3r	EA	5.517	5,750	530		6,280
1560	NEMA 4	"	5.755	13,510	550		14,060
1580	NEMA 12	"	7.018	4,400	670		5,070
1590	600a						
1600	NEMA 1	EA	8.000	5,180	770		5,950
1620	NEMA 3r	"	8.000	7,690	770		8,460
1640	NEMA 4	"	8.989	12,470	860		13,330
1660	NEMA 12	"	12.308	8,100	1,180		9,280
2000	Non-fused, 240-600v, heavy duty, 3 phase, 30 amp						
2020	NEMA 1	EA	1.143	240	110		350
2040	NEMA 3r	"	1.143	400	110		510
2060	NEMA 4	"	1.739	1,560	170		1,730
2080	NEMA 12	"	1.739	470	170		640
2090	60a						
2100	NEMA1	EA	1.143	330	110		440
2120	NEMA 3r	"	1.143	590	110		700
2140	NEMA 4	"	1.739	1,700	170		1,870
2160	NEMA 12	"	1.739	570	170		740
2170	100a						
2180	NEMA 1	EA	1.739	530	170		700
2200	NEMA 3r	"	1.739	830	170		1,000
2220	NEMA 4	"	2.500	3,430	240		3,670
2240	NEMA 12	"	2.500	800	240		1,040
2260	200a, NEMA 1	"	2.500	810	240		1,050
2680	600a, NEMA 12	"	12.308	4,510	1,180		5,690
3000	Bolt-on hubs						
3010	3/4" - 1-1/2"	EA	0.250	28.00	24.00		52.00
3020	2"	"	0.296	51.00	28.50		80.00
3022	2-1/2"	"	0.296	80.00	28.50		110
3024	3"	"	0.348	150	33.50		180
3030	3-1/2"	"	0.400	220	38.50		260
3040	4"	"	0.400	280	38.50		320
3045	Watertight hubs						
3050	1/2"	EA	0.250	22.50	24.00		46.50
3060	3/4"	"	0.296	33.25	28.50		62.00
3070	1"	"	0.400	34.50	38.50		73.00
3080	1-1/4"	"	0.444	39.50	42.75		82.00
3090	1-1/2"	"	0.471	58.00	45.25		100

BASIC MATERIALS

ID Code	Description	Output		Unit Costs			
	Component Descriptions	Unit of Meas.	Manhr / Unit	Material Cost	Labor Cost	Equipment Cost	Total Cost
26 - 28166	**SAFETY SWITCHES, Cont'd...**						**26 - 28166**
3100	2"	EA	0.500	86.00	48.00		130
3110	2-1/2"	"	0.533	110	51.00		160
3120	3"	"	0.615	130	59.00		190
3130	3-1/2"	"	0.800	200	77.00		280
3140	4"	"	0.851	280	82.00		360
3145	Non-fused, 600v, 3 pole, NEMA 7						
3150	60a	EA	2.222	2,260	210		2,470
3160	100a	"	3.200	2,830	310		3,140
3170	225a	"	4.000	6,160	380		6,540
3175	NEMA 9						
3180	60a	EA	2.500	1,900	240		2,140
3190	100a	"	3.333	2,390	320		2,710
3200	225a	"	4.211	5,330	400		5,730
3205	Fusible bolted pressure switches, 600v/3 pole, NEMA						
3210	800a	EA	16.000	14,490	1,540		16,030
3220	1200a	"	21.978	17,550	2,110		19,660
3230	1600a	"	25.000	18,940	2,400		21,340
3240	2000a	"	29.963	19,490	2,880		22,370
3250	2500a	"	34.934	22,280	3,350		25,630
3260	3000a	"	44.944	30,100	4,310		34,410
3270	4000a	"	51.948	40,140	4,990		45,130
3275	Non-fusible						
3280	800a	EA	14.493	13,920	1,390		15,310
3290	1200a	"	20.000	15,610	1,920		17,530
3300	1600a	"	22.989	16,720	2,210		18,930
3310	2000a	"	27.972	17,720	2,690		20,410
3320	2500a	"	34.934	20,340	3,350		23,690
3330	3000a	"	44.944	28,990	4,310		33,300
3340	4000a	"	51.948	39,030	4,990		44,020
3400	Fusible load interrupter switches, 4.16 kv, NEMA 1						
3410	200a	EA	29.963	13,910	2,880		16,790
3420	600a	"	70.175	17,530	6,740		24,270
3425	Fusible load interrupter switch, 13.8 kv						
3430	NEMA 1, 600a	EA	100.000	19,450	9,600		29,050
3435	NEMA 3R, 600a	"	100.000	23,340	9,600		32,940
3438	4.16 kv, NEMA 3R						
3440	200a	EA	29.963	12,150	2,880		15,030
3450	600a	"	70.175	21,890	6,740		28,630

BASIC MATERIALS

ID Code	Description — Component Descriptions	Output — Unit of Meas.	Output — Manhr / Unit	Unit Costs — Material Cost	Unit Costs — Labor Cost	Unit Costs — Equipment Cost	Unit Costs — Total Cost
26 - 28166	**SAFETY SWITCHES, Cont'd...**						**26 - 28166**
3460	Non-fused load interrupter switch, 4.16 kv, NEMA 1						
3470	200a	EA	29.963	9,710	2,880		12,590
3480	600a	"	70.175	16,760	6,740		23,500
3490	13.8 kv, NEMA 1, 600a	"	100.000	17,980	9,600		27,580
3495	4.16 kv, NEMA 3R						
3500	200a	EA	29.963	10,210	2,880		13,090
3510	600a	"	70.175	20,450	6,740		27,190
3520	13.8 kv, NEMA 3R, 600a	"	100.000	21,890	9,600		31,490
3530	Interrupter switch accessories, strip heater	"					1,320
3540	Cable lugs	"					240
3550	Key interlock	"					1,690
3560	Auxiliary switch	"					950
3565	Lightning arrester						
3570	5 kva	EA					9,020
3580	15 kv	"					9,860
26 - 28168	**SAFETY SWITCHES, HEAVY DUTY**						**26 - 28168**
0980	Safety switch, 600v, 3 pole, heavy duty, NEMA-1						
1000	30a	EA	1.000	300	96.00		400
1020	60a	"	1.143	410	110		520
1040	100a	"	1.600	780	150		930
1100	200a	"	2.500	1,220	240		1,460
1120	400a	"	5.517	3,130	530		3,660
1140	600a	"	8.000	5,560	770		6,330
1160	800a	"	10.526	12,590	1,010		13,600
1200	1200a	"	14.286	15,640	1,370		17,010
26 - 28169	**TRANSFER SWITCHES**						**26 - 28169**
0980	Automatic transfer switch 600v, 3 pole						
1000	30a	EA	3.478	3,920	330		4,250
1020	60a	"	3.478	4,740	330		5,070
1040	100a	"	4.762	5,190	460		5,650
1080	150a	"	6.015	6,930	580		7,510
1100	225a	"	8.000	8,660	770		9,430
1120	260a	"	8.000	9,450	770		10,220
1140	400a	"	10.000	11,720	960		12,680
1160	600a	"	15.094	17,050	1,450		18,500
1180	800a	"	18.182	21,580	1,750		23,330
1200	1000a	"	21.053	30,770	2,020		32,790

BASIC MATERIALS

ID Code	Description / Component Descriptions	Unit of Meas.	Manhr / Unit	Material Cost	Labor Cost	Equipment Cost	Total Cost
26 - 28169	**TRANSFER SWITCHES, Cont'd...**						**26 - 28169**
1220	1200a	EA	22.857	35,560	2,190		37,750
1240	1600a	"	25.000	44,880	2,400		47,280
1260	2000a	"	29.630	45,410	2,840		48,250
1280	2600a	"	42.105	93,190	4,040		97,230
1300	3000a	"	50.000	146,440	4,800		151,240

POWER GENERATION

ID Code	Description / Component Descriptions	Unit of Meas.	Manhr / Unit	Material Cost	Labor Cost	Equipment Cost	Total Cost
26 - 32001	**GENERATORS**						**26 - 32001**
0980	Diesel generator, with auto transfer switch						
1000	30kw	EA	30.769	37,600	2,950		40,550
1040	50kw	"	30.769	47,610	2,950		50,560
1080	75kw	"	42.105	60,450	4,040		64,490
1200	100kw	"	47.059	67,080	4,520		71,600
1220	125kw	"	50.000	71,690	4,800		76,490
1240	150kw	"	57.143	82,650	5,490		88,140
1260	175kw	"	66.667	86,130	6,400		92,530
1280	200kw	"	80.000	89,340	7,680		97,020
1300	250kw	"	88.889	97,640	8,530		106,170
1320	300kw	"	100.000	117,430	9,600		127,030
1340	350kw	"	114.286	125,720	10,970		136,690
1360	400kw	"	133.333	154,340	12,800		167,140
1380	450kw	"	145.455	165,310	13,960		179,270
1400	500kw	"	160.000	179,080	15,360		194,440
1420	600kw	"	200.000	245,280	19,200		264,480
1500	750kw	"	200.000	342,390	19,200		361,590

BATTERY EQUIPMENT

ID Code	Description / Component Descriptions	Unit of Meas.	Manhr / Unit	Material Cost	Labor Cost	Equipment Cost	Total Cost
26 - 33530	**UNINTERRUPTIBLE POWER**						**26 - 33530**
1010	Uninterruptible power systems, (UPS), 3 kva	EA	8.000	11,340	770		12,110
1020	5 kva	"	11.004	12,720	1,060		13,780
1030	7.5 kva	"	16.000	15,260	1,540		16,800
1040	10 kva	"	21.978	19,080	2,110		21,190
1050	15 kva	"	22.857	22,900	2,190		25,090
1060	20 kva	"	24.024	31,810	2,310		34,120
1070	25 kva	"	25.000	40,710	2,400		43,110
1080	30 kva	"	25.974	41,980	2,490		44,470

BATTERY EQUIPMENT

ID Code	Description		Output		Unit Costs			
	Component Descriptions	Unit of Meas.	Manhr / Unit	Material Cost	Labor Cost	Equipment Cost	Total Cost
26 - 33530	**UNINTERRUPTIBLE POWER, Cont'd...**						**26 - 33530**
1090	35 kva	EA	27.027	44,530	2,590		47,120
1100	40 kva	"	27.972	48,350	2,690		51,040
1110	45 kva	"	28.986	50,890	2,780		53,670
1120	50 kva	"	29.963	54,710	2,880		57,590
1130	62.5 kva	"	32.000	64,890	3,070		67,960
1140	75 kva	"	34.934	75,060	3,350		78,410
1150	100 kva	"	36.036	100,510	3,460		103,970
1160	150 kva	"	50.000	152,680	4,800		157,480
1170	200 kva	"	55.172	203,570	5,300		208,870
1180	300 kva	"	74.766	305,360	7,180		312,540
1190	400 kva	"	89.888	456,820	8,630		465,450
1200	500 kva	"	109.589	571,020	10,520		581,540

POWER FILTERS AND CONDITIONERS

ID Code	Description		Output		Unit Costs			
26 - 35002	**CENTRAL INVERTER SYSTEMS**						**26 - 35002**
1000	Central inverter systems						
1010	500va	EA	2.963	15,830	280		16,110
1020	1000va	"	4.000	17,300	380		17,680
1030	1500va	"	5.333	20,240	510		20,750
1040	2400va	"	6.667	26,110	640		26,750
1050	3000va	"	8.502	27,580	820		28,400
1060	4500va	"	10.000	41,960	960		42,920
1070	6000va	"	11.004	52,810	1,060		53,870
1080	7500va	"	14.011	61,030	1,350		62,380
1090	10,000va	"	16.000	70,420	1,540		71,960
1100	16,600va	"	22.989	122,060	2,210		124,270
1110	25,000va	"	34.934	145,530	3,350		148,880
26 - 35009	**POWER LINE FILTERS**						**26 - 35009**
0980	Heavy duty power line filter, 240v						
1000	100a	EA	10.000	7,260	960		8,220
1020	300a	"	16.000	24,100	1,540		25,640
1040	600a	"	24.242	33,400	2,330		35,730

POWER FILTERS AND CONDITIONERS

ID Code	Description — Component Descriptions	Output — Unit of Meas.	Output — Manhr / Unit	Unit Costs — Material Cost	Unit Costs — Labor Cost	Unit Costs — Equipment Cost	Unit Costs — Total Cost
26 - 35130		**CAPACITORS**					**26 - 35130**
1800	Three phase capacitors						
1805	240v						
1810	1.5 kvar	EA	2.500	830	240		1,070
1820	2.5 kvar	"	3.200	920	310		1,230
1830	3.0 kvar	"	4.000	1,030	380		1,410
1840	4 kvar	"	5.000	1,120	480		1,600
1850	5 kvar	"	5.333	1,190	510		1,700
1860	6 kvar	"	5.714	1,330	550		1,880
1870	7.5 kvar	"	6.154	1,470	590		2,060
1880	10 kvar	"	8.000	1,630	770		2,400
1890	15 kvar	"	9.501	2,220	910		3,130
1900	20 kvar	"	11.994	2,700	1,150		3,850
1910	25 kvar	"	13.008	3,160	1,250		4,410
1920	40 kvar	"	18.018	5,630	1,730		7,360
1930	50 kvar	"	20.997	6,820	2,020		8,840
1940	60 kvar	"	21.505	8,590	2,060		10,650
1950	75 kvar	"	25.000	9,780	2,400		12,180
1960	100 kvar	"	29.963	11,570	2,880		14,450
1965	480v						
1970	1.5 kvar	EA	2.500	710	240		950
1980	2.5 kvar	"	3.200	750	310		1,060
1990	3 kvar	"	4.000	800	380		1,180
2000	4 kvar	"	5.000	870	480		1,350
2010	5 kvar	"	5.333	940	510		1,450
2020	6 kvar	"	5.714	990	550		1,540
2030	7.5 kvar	"	6.154	1,010	590		1,600
2040	10 kvar	"	8.000	1,150	770		1,920
2050	12.5 kvar	"	9.501	1,300	910		2,210
2060	15 kvar	"	11.994	1,420	1,150		2,570
2070	18 kvar	"	12.500	1,470	1,200		2,670
2080	20 kvar	"	13.008	1,540	1,250		2,790
2090	22.5 kvar	"	13.491	1,600	1,300		2,900
2100	25 kvar	"	14.842	1,720	1,420		3,140
2110	30 kvar	"	14.842	2,020	1,420		3,440
2120	35 kvar	"	16.000	2,270	1,540		3,810
2130	40 kvar	"	18.018	2,560	1,730		4,290
2140	45 kvar	"	20.000	2,660	1,920		4,580
2150	50 kvar	"	20.997	3,410	2,020		5,430

POWER FILTERS AND CONDITIONERS

ID Code	Description / Component Descriptions	Output Unit of Meas.	Manhr / Unit	Material Cost	Labor Cost	Equipment Cost	Total Cost
26 - 35130	**CAPACITORS, Cont'd...**						**26 - 35130**
2160	60 kvar	EA	21.978	3,990	2,110		6,100
2170	70 kvar	"	24.024	3,670	2,310		5,980
2180	75 kvar	"	25.000	3,800	2,400		6,200
2190	80 kvar	"	27.027	4,740	2,590		7,330
2200	90 kvar	"	28.986	5,040	2,780		7,820
2210	100 kvar	"	29.963	5,340	2,880		8,220
2220	125 kvar	"	33.058	6,820	3,170		9,990
2230	150 kvar	"	37.037	8,010	3,560		11,570

FACILITY LIGHTNING PROTECTION

ID Code	Component Descriptions	Unit of Meas.	Manhr / Unit	Material Cost	Labor Cost	Equipment Cost	Total Cost
26 - 41130	**LIGHTNING PROTECTION**						**26 - 41130**
0100	Lightning protection						
0980	Copper point, nickel plated, 12'						
1000	1/2" dia.	EA	1.000	68.00	96.00		160
1020	5/8" dia.	"	1.000	76.00	96.00		170

LIGHTING

ID Code	Component Descriptions	Unit of Meas.	Manhr / Unit	Material Cost	Labor Cost	Equipment Cost	Total Cost
26 - 51101	**INTERIOR LIGHTING**						**26 - 51101**
0010	Recessed fluorescent fixtures, 2'x2'						
0015	2 lamp	EA	0.727	110	70.00		180
0020	4 lamp	"	0.727	140	70.00		210
0030	2 lamp w/flange	"	1.000	130	96.00		230
0040	4 lamp w/flange	"	1.000	160	96.00		260
0045	1'x4'						
0050	2 lamp	EA	0.667	110	64.00		170
0060	3 lamp	"	0.667	150	64.00		210
0070	2 lamp w/flange	"	0.727	130	70.00		200
0080	3 lamp w/flange	"	0.727	180	70.00		250
0085	2'x4'						
0090	2 lamp	EA	0.727	130	70.00		200
0100	3 lamp	"	0.727	160	70.00		230
0110	4 lamp	"	0.727	150	70.00		220
0120	2 lamp w/flange	"	1.000	160	96.00		260
0130	3 lamp w/flange	"	1.000	180	96.00		280
0140	4 lamp w/flange	"	1.000	180	96.00		280
0145	4'x4'						

LIGHTING

ID Code	Component Descriptions	Unit of Meas.	Manhr / Unit	Material Cost	Labor Cost	Equipment Cost	Total Cost
	Description	**Output**		**Unit Costs**			

26 - 51101 — **INTERIOR LIGHTING, Cont'd...** — **26 - 51101**

ID Code	Component Descriptions	Unit of Meas.	Manhr / Unit	Material Cost	Labor Cost	Equipment Cost	Total Cost
0150	4 lamp	EA	1.000	530	96.00		630
0160	6 lamp	"	1.000	630	96.00		730
0170	8 lamp	"	1.000	680	96.00		780
0180	4 lamp w/flange	"	1.509	660	140		800
0190	6 lamp w/flange	"	1.509	820	140		960
0200	8 lamp, w/flange	"	1.509	910	140		1,050
0205	Surface mounted incandescent fixtures						
0210	40w	EA	0.667	160	64.00		220
0220	75w	"	0.667	170	64.00		230
0230	100w	"	0.667	180	64.00		240
0240	150w	"	0.667	240	64.00		300
0245	Pendant						
0250	40w	EA	0.800	130	77.00		210
0260	75w	"	0.800	150	77.00		230
0270	100w	"	0.800	170	77.00		250
0280	150w	"	0.800	190	77.00		270
0281	Contractor grade recessed down lights						
0282	100 watt housing only	EA	1.000	110	96.00		210
0283	150 watt housing only	"	1.000	160	96.00		260
0284	100 watt trim	"	0.500	89.00	48.00		140
0285	150 watt trim	"	0.500	140	48.00		190
0287	Recessed incandescent fixtures						
0290	40w	EA	1.509	220	140		360
0300	75w	"	1.509	240	140		380
0310	100w	"	1.509	260	140		400
0320	150w	"	1.509	280	140		420
0325	Exit lights, 120v						
0330	Recessed	EA	1.250	74.00	120		190
0340	Back mount	"	0.727	120	70.00		190
0350	Universal mount	"	0.727	130	70.00		200
0360	Emergency battery units, 6v-120v, 50 unit	"	1.509	260	140		400
0370	With 1 head	"	1.509	300	140		440
0380	With 2 heads	"	1.509	330	140		470
0390	Mounting bucket	"	0.727	52.00	70.00		120
0395	Light track single circuit						
0400	2'	EA	0.500	65.00	48.00		110
0410	4'	"	0.500	77.00	48.00		130
0420	8'	"	1.000	100	96.00		200

LIGHTING

ID Code	Component Descriptions	Unit of Meas.	Manhr / Unit	Material Cost	Labor Cost	Equipment Cost	Total Cost
	Description	**Output**		**Unit Costs**			
26 - 51101	**INTERIOR LIGHTING, Cont'd...**					**26 - 51101**	
0430	12'	EA	1.509	150	140		290
0435	Fittings and accessories						
0440	Dead end	EA	0.145	25.75	14.00		39.75
0450	Starter kit	"	0.250	34.25	24.00		58.00
0460	Conduit feed	"	0.145	33.25	14.00		47.25
0470	Straight connector	"	0.145	29.50	14.00		43.50
0480	Center feed	"	0.145	47.00	14.00		61.00
0490	L-connector	"	0.145	33.25	14.00		47.25
0501	T-connector	"	0.145	44.75	14.00		59.00
0510	X-connector	"	0.200	54.00	19.25		73.00
0520	Cord and plug	"	0.100	54.00	9.60		64.00
0530	Rigid corner	"	0.145	71.00	14.00		85.00
0540	Flex connector	"	0.145	56.00	14.00		70.00
0550	2 way connector	"	0.200	160	19.25		180
0560	Spacer clip	"	0.050	2.40	4.80		7.20
0570	Grid box	"	0.145	13.50	14.00		27.50
0580	T-bar clip	"	0.050	3.59	4.80		8.39
0590	Utility hook	"	0.145	10.50	14.00		24.50
0595	Fixtures, square						
0600	R-20	EA	0.145	66.00	14.00		80.00
0610	R-30	"	0.145	100	14.00		110
0620	40w flood	"	0.145	170	14.00		180
0630	40w spot	"	0.145	170	14.00		180
0640	100w flood	"	0.145	190	14.00		200
0650	100w spot	"	0.145	150	14.00		160
0660	Mini spot	"	0.145	63.00	14.00		77.00
0670	Mini flood	"	0.145	140	14.00		150
0680	Quartz, 500w	"	0.145	370	14.00		380
0690	R-20 sphere	"	0.145	110	14.00		120
0700	R-30 sphere	"	0.145	58.00	14.00		72.00
0710	R-20 cylinder	"	0.145	78.00	14.00		92.00
0720	R-30 cylinder	"	0.145	90.00	14.00		100
0730	R-40 cylinder	"	0.145	92.00	14.00		110
0740	R-30 wall wash	"	0.145	140	14.00		150
0750	R-40 wall wash	"	0.145	180	14.00		190
0755	Explosion proof, incan., surface mounted						
0760	100w - 200w	EA	1.739	860	170		1,030
0770	300w	"	1.739	1,210	170		1,380

LIGHTING

ID Code	Component Descriptions	Unit of Meas.	Manhr / Unit	Material Cost	Labor Cost	Equipment Cost	Total Cost
	Description	**Output**		**Unit Costs**			
26 - 51101		**INTERIOR LIGHTING, Cont'd...**					**26 - 51101**
0780	500w	EA	1.739	1,780	170		1,950
0785	With guard						
0790	100w-200w	EA	2.222	820	210		1,030
0800	300w	"	2.222	1,140	210		1,350
0810	500w	"	2.222	1,670	210		1,880
0820	Reflectors for incan. light fixtures, dome	"	0.250	150	24.00		170
0830	Angle	"	0.250	170	24.00		190
0840	High-bay	"	0.296	310	28.50		340
0845	Explosion proof fluor. fixtures, 800 ms.						
0850	1 lamp	EA	2.222	3,320	210		3,530
0860	2 lamp	"	2.667	4,110	260		4,370
0870	3 lamp	"	2.963	6,160	280		6,440
0880	4 lamp	"	3.200	7,970	310		8,280
0885	Explosion proof hp sodium fixtures						
0890	50w-70w	EA	2.222	2,250	210		2,460
0900	100w	"	2.222	2,300	210		2,510
0910	150w	"	2.500	2,400	240		2,640
0920	200w	"	2.500	2,430	240		2,670
0930	250w	"	2.500	2,550	240		2,790
0940	310w	"	2.500	2,630	240		2,870
0950	400w	"	2.500	3,960	240		4,200
0955	With guard						
0960	50w-70w	EA	2.500	2,270	240		2,510
0970	100w	"	2.500	2,320	240		2,560
0980	150w	"	2.759	2,500	260		2,760
0990	200w	"	2.759	2,550	260		2,810
1000	250w	"	2.759	2,630	260		2,890
1010	310w	"	2.759	2,730	260		2,990
1020	400w	"	2.759	4,110	260		4,370
1025	Explosion proof metal halide fixtures						
1030	175w	EA	2.500	1,940	240		2,180
1040	250w	"	2.500	2,120	240		2,360
1050	400w	"	2.500	3,140	240		3,380
1060	With guard, 175w	"	2.759	1,990	260		2,250
1070	250w	"	2.759	2,170	260		2,430
1080	400w	"	2.759	3,270	260		3,530
1085	Energy saving rapid start fluor. lamps						
1090	F30 cw	EA	0.100	11.25	9.60		20.75

LIGHTING

ID Code	Component Descriptions	Unit of Meas.	Manhr / Unit	Material Cost	Labor Cost	Equipment Cost	Total Cost

ID Code	Component Descriptions	Unit of Meas.	Manhr / Unit	Material Cost	Labor Cost	Equipment Cost	Total Cost
1100	F40 cw	EA	0.100	11.25	9.60		20.75
1110	F40 cwx	"	0.100	12.50	9.60		22.00
1120	F30 ww	"	0.100	17.00	9.60		26.50
1130	F40 ww	"	0.100	9.24	9.60		18.75
1140	F40 wwx	"	0.100	12.50	9.60		22.00
1145	Slimline						
1150	F48 cw	EA	0.145	18.75	14.00		32.75
1170	F96 cwx	"	0.145	33.50	14.00		47.50
1180	F48 ww	"	0.145	25.00	14.00		39.00
1190	F96 ww	"	0.145	23.00	14.00		37.00
1200	F96 wwx	"	0.145	21.50	14.00		35.50
1210	High output	"	0.145	14.25	14.00		28.25
1220	F96 cwx	"	0.145	21.00	14.00		35.00
1230	F96 cw	"	0.145	15.75	14.00		29.75
1240	Power groove, F48 cw	"	0.145	36.00	14.00		50.00
1245	Circle						
1260	Fc6 cw	EA	0.100	15.50	9.60		25.00
1270	Fc8 cw	"	0.100	13.00	9.60		22.50
1280	Fc12 cw	"	0.100	14.00	9.60		23.50
1290	Fc16 cw	"	0.100	20.50	9.60		30.00
1300	Fc6 ww	"	0.100	16.50	9.60		26.00
1310	Fc8 ww	"	0.100	14.00	9.60		23.50
1320	Fc12 ww	"	0.100	17.00	9.60		26.50
1330	Fc16 ww	"	0.100	24.75	9.60		34.25
1335	Incandescent lamps						
1340	200w	EA	0.100	7.34	9.60		17.00
1350	300w	"	0.100	8.01	9.60		17.50
1360	500w	"	0.100	18.00	9.60		27.50
1370	750w	"	0.145	43.75	14.00		58.00
1380	1000w	"	0.200	47.75	19.25		67.00
1390	1500w	"	0.200	70.00	19.25		89.00
1395	Energy saving reflector floodlight lamps						
1400	25w	EA	0.100	12.75	9.60		22.25
1410	30w	"	0.100	13.75	9.60		23.25
1420	50w	"	0.100	13.75	9.60		23.25
1430	75w	"	0.100	16.00	9.60		25.50
1440	120w	"	0.100	16.00	9.60		25.50
1450	150w	"	0.100	22.00	9.60		31.50

LIGHTING

ID Code	Component Descriptions	Unit of Meas.	Manhr / Unit	Material Cost	Labor Cost	Equipment Cost	Total Cost
	Description	**Output**		**Unit Costs**			
26 - 51101	**INTERIOR LIGHTING, Cont'd...**						**26 - 51101**
1460	200w	EA	0.100	23.00	9.60		32.50
1470	300w	"	0.100	31.00	9.60		40.50
1480	500w	"	0.145	51.00	14.00		65.00
1490	750w	"	0.145	67.00	14.00		81.00
1495	Reflector spotlight						
1500	75w	EA	0.100	12.00	9.60		21.50
1510	100w	"	0.100	14.25	9.60		23.75
1520	125w	"	0.100	24.75	9.60		34.25
1530	150w	"	0.100	26.00	9.60		35.50
1540	250w	"	0.100	43.25	9.60		53.00
1550	300w	"	0.100	69.00	9.60		79.00
1560	400w	"	0.145	76.00	14.00		90.00
1570	500w	"	0.145	95.00	14.00		110
1580	1000w	"	0.200	200	19.25		220
1585	Medium par flood lamps						
1590	75w	EA	0.100	11.75	9.60		21.25
1600	100w	"	0.100	13.25	9.60		22.75
1610	150w	"	0.100	14.25	9.60		23.75
1620	200w	"	0.100	55.00	9.60		65.00
1630	300w	"	0.100	61.00	9.60		71.00
1640	500w	"	0.145	120	14.00		130
1645	Medium par spot lamps						
1650	75w	EA	0.100	14.75	9.60		24.25
1660	120w	"	0.100	15.25	9.60		24.75
1670	150w	"	0.100	15.75	9.60		25.25
1675	Tubular quartz lamps						
1680	100w	EA	0.145	98.00	14.00		110
1690	150w	"	0.145	98.00	14.00		110
1700	200w	"	0.145	110	14.00		120
1710	400w	"	0.145	120	14.00		130
1720	500w	"	0.200	120	19.25		140
1730	750w	"	0.200	120	19.25		140
1740	1000w	"	0.250	150	24.00		170
1750	1250w	"	0.250	160	24.00		180
1760	1500w	"	0.250	200	24.00		220
1765	Ballast replacements rapid start fluor						
1770	1f-40-120v	EA	0.727	32.00	70.00		100
1780	1f-40-277v	"	0.727	54.00	70.00		120

LIGHTING

ID Code	Component Descriptions	Unit of Meas.	Manhr / Unit	Material Cost	Labor Cost	Equipment Cost	Total Cost
	Description	**Output**		**Unit Costs**			

ID Code	Component Descriptions	Unit of Meas.	Manhr / Unit	Material Cost	Labor Cost	Equipment Cost	Total Cost
26 - 51101	**INTERIOR LIGHTING, Cont'd...**						**26 - 51101**
1790	1f-96-120v	EA	0.727	170	70.00		240
1800	1f-96-277v	"	0.727	200	70.00		270
1810	2f-40-120v	"	0.727	45.50	70.00		120
1820	2f-40-277v	"	0.727	54.00	70.00		120
1830	2f-96-120v	"	0.727	140	70.00		210
1840	2f-96-277v	"	0.727	150	70.00		220
1850	Circline, 1fc6-1fc16	"	0.727	42.00	70.00		110
1855	Very high output, 1500ma						
1860	1f48-120v	EA	0.727	240	70.00		310
1870	1f48-277v	"	0.727	240	70.00		310
1880	1f96-120v	"	0.727	210	70.00		280
1890	1f96-277v	"	0.727	240	70.00		310
1900	2f48-120v	"	0.727	210	70.00		280
1910	2f48-277v	"	0.727	240	70.00		310
1920	2f96-120v	"	0.727	240	70.00		310
1930	2f96-277v	"	0.727	270	70.00		340
1935	Mercury, multi tap						
1940	475w	EA	1.000	200	96.00		300
1950	100w	"	1.000	200	96.00		300
1960	175w	"	1.000	220	96.00		320
1970	250w	"	1.000	270	96.00		370
1980	400w	"	1.000	310	96.00		410
1990	1000w	"	1.000	760	96.00		860
1995	Metal halide, multi tap						
2000	175w	EA	1.000	180	96.00		280
2010	250w	"	1.000	230	96.00		330
2020	400w	"	1.000	290	96.00		390
2030	1000w	"	1.000	510	96.00		610
2035	1500w	"	1.000	670	96.00		770
2040	High pressure sodium						
2050	70w	EA	1.000	300	96.00		400
2060	100w	"	1.000	320	96.00		420
2070	150w	"	1.000	340	96.00		440
2080	250w	"	1.000	510	96.00		610
2090	400w	"	1.000	580	96.00		680
2100	1000w	"	1.000	790	96.00		890

LIGHTING

ID Code	Description		Output		Unit Costs			
	Component Descriptions		Unit of Meas.	Manhr / Unit	Material Cost	Labor Cost	Equipment Cost	Total Cost
26 - 51201	**ENERGY EFFICIENT INTERIOR LIGHTING**						**26 - 51201**	
1000	Ballast							
1010	Fluorescent, 12 VDC							
1020	Minimum		EA	0.667	160	64.00		220
1030	Average		"	0.667	190	64.00		250
1040	Maximum		"	0.667	230	64.00		290
1050	24 VDC							
1060	Minimum		EA	0.667	170	64.00		230
1070	Average		"	0.667	210	64.00		270
1080	Maximum		"	0.667	240	64.00		300
1090	Pressure sodium, 12 VDC							
1100	Minimum		EA	0.667	190	64.00		250
1110	Average		"	0.667	220	64.00		280
1120	Maximum		"	0.667	240	64.00		300
1130	24 VDC							
1140	Minimum		EA	0.667	210	64.00		270
1150	Average		"	0.667	230	64.00		290
1160	Maximum		"	0.667	240	64.00		300
3000	Lamps							
3010	Photovoltaic source, Fluorescent, 12 VDC, 7 Watt							
3020	Minimum		EA	0.267	35.50	25.50		61.00
3030	Average		"	0.267	42.50	25.50		68.00
3040	Maximum		"	0.267	49.50	25.50		75.00
3050	11 Watt							
3060	Minimum		EA	0.267	39.00	25.50		65.00
3070	Average		"	0.267	46.00	25.50		72.00
3080	Maximum		"	0.267	53.00	25.50		79.00
3090	15 Watt							
3100	Minimum		EA	0.267	42.50	25.50		68.00
3110	Average		"	0.267	49.50	25.50		75.00
3120	Maximum		"	0.267	57.00	25.50		83.00
3130	25 Watt							
3140	Minimum		EA	0.267	47.75	25.50		73.00
3150	Average		"	0.267	53.00	25.50		79.00
3160	Maximum		"	0.267	62.00	25.50		88.00
3170	30 Watt							
3180	Minimum		EA	0.267	53.00	25.50		79.00
3190	Average		"	0.267	60.00	25.50		86.00
3200	Maximum		"	0.267	67.00	25.50		93.00

LIGHTING

ID Code	Component Descriptions	Unit of Meas.	Manhr / Unit	Material Cost	Labor Cost	Equipment Cost	Total Cost
	Description	**Output**		**Unit Costs**			

26 - 51201   **ENERGY EFFICIENT INTERIOR LIGHTING, Cont'd...**   **26 - 51201**

ID Code	Component Descriptions	Unit of Meas.	Manhr / Unit	Material Cost	Labor Cost	Equipment Cost	Total Cost
3210	LED, 85-265 V, 300 lumens						
3220	Minimum	EA	0.267	76.00	25.50		100
3230	Average	"	0.267	83.00	25.50		110
3240	Maximum	"	0.267	89.00	25.50		110
3250	600 lumens						
3260	Minimum	EA	0.267	190	25.50		220
3270	Average	"	0.267	410	25.50		440
3280	Maximum	"	0.267	210	25.50		240
3290	12-24 V, 2500 lumens						
3300	Minimum	EA	0.667	770	64.00		830
3310	Average	"	0.667	790	64.00		850
3320	Maximum	"	0.667	800	64.00		860
3330	Compact fluorescent, 13 Watt, 900 lumens						
3340	Minimum	EA	0.267	57.00	25.50		83.00
3350	Average	"	0.267	64.00	25.50		90.00
3360	Maximum	"	0.267	71.00	25.50		97.00
3370	36 Watt, 2800 lumens						
3380	Minimum	EA	0.267	44.25	25.50		70.00
3390	Average	"	0.267	49.50	25.50		75.00
3400	Maximum	"	0.267	53.00	25.50		79.00
3410	40 Watt, 3150 lumens						
3420	Minimum	EA	0.267	47.75	25.50		73.00
3430	Average	"	0.267	53.00	25.50		79.00
3440	Maximum	"	0.267	57.00	25.50		83.00
3450	Low Pressure Sodium, 18 Watt, 1800 lumens						
3460	Minimum	EA	0.267	110	25.50		140
3470	Average	"	0.267	120	25.50		150
3480	Maximum	"	0.267	120	25.50		150
3490	35 Watt, 5000 lumens						
3500	Minimum	EA	0.267	140	25.50		170
3510	Average	"	0.267	150	25.50		180
3520	Maximum	"	0.267	150	25.50		180
3530	40 Watt, 3150 lumens						
3540	Minimum	EA	0.267	47.75	25.50		73.00
3550	Average	"	0.267	53.00	25.50		79.00
3560	Maximum	"	0.267	57.00	25.50		83.00

LIGHTING

ID Code	Component Descriptions	Unit of Meas.	Manhr / Unit	Material Cost	Labor Cost	Equipment Cost	Total Cost
	Description	**Output**		**Unit Costs**			

26 - 51401	**INDUSTRIAL LIGHTING**					**26 - 51401**	
0100	Surface mounted fluorescent, wrap around lens						
0110	1 lamp	EA	0.800	130	77.00		210
0120	2 lamps	"	0.889	190	85.00		280
0140	4 lamps	"	1.000	200	96.00		300
0250	Wall mounted fluorescent						
0300	2-20w lamps	EA	0.500	140	48.00		190
0320	2-30w lamps	"	0.500	150	48.00		200
0340	2-40w lamps	"	0.667	150	64.00		210
0350	Indirect, with wood shielding, 2049w lamps						
0360	4'	EA	1.000	160	96.00		260
0380	8'	"	1.600	200	150		350
0390	Industrial fluorescent, 2 lamp						
0400	4'	EA	0.727	110	70.00		180
0420	8'	"	1.333	170	130		300
0490	Strip fluorescent						
0510	4'						
0520	1 lamp	EA	0.667	67.00	64.00		130
0540	2 lamps	"	0.667	81.00	64.00		140
0550	8'						
0560	1 lamp	EA	0.727	97.00	70.00		170
0580	2 lamps	"	0.889	150	85.00		240
0640	Wire guard for strip fixture, 4' long	"	0.348	15.00	33.50		48.50
0660	Strip fluorescent, 8' long, two 4' lamps	"	1.333	200	130		330
0680	With four 4' lamps	"	1.600	250	150		400
0690	Wet location fluorescent, plastic housing						
0695	4' long						
0700	1 lamp	EA	1.000	160	96.00		260
0720	2 lamps	"	1.333	240	130		370
0730	8' long						
0740	2 lamps	EA	1.600	410	150		560
0760	4 lamps	"	1.739	540	170		710
1000	Parabolic troffer, 2'x2'						
1020	With 2 U lamps	EA	1.000	190	96.00		290
1060	With 3 U lamps	"	1.143	220	110		330
1080	2'x4'						
1100	With 2 40w lamps	EA	1.143	220	110		330
1120	With 3 40w lamps	"	1.333	220	130		350
1140	With 4 40w lamps	"	1.333	230	130		360

LIGHTING

ID Code	Description / Component Descriptions	Output / Unit of Meas.	Manhr / Unit	Unit Costs / Material Cost	Labor Cost	Equipment Cost	Total Cost
26 - 51401	**INDUSTRIAL LIGHTING, Cont'd...**						**26 - 51401**
1180	1'x4'						
1220	With 1 T-12 lamp, 9 cell	EA	0.727	110	70.00		180
1240	With 2 T-12 lamps	"	0.889	130	85.00		210
1260	With 1 T-12 lamp, 20 cell	"	0.727	130	70.00		200
1280	With 2 T-12 lamps	"	0.889	140	85.00		230
1480	Steel sided surface fluorescent, 2'x4'						
1500	3 lamps	EA	1.333	220	130		350
1520	4 lamps	"	1.333	250	130		380
2100	Outdoor sign fluor., 1 lamp, remote ballast						
2120	4' long	EA	6.015	4,630	580		5,210
2140	6' long	"	8.000	5,560	770		6,330
2620	Recess mounted, commercial, 2'x2', 13" high						
2640	100w	EA	4.000	1,540	380		1,920
2660	250w	"	4.494	1,690	430		2,120
3120	High pressure sodium, hi-bay open						
3140	400w	EA	1.739	670	170		840
3160	1000w	"	2.424	1,150	230		1,380
3170	Enclosed						
3180	400w	EA	2.424	1,080	230		1,310
3200	1000w	"	2.963	1,500	280		1,780
3210	Metal halide hi-bay, open						
3220	400w	EA	1.739	410	170		580
3240	1000w	"	2.424	850	230		1,080
3250	Enclosed						
3260	400w	EA	2.424	930	230		1,160
3280	1000w	"	2.963	900	280		1,180
3500	High pressure sodium, low-bay surface mounted						
3520	100w	EA	1.000	340	96.00		440
3540	150w	"	1.143	370	110		480
3560	250w	"	1.333	420	130		550
3580	400w	"	1.600	530	150		680
3590	Metal halide, low-bay, pendant mounted						
3600	175w	EA	1.333	540	130		670
3620	250w	"	1.600	740	150		890
3660	400w	"	2.222	800	210		1,010
4000	Indirect luminaire, square, metal halide, freestanding						
4020	175w	EA	1.000	670	96.00		770
4040	250w	"	1.000	730	96.00		830

LIGHTING

ID Code	Description — Component Descriptions	Output — Unit of Meas.	Output — Manhr / Unit	Unit Costs — Material Cost	Unit Costs — Labor Cost	Unit Costs — Equipment Cost	Unit Costs — Total Cost
26 - 51401	**INDUSTRIAL LIGHTING, Cont'd...**						**26 - 51401**
4060	400w	EA	1.000	750	96.00		850
4070	High pressure sodium						
4080	150w	EA	1.000	1,220	96.00		1,320
4100	250w	"	1.000	1,300	96.00		1,400
4120	400w	"	1.000	1,440	96.00		1,540
4125	Round, metal halide						
4140	175w	EA	1.000	1,400	96.00		1,500
4160	250w	"	1.000	1,450	96.00		1,550
4180	400w	"	1.000	1,520	96.00		1,620
4190	High pressure sodium						
4200	150w	EA	1.000	1,340	96.00		1,440
4220	250w	"	1.000	1,560	96.00		1,660
4240	400w	"	1.000	1,620	96.00		1,720
4250	Wall mounted, metal halide						
4260	175w	EA	2.500	610	240		850
4280	250w	"	2.500	600	240		840
4300	400w	"	3.200	660	310		970
4310	High pressure sodium						
4320	150w	EA	2.500	570	240		810
4340	250w	"	2.500	600	240		840
4360	400w	"	3.200	620	310		930
4480	Wall pack lithonia, high pressure sodium						
4500	35w	EA	0.889	87.00	85.00		170
4520	55w	"	1.000	110	96.00		210
4540	150w	"	1.600	280	150		430
4560	250w	"	1.739	300	170		470
4570	Low pressure sodium						
4580	35w	EA	1.739	470	170		640
4600	55w	"	2.000	640	190		830
4610	Wall pack hubbell, high pressure sodium						
4620	35w	EA	0.889	390	85.00		480
4640	150w	"	1.600	490	150		640
4660	250w	"	1.739	630	170		800
4700	Compact fluorescent						
4720	2-7w	EA	1.000	240	96.00		340
4740	2-13w	"	1.333	270	130		400
4760	1-18w	"	1.333	320	130		450
6000	Handball & racquet ball court, 2'x2', metal halide						

LIGHTING

ID Code	Description		Output		Unit Costs			
	Component Descriptions		Unit of Meas.	Manhr / Unit	Material Cost	Labor Cost	Equipment Cost	Total Cost
26 - 51401		**INDUSTRIAL LIGHTING, Cont'd...**					**26 - 51401**	
6020	250w		EA	2.500	850	240		1,090
6040	400w		"	2.759	1,010	260		1,270
6060	High pressure sodium							
6080	250w		EA	2.500	930	240		1,170
6100	400w		"	2.759	1,010	260		1,270
6120	Bollard light, 42" w/found., high pressure sodium							
6160	70w		EA	2.581	1,420	250		1,670
6180	100w		"	2.581	1,450	250		1,700
6200	150w		"	2.581	1,470	250		1,720
8000	Light fixture lamps							
8010	Lamp							
8020	20w med. bi-pin base, cool white, 24"		EA	0.145	11.50	14.00		25.50
8040	30w cool white, rapid start, 36"		"	0.145	14.50	14.00		28.50
8060	40w cool white U, 3"		"	0.145	31.50	14.00		45.50
8080	40w cool white, rapid start, 48"		"	0.145	13.50	14.00		27.50
8100	70w high pressure sodium, mogul base		"	0.200	100	19.25		120
8120	75w slimline, 96"		"	0.200	31.00	19.25		50.00
8130	100w							
8140	Incandescent, 100a, inside frost		EA	0.100	5.40	9.60		15.00
8160	Mercury vapor, clear, mogul base		"	0.200	94.00	19.25		110
8180	High pressure sodium, mogul base		"	0.200	140	19.25		160
8190	150w							
8200	Par 38 flood or spot, incandescent		EA	0.100	31.75	9.60		41.25
8220	High pressure sodium, 1/2 mogul base		"	0.200	130	19.25		150
8230	175w							
8240	Mercury vapor, clear, mogul base		EA	0.200	57.00	19.25		76.00
8260	Metal halide, clear, mogul base		"	0.200	110	19.25		130
8270	High pressure sodium, mogul base		"	0.200	130	19.25		150
8530	250w							
8540	Mercury vapor, clear, mogul base		EA	0.200	79.00	19.25		98.00
8560	Metal halide, clear, mogul base		"	0.200	110	19.25		130
8580	High pressure sodium, mogul base		"	0.200	140	19.25		160
8590	400w							
8600	Mercury vapor, clear, mogul base		EA	0.200	87.00	19.25		110
8620	Metal halide, clear, mogul base		"	0.200	110	19.25		130
8640	High pressure sodium, mogul base		"	0.200	140	19.25		160
8650	1000w							
8660	Mercury vapor, clear, mogul base		EA	0.250	210	24.00		230

LIGHTING

ID Code	Component Descriptions	Unit of Meas.	Manhr / Unit	Material Cost	Labor Cost	Equipment Cost	Total Cost
	Description	**Output**		**Unit Costs**			
26 - 51401	**INDUSTRIAL LIGHTING, Cont'd...**					**26 - 51401**	
8680	High pressure sodium, mogul base	EA	0.250	370	24.00		390
26 - 56003	**EXTERIOR LIGHTING**					**26 - 56003**	
1200	Exterior light fixtures						
1210	Rectangle, high pressure sodium						
1220	70w	EA	2.500	480	240		720
1240	100w	"	2.581	500	250		750
1260	150w	"	2.581	530	250		780
1280	250w	"	2.759	720	260		980
1300	400w	"	3.478	810	330		1,140
1310	Flood, rectangular, high pressure sodium						
1320	70w	EA	2.500	350	240		590
1340	100w	"	2.581	410	250		660
1360	150w	"	2.581	380	250		630
1400	400w	"	3.478	470	330		800
1420	1000w	"	4.494	790	430		1,220
1430	Round						
1440	400w	EA	3.478	890	330		1,220
1460	1000w	"	4.494	1,380	430		1,810
1470	Round, metal halide						
1480	400w	EA	3.478	970	330		1,300
1500	1000w	"	4.494	1,450	430		1,880
1980	Light fixture arms, cobra head, 6', high press. sodium						
2000	100w	EA	2.000	530	190		720
2021	150w	"	2.500	840	240		1,080
2060	250w	"	2.500	880	240		1,120
2080	400w	"	2.963	900	280		1,180
2090	Flood, metal halide						
2100	400w	EA	3.478	880	330		1,210
2120	1000w	"	4.494	1,200	430		1,630
6260	1500w	"	6.015	1,510	580		2,090
6270	Mercury vapor						
6280	250w	EA	2.759	580	260		840
6300	400w	"	3.478	650	330		980
6360	Incandescent						
6380	300w	EA	1.739	130	170		300
6410	500w	"	2.000	240	190		430
6420	1000w	"	3.200	270	310		580

LIGHTING

Description		Output		Unit Costs			
ID Code	Component Descriptions	Unit of Meas.	Manhr / Unit	Material Cost	Labor Cost	Equipment Cost	Total Cost
26 - 56004	**ENERGY EFFICIENT EXTERIOR LIGHTING**						**26 - 56004**
1000	Solar Powered, LED area light, 100 Watt, Zone 4						
1010	Minimum	EA	1.333	1,740	130		1,870
1020	Average	"	1.600	1,950	150		2,100
1030	Maximum	"	2.000	2,130	190		2,320
1040	Zone 2						
1050	Minimum	EA	1.333	2,290	130		2,420
1060	Average	"	1.600	2,390	150		2,540
1070	Maximum	"	2.000	2,480	190		2,670
1080	Zone 4DD						
1090	Minimum	EA	1.333	2,630	130		2,760
1100	Average	"	1.600	2,820	150		2,970
1110	Maximum	"	2.000	3,010	190		3,200
1120	Zone 2DD						
1130	Minimum	EA	1.333	2,910	130		3,040
1140	Average	"	1.600	3,140	150		3,290
1150	Maximum	"	2.000	3,370	190		3,560

DIVISION 27
COMMUNICATIONS

COMMUNICATIONS

	Description		Output		Unit Costs			
ID Code	Component Descriptions		Unit of Meas.	Manhr / Unit	Material Cost	Labor Cost	Equipment Cost	Total Cost
27 - 32001		**TELEPHONE SYSTEMS**						**27 - 32001**
0480	Communication cable							
0490	25 pair		LF	0.026	1.43	2.47		3.90
0520	100 pair		"	0.029	6.84	2.74		9.58
0530	150 pair		"	0.033	10.50	3.20		13.75
0540	200 pair		"	0.040	14.00	3.84		17.75
0550	300 pair		"	0.042	17.75	4.04		21.75
0560	400 pair		"	0.044	24.25	4.26		28.50
0700	Cable tap in manhole or junction box							
0800	25 pair cable		EA	3.810	9.97	370		380
0900	50 pair cable		"	7.547	20.25	720		740
1000	75 pair cable		"	11.268	30.25	1,080		1,110
1020	100 pair cable		"	15.094	40.25	1,450		1,490
1040	150 pair cable		"	22.222	60.00	2,130		2,190
1060	200 pair cable		"	29.630	81.00	2,840		2,920
1080	300 pair cable		"	44.444	120	4,270		4,390
1100	400 pair cable		"	61.538	160	5,910		6,070
2000	Cable terminations, manhole or junction box							
2020	25 pair cable		EA	3.756	9.97	360		370
2040	50 pair cable		"	7.477	20.25	720		740
2060	100 pair cable		"	15.094	40.25	1,450		1,490
2080	150 pair cable		"	22.222	60.00	2,130		2,190
2100	200 pair cable		"	29.630	81.00	2,840		2,920
2120	300 pair cable		"	44.444	120	4,270		4,390
2140	400 pair cable		"	61.538	130	5,910		6,040
3000	Telephones, standard							
3010	1 button		EA	2.963	190	280		470
3020	2 button		"	3.478	270	330		600
3030	6 button		"	5.333	410	510		920
3040	12 button		"	7.619	1,030	730		1,760
3050	18 button		"	8.889	1,080	850		1,930
3055	Hazardous area							
3060	Desk		EA	7.273	3,090	700		3,790
3070	Wall		"	5.000	1,440	480		1,920
3075	Accessories							
3080	Standard ground		EA	1.600	51.00	150		200
3090	Push button		"	1.600	52.00	150		200
3100	Buzzer		"	1.600	54.00	150		200
3110	Interface device		"	0.800	29.00	77.00		110

COMMUNICATIONS

ID Code	Component Descriptions	Unit of Meas.	Manhr / Unit	Material Cost	Labor Cost	Equipment Cost	Total Cost
	Description	**Output**		**Unit Costs**			

27 - 32001 — **TELEPHONE SYSTEMS, Cont'd...** — **27 - 32001**

ID Code	Component Descriptions	Unit of Meas.	Manhr / Unit	Material Cost	Labor Cost	Equipment Cost	Total Cost
3120	Long cord	EA	0.800	30.50	77.00		110
3130	Interior jack	"	0.400	18.75	38.50		57.00
3140	Exterior jack	"	0.615	37.25	59.00		96.00
3145	Hazardous area						
3150	Selector switch	EA	3.200	300	310		610
3160	Bell	"	3.200	460	310		770
3170	Horn	"	4.211	690	400		1,090
3180	Horn relay	"	3.077	570	300		870

AUDIO-VIDEO SYSTEMS

27 - 41005 — **TELEVISION SYSTEMS** — **27 - 41005**

ID Code	Component Descriptions	Unit of Meas.	Manhr / Unit	Material Cost	Labor Cost	Equipment Cost	Total Cost
0100	TV outlet, self terminating, w/cover plate	EA	0.308	8.72	29.50		38.25
0120	Thru splitter	"	1.600	19.00	150		170
0140	End of line	"	1.333	15.75	130		150
0480	In line splitter multitap						
0490	4 way	EA	1.818	31.75	170		200
0520	2 way	"	1.702	23.75	160		180
1000	Equipment cabinet	"	1.600	79.00	150		230
1010	Antenna						
1020	Broad band UHF	EA	3.478	160	330		490
1040	Lightning arrester	"	0.727	48.25	70.00		120
1060	TV cable	LF	0.005	0.72	0.48		1.20
1070	Coaxial cable rg	"	0.005	0.49	0.48		0.97
1080	Cable drill, with replacement tip	EA	0.500	7.94	48.00		56.00
1100	Cable blocks for in-line taps	"	0.727	15.75	70.00		86.00
1120	In-line taps PTU-series 36 TV system	"	1.143	19.00	110		130
2010	Control receptacles	"	0.449	12.50	43.25		56.00
2020	Coupler	"	2.424	23.75	230		250
2030	Head end equipment	"	6.667	3,000	640		3,640
2040	TV camera	"	1.667	1,620	160		1,780
2050	TV power bracket	"	0.800	140	77.00		220
2060	TV monitor	"	1.455	1,240	140		1,380
2070	Video recorder	"	2.105	2,410	200		2,610
2080	Console	"	8.502	4,990	820		5,810
2090	Selector switch	"	1.379	780	130		910
2100	TV controller	"	1.404	370	130		500

AUDIO-VIDEO SYSTEMS

ID Code	Component Descriptions	Unit of Meas.	Manhr / Unit	Material Cost	Labor Cost	Equipment Cost	Total Cost
		Description	**Output**		**Unit Costs**		
27 - 41010	**CLOSED CIRCUIT TV SYSTEMS**						**27 - 41010**
1000	Camera color, minimum	EA	1.600	420	150		570
1020	Maximum	"	1.600	3,080	150		3,230
1040	Black and white, minimum	"	1.600	340	150		490
1060	Maximum	"	1.600	2,150	150		2,300
1080	Lens, auto tris, minimum	"	1.000	1,390	96.00		1,490
1100	Maximum	"	1.000	2,110	96.00		2,210
1120	Zoom, minimum	"	1.000	4,060	96.00		4,160
1140	Maximum	"	1.000	14,830	96.00		14,930
1160	Fixed, minimum	"	1.000	75.00	96.00		170
1180	Maximum	"	1.000	160	96.00		260
1200	Pinhole, minimum	"	1.000	840	96.00		940
1220	Maximum	"	1.000	1,630	96.00		1,730
1240	Low light level, minimum	"	1.000	870	96.00		970
1260	Maximum	"	1.000	2,130	96.00		2,230
1280	Enclosure indoor, minimum	"	1.501	360	140		500
1300	Maximum	"	3.509	5,500	340		5,840
1320	Outdoor, minimum	"	2.500	630	240		870
1340	Maximum	"	5.517	2,870	530		3,400
1360	Wiper kit	"	0.750	43.75	72.00		120
1380	Lube kit	"	0.500	170	48.00		220
1400	O-Ring kit	"	1.000	46.25	96.00		140
1420	Recharge kit	"	2.500	1,170	240		1,410
1440	Automatic washer/wiper	"	3.509	1,360	340		1,700
1460	Mount, minimum	"	1.000	51.00	96.00		150
1480	Maximum	"	8.000	850	770		1,620
1500	Defroster, minimum	"	1.000	100	96.00		200
1520	Maximum	"	2.000	550	190		740
1540	Alarm thermostat kit	"	1.501	890	140		1,030
1560	Shroud, minimum	"	0.750	84.00	72.00		160
1580	Maximum	"	1.501	340	140		480
1600	Heater, maximum	"	1.751	530	170		700
1620	Air funnel	"	1.501	620	140		760
1640	Visor, minimum	"	0.500	100	48.00		150
1660	Maximum	"	0.500	200	48.00		250
2000	Camera enclosure dome indoor, minimum	"	1.751	360	170		530
2020	Maximum	"	3.756	2,120	360		2,480
2040	Outdoor, minimum	"	2.759	750	260		1,010
2060	Maximum	"	6.504	1,810	620		2,430

AUDIO-VIDEO SYSTEMS

	Description	Output		Unit Costs			
ID Code	Component Descriptions	Unit of Meas.	Manhr / Unit	Material Cost	Labor Cost	Equipment Cost	Total Cost
27 - 41010	**CLOSED CIRCUIT TV SYSTEMS, Cont'd...**					**27 - 41010**	
2080	Pan/tilt indoor, minimum	EA	4.520	640	430		1,070
2100	Maximum	"	7.767	2,550	750		3,300
2120	Outdoor, minimum	"	5.000	640	480		1,120
2140	Maximum	"	9.756	4,290	940		5,230
2160	Insulation, minimum	"	1.000	84.00	96.00		180
2180	Maximum	"	2.500	810	240		1,050
2200	Dome tamper switch	"	0.500	84.00	48.00		130
2220	Blower, minimum	"	1.000	180	96.00		280
2240	Maximum	"	1.501	320	140		460
2260	Heater, minimum	"	0.500	130	48.00		180
2280	Ceiling adapter	"	0.500	81.00	48.00		130
2290	Lock, minimum	"	0.350	33.00	33.50		67.00
2300	Maximum	"	0.750	190	72.00		260
4000	Monitor color, minimum	"	1.501	440	140		580
4020	Maximum	"	2.000	3,550	190		3,740
4040	Black and white, minimum	"	1.501	570	140		710
4060	Maximum	"	2.000	850	190		1,040
4080	Rack, minimum	"	1.000	290	96.00		390
4100	Maximum	"	1.751	700	170		870
4120	Time lapse recorder, minimum	"	1.000	440	96.00		540
4140	Maximum	"	4.000	4,360	380		4,740
4160	Scanner, minimum	"	1.000	670	96.00		770
4180	Maximum	"	1.501	1,960	140		2,100
4200	Camera switcher, minimum	"	1.501	770	140		910
4220	Maximum	"	4.520	9,830	430		10,260
4240	Control, minimum	"	0.750	240	72.00		310
4260	Maximum	"	4.520	9,210	430		9,640
4280	Seq. processor	"	2.500	500	240		740
4300	Screen splitter	"	1.501	1,130	140		1,270
5000	Wireless video, minimum	"	6.504	1,980	620		2,600
5020	Maximum	"	12.121	8,810	1,160		9,970
5040	Control equalizing amp., minimum	"	1.751	780	170		950
5060	Maximum	"	2.500	870	240		1,110
5080	Relay box	"	1.501	1,450	140		1,590
5100	Generator	"	2.000	1,270	190		1,460
5120	Transformer	"	1.000	89.00	96.00		190
5140	Camera power supply	"	1.501	500	140		640
5160	Motion detector	"	2.000	190	190		380

AUDIO-VIDEO SYSTEMS

ID Code	Description — Component Descriptions	Output — Unit of Meas.	Manhr / Unit	Unit Costs — Material Cost	Labor Cost	Equipment Cost	Total Cost
27 - 41010	**CLOSED CIRCUIT TV SYSTEMS, Cont'd...**						**27 - 41010**
5180	Control test board	EA	0.750	320	72.00		390
5200	Receiver, minimum	"	1.000	650	96.00		750
5220	Maximum	"	2.759	2,470	260		2,730
5240	Keyboard control, minimum	"	2.000	820	190		1,010
5260	Maximum	"	3.008	2,420	290		2,710
5280	Matrix CPU, minimum	"	2.000	3,050	190		3,240
5300	Maximum	"	8.000	10,910	770		11,680
5320	Relay interface, minimum	"	1.000	1,460	96.00		1,560
5340	Maximum	"	2.500	2,790	240		3,030
5360	Alarm interface, minimum	"	1.000	630	96.00		730
5380	Maximum	"	2.000	1,690	190		1,880
5400	Video input card, minimum	"	1.000	440	96.00		540
5420	Maximum	"	2.500	3,560	240		3,800
5440	Transmitter/CPU, minimum	"	8.000	4,070	770		4,840
5460	Maximum	"	12.121	8,650	1,160		9,810
5480	Output card	"	1.000	480	96.00		580
5500	Input card	"	1.000	610	96.00		710
5520	Interface card	"	1.000	700	96.00		800
5540	Auto/random scan	"	0.500	350	48.00		400
5560	Pretested cable, minimum	"	0.500	170	48.00		220
5580	Maximum	"	0.750	410	72.00		480
5590	Wiring harness	"	0.350	12.75	33.50		46.25

DISTRIBUTED AUDIO-VIDEO COMMUNICATIONS SYSTEMS

ID Code	Description — Component Descriptions	Output — Unit of Meas.	Manhr / Unit	Material Cost	Labor Cost	Equipment Cost	Total Cost
27 - 51002	**SIGNALING SYSTEMS**						**27 - 51002**
1000	Signaling systems						
1010	4" bell	EA	0.602	180	58.00		240
1020	6" bell	"	0.650	210	62.00		270
1030	10" bell	"	0.748	240	72.00		310
1035	Buzzer						
1040	Size 0	EA	0.444	46.75	42.75		90.00
1050	Size 1	"	0.444	50.00	42.75		93.00
1060	Size 2	"	0.500	53.00	48.00		100
1070	Size 3	"	0.533	56.00	51.00		110
1080	Horn	"	0.615	160	59.00		220
1090	Chime	"	0.533	240	51.00		290
1095	Push button						

DISTRIBUTED AUDIO-VIDEO COMMUNICATIONS SYSTEMS

		Output		Unit Costs			
	Description						
ID Code	Component Descriptions	Unit of Meas.	Manhr / Unit	Material Cost	Labor Cost	Equipment Cost	Total Cost
27 - 51002			**SIGNALING SYSTEMS, Cont'd...**			**27 - 51002**	
1100	Standard	EA	0.400	53.00	38.50		92.00
1110	Weatherproof	"	0.500	80.00	48.00		130
1115	Door opener						
1120	Mortise	EA	0.500	56.00	48.00		100
1130	Rim	"	0.400	79.00	38.50		120
1140	Transformer	"	0.444	28.25	42.75		71.00
2000	Contractor grade doorbell chime kit						
2020	Chime	EA	1.000	54.00	96.00		150
2040	Doorbutton	"	0.320	7.48	30.75		38.25
2050	Transformer	"	0.500	24.00	48.00		72.00
27 - 51003			**SOUND SYSTEMS**			**27 - 51003**	
1010	Power amplifiers	EA	3.478	1,320	330		1,650
1020	Pre-amplifiers	"	2.759	1,050	260		1,310
1030	Tuner	"	1.455	670	140		810
1040	Equalizer	"	1.600	1,710	150		1,860
1050	Mixer	"	2.222	700	210		910
1060	Tape recorder	"	1.860	2,230	180		2,410
1065	Microphone	"	1.000	190	96.00		290
1070	Cassette Player	"	2.162	1,130	210		1,340
1080	Record player	"	1.905	99.00	180		280
1090	Equipment rack	"	1.290	130	120		250
1095	Speaker						
1110	Wall	EA	4.000	700	380		1,080
1120	Paging	"	0.800	250	77.00		330
1130	Column	"	0.533	370	51.00		420
1140	Single	"	0.615	85.00	59.00		140
1150	Double	"	4.444	260	430		690
1160	Volume control	"	0.533	85.00	51.00		140
1170	Plug-in	"	0.800	260	77.00		340
1180	Desk	"	0.400	210	38.50		250
1190	Outlet	"	0.400	42.25	38.50		81.00
1200	Stand	"	0.296	85.00	28.50		110
1210	Console	"	8.000	19,920	770		20,690
1220	Power supply	"	1.290	370	120		490

HEALTHCARE COMMUNICATIONS AND MONITORING SYSTEMS

ID Code	Description / Component Descriptions	Output Unit of Meas.	Manhr / Unit	Material Cost	Labor Cost	Equipment Cost	Total Cost
27 - 52230	**CALL SYSTEMS**						**27 - 52230**
1010	Call systems, single bed station	EA	0.533	290	51.00		340
1020	Double bed station	"	0.727	530	70.00		600
1030	Call-in cord	"	0.200	100	19.25		120
1040	Pull cord	"	0.200	160	19.25		180
1050	Pillow speaker	"	0.276	320	26.50		350
1060	Dome light	"	0.533	85.00	51.00		140
1070	Zone light	"	0.533	74.00	51.00		130
1080	Stake station	"	0.615	230	59.00		290
1090	Duty station	"	0.500	290	48.00		340
1100	Utility station	"	0.615	230	59.00		290
1110	Nurses station	"	0.533	240	51.00		290
1120	Surgical station	"	0.727	550	70.00		620
1130	Master station	"	2.500	5,730	240		5,970
1140	Control station	"	8.000	2,010	770		2,780
1150	Annunciator	"	2.000	760	190		950
1160	Power supply	"	1.538	560	150		710
1170	Speakers	"	0.800	85.00	77.00		160
1180	Foot switch	"	0.296	100	28.50		130
3000	Code blue systems						
3010	Bed station	EA	0.727	550	70.00		620
3020	Dome light	"	0.667	85.00	64.00		150
3030	Zone light	"	0.727	150	70.00		220
3040	Pull cord	"	0.250	160	24.00		180
3050	Nurses station	"	0.533	240	51.00		290
3060	Annunciator	"	2.000	760	190		950
3070	Power supply	"	1.455	560	140		700
3075	Nurse station indicator, alarm annunciators, flush						
3080	4 circuit	EA	4.000	4,890	380		5,270
3090	6 circuit	"	8.000	5,730	770		6,500
3100	12 circuit	"	12.012	9,760	1,150		10,910
3105	Desktop						
3110	4 circuit	EA	3.478	5,390	330		5,720
3120	6 circuit	"	3.478	6,730	330		7,060
3130	12 circuit	"	5.000	10,430	480		10,910

DISTRIBUTED SYSTEMS

ID Code	Description Component Descriptions	Output Unit of Meas.	Output Manhr / Unit	Unit Costs Material Cost	Unit Costs Labor Cost	Unit Costs Equipment Cost	Unit Costs Total Cost
27 - 53130	**CLOCK SYSTEMS**						**27 - 53130**
1000	Clock systems						
1010	Single face	EA	0.800	190	77.00		270
1020	Double face	"	0.800	550	77.00		630
1030	Skeleton	"	2.759	420	260		680
1040	Master	"	5.000	4,400	480		4,880
1050	Signal generator	"	4.000	3,800	380		4,180
1060	Elapsed time indicator	"	0.800	750	77.00		830
1070	Controller	"	0.533	170	51.00		220
1080	Clock and speaker	"	1.096	320	110		430
1085	Bell						
1090	Standard	EA	0.533	150	51.00		200
1100	Weatherproof	"	0.800	180	77.00		260
1105	Horn						
1110	Standard	EA	0.727	96.00	70.00		170
1120	Weatherproof	"	0.952	120	91.00		210
1130	Chime	"	0.533	110	51.00		160
1140	Buzzer	"	0.533	38.50	51.00		90.00
1150	Flasher	"	0.615	150	59.00		210
1160	Control Board	"	3.478	550	330		880
1170	Program unit	"	5.000	590	480		1,070
1180	Clock back box	"	0.500	33.50	48.00		82.00
1190	Double clock back box	"	0.667	74.00	64.00		140
1200	Wire guard	"	0.200	22.25	19.25		41.50

DIVISION 28
SAFETY & SECURITY

ELECTRONIC

ID Code	Description Component Descriptions	Output Unit of Meas.	Output Manhr / Unit	Unit Costs Material Cost	Unit Costs Labor Cost	Unit Costs Equipment Cost	Unit Costs Total Cost
28 - 16005	**SECURITY SYSTEMS**						**28 - 16005**
1000	Sensors						
1020	Balanced magnetic door switch, surface mounted	EA	0.500	260	48.00		310
1040	With remote test	"	1.000	330	96.00		430
1060	Flush mounted	"	1.860	240	180		420
1080	Mounted bracket	"	0.348	18.50	33.50		52.00
1100	Mounted bracket spacer	"	0.348	16.75	33.50		50.00
1120	Photoelectric sensor, for fence						
1140	6 beam	EA	2.759	27,150	260		27,410
1160	9 beam	"	4.255	33,190	410		33,600
1170	Photoelectric sensor, 12 volt dc						
1180	500' range	EA	1.600	800	150		950
1190	800' range	"	2.000	890	190		1,080
1195	Capacitance wire grid kit						
1200	Surface	EA	1.000	220	96.00		320
1220	Duct	"	1.600	170	150		320
1240	Tube grid kit	"	0.500	280	48.00		330
1260	Vibration sensor, 30 max per zone	"	0.500	330	48.00		380
1280	Audio sensor, 30 max per zone	"	0.500	350	48.00		400
1290	Inertia sensor						
1300	Outdoor	EA	0.727	260	70.00		330
1320	Indoor	"	0.500	170	48.00		220
2000	Ultrasonic transmitter, 20 max per zone						
2020	Omni-directional	EA	1.600	190	150		340
2040	Directional	"	1.333	200	130		330
2050	Transceiver						
2060	Omni-directional	EA	1.000	200	96.00		300
2080	Directional	"	1.000	220	96.00		320
2560	Passive infrared sensor, 20 max per zone	"	1.600	1,430	150		1,580
3020	Access/secure unit, balanced magnetic switch	"	1.600	890	150		1,040
3040	Photoelectric sensor	"	1.600	1,470	150		1,620
3060	Photoelectric fence sensor	"	1.600	1,510	150		1,660
3080	Capacitance sensor	"	1.739	1,750	170		1,920
3100	Audio and vibration sensor	"	1.600	1,540	150		1,690
3120	Inertia sensor	"	1.600	2,060	150		2,210
3160	Ultrasonic sensor	"	1.739	2,360	170		2,530
3200	infrared sensor	"	2.000	1,470	190		1,660
4020	Monitor panel, with access/secure tone, standard	"	1.739	1,000	170		1,170
4040	High security	"	2.000	1,470	190		1,660

ELECTRONIC

ID Code	Component Descriptions	Unit of Meas.	Manhr / Unit	Material Cost	Labor Cost	Equipment Cost	Total Cost
	Description	**Output**		**Unit Costs**			

ID Code	Component Descriptions	Unit of Meas.	Manhr / Unit	Material Cost	Labor Cost	Equipment Cost	Total Cost
28 - 16005	**SECURITY SYSTEMS, Cont'd...**						**28 - 16005**
4060	Emergency power indicator	EA	0.500	600	48.00		650
4070	Monitor rack with 115v power supply						
4080	1 zone	EA	1.000	840	96.00		940
4100	10 zone	"	2.500	4,260	240		4,500
4120	Monitor cabinet, wall mounted						
4140	1 zone	EA	1.000	1,280	96.00		1,380
4160	5 zone	"	1.600	2,120	150		2,270
4180	10 zone	"	1.739	4,630	170		4,800
4200	20 zone	"	2.000	6,450	190		6,640
4240	Floor mounted, 50 zone	"	4.000	6,710	380		7,090
5000	Security system accessories						
5020	Tamper assembly for monitor cabinet	EA	0.444	170	42.75		210
5040	Monitor panel blank	"	0.348	22.25	33.50		56.00
5060	Audible alarm	"	0.500	190	48.00		240
5080	Audible alarm control	"	0.348	780	33.50		810
5090	Termination screw, terminal cabinet						
5100	25 pair	EA	1.600	560	150		710
5120	50 pair	"	2.500	890	240		1,130
5140	150 pair	"	5.000	1,450	480		1,930
5150	Universal termination, cabinets & panel						
5160	Remote test	EA	1.739	130	170		300
5180	No remote test	"	0.727	93.00	70.00		160
5220	High security line supervision termination	"	1.000	650	96.00		750
5240	Door cord for capacitance sensor, 12"	"	0.500	22.25	48.00		70.00
5260	Insulation block kit for capacitance sensor	"	0.348	100	33.50		130
5280	Termination block for capacitance sensor	"	0.348	22.50	33.50		56.00
5360	Guard alert display	"	0.615	2,360	59.00		2,420
5380	Uninterrupted power supply	"	8.000	2,400	770		3,170
5390	Plug-in 40kva transformer						
5400	12 volt	EA	0.348	95.00	33.50		130
5420	18 volt	"	0.348	63.00	33.50		97.00
5440	24 volt	"	0.348	44.50	33.50		78.00
5520	Test relay	"	0.348	140	33.50		170
5580	Coaxial cable, 50 ohm	LF	0.006	0.65	0.57		1.22
6010	Door openers	EA	0.500	170	48.00		220
6015	Push buttons						
6020	Standard	EA	0.348	37.00	33.50		71.00
6030	Weatherproof	"	0.444	56.00	42.75		99.00

ELECTRONIC

	Description	Output		Unit Costs			
ID Code	Component Descriptions	Unit of Meas.	Manhr / Unit	Material Cost	Labor Cost	Equipment Cost	Total Cost
28 - 16005	**SECURITY SYSTEMS, Cont'd...**					**28 - 16005**	
6040	Bells	EA	0.727	140	70.00		210
6045	Horns						
6050	Standard	EA	1.000	150	96.00		250
6060	Weatherproof	"	1.250	280	120		400
6070	Chimes	"	0.667	220	64.00		280
6080	Flasher	"	0.615	160	59.00		220
6090	Motion detectors	"	1.509	650	140		790
6100	Intercom units	"	0.727	150	70.00		220
6110	Remote annunciator	"	5.000	5,050	480		5,530
6500	Control/communications, panels, minimum	"	6.015	2,160	580		2,740
6520	Maximum	"	10.000	5,100	960		6,060
6540	CPU w/shield, minimum	"	4.000	770	380		1,150
6560	Maximum	"	8.000	3,700	770		4,470
6580	Attack resistant pkg.	"	5.000	2,560	480		3,040
6600	Fire/burglar pkg.	"	7.018	2,860	670		3,530
6640	Standard burglar	"	6.015	2,390	580		2,970
6680	Fire	"	5.000	1,830	480		2,310
6700	Input/Output board	"	2.000	1,410	190		1,600
6720	Panel accessory flush mount kit, brass	"	0.500	120	48.00		170
6740	Stainless steel	"	0.500	88.00	48.00		140
6760	Desk stand	"	0.500	130	48.00		180
6780	Conduit box	"	0.500	60.00	48.00		110
6800	Fire alarm kit	"	0.750	76.00	72.00		150
6820	Lock set and key	"	0.350	17.75	33.50		51.00
6840	Tamper switch	"	0.350	22.75	33.50		56.00
6860	Dual battery harness	"	0.500	45.25	48.00		93.00
6880	Powered loop interface	"	0.750	210	72.00		280
6900	Battery 12v DC, 6 amp	"	0.500	140	48.00		190
6920	Reversing, relay module	"	1.000	350	96.00		450
6940	Dual phone line switcher	"	1.000	390	96.00		490
6960	Class A circuit module	"	1.000	400	96.00		500
6980	Plug-in relay	"	0.500	27.75	48.00		76.00
7000	Mounting bracket	"	0.750	53.00	72.00		130
7020	Telephone cord 7'-8 conductor	"	0.500	35.25	48.00		83.00
7040	Listen-in amplifier	"	2.000	290	190		480
7060	Listen-in pickup	"	0.500	130	48.00		180
7080	Bell supervision module	"	1.000	380	96.00		480
7100	Zone control module	"	1.501	470	140		610

ELECTRONIC

ID Code	Component Descriptions	Unit of Meas.	Manhr / Unit	Material Cost	Labor Cost	Equipment Cost	Total Cost
28 - 16005	**SECURITY SYSTEMS, Cont'd...**						**28 - 16005**
7120	Independ. control keypad	EA	1.000	310	96.00		410
7140	Independ. zone control	"	2.500	540	240		780
7160	Command center	"	4.520	860	430		1,290
7180	Control/commun. panel, attack resistant enclosure	"	1.501	390	140		530
7200	Fire enclosure	"	1.000	250	96.00		350
7220	Serial output module	"	1.000	310	96.00		410
7240	Bell noise filter	"	0.350	22.75	33.50		56.00
7260	Release module	"	1.000	360	96.00		460
7280	Printer/ctr. interface	"	1.751	670	170		840
7300	Battery charger module	"	1.501	510	140		650
7320	Cable ribbon	"	0.500	160	48.00		210
7340	50' printer cable	"	0.500	240	48.00		290
7360	Local security printer	"	1.000	2,940	96.00		3,040
7380	Printer paper	"	0.500	150	48.00		200
7400	Inert reader system card reader insert	"	2.500	1,200	240		1,440
7420	Magnetic cards	"	0.350	12.50	33.50		46.00
7440	Photo ID cards	"	0.350	17.75	33.50		51.00
7460	Proximity card reader	"	2.500	1,240	240		1,480
7480	Decoder	"	2.000	1,310	190		1,500
7500	Cards	"	0.500	37.75	48.00		86.00
7520	Wiegand, swipe reader interface	"	3.008	1,370	290		1,660
7540	Card reader	"	3.008	1,210	290		1,500
7560	Swipe cards	"	0.500	240	48.00		290
7580	Photo cards	"	0.500	290	48.00		340
7600	System command, zone expans. ctr. control comm.	"	4.520	1,320	430		1,750
7620	Command center	"	2.500	420	240		660
7640	Center	"	2.500	390	240		630
7660	Transformer	"	0.500	50.00	48.00		98.00
7680	Enclosure	"	0.750	130	72.00		200
7700	Skirt	"	0.500	91.00	48.00		140
7720	Bar code programmer	"	8.000	2,080	770		2,850
7740	Wand assembly	"	4.000	570	380		950
7760	Wand tip	"	0.350	22.75	33.50		56.00
7780	Connector cord	"	0.500	45.25	48.00		93.00
7800	Adapter	"	0.500	76.00	48.00		120
7820	Carry case	"	0.350	150	33.50		180
7900	Video - Intro	"	0.500	380	48.00		430
7920	Central station receiver	"	24.242	20,160	2,330		22,490

ELECTRONIC

ID Code	Component Descriptions	Unit of Meas.	Manhr / Unit	Material Cost	Labor Cost	Equipment Cost	Total Cost
	Description	**Output**		**Unit Costs**			

28 - 16005   SECURITY SYSTEMS, Cont'd...   28 - 16005

ID Code	Component Descriptions	Unit of Meas.	Manhr / Unit	Material Cost	Labor Cost	Equipment Cost	Total Cost
7940	Spares pkg.	EA	16.000	7,560	1,540		9,100
7960	Printer paper	"	0.500	60.00	48.00		110
7980	Main processing unit	"	12.121	5,040	1,160		6,200
8000	Terminator card	"	1.751	500	170		670
8020	Power supply card	"	2.000	1,510	190		1,700
8040	Receiver line card	"	2.000	1,510	190		1,700
8060	Telco terminator card	"	1.501	250	140		390
8080	Printer	"	2.500	3,020	240		3,260
8100	Printer terminator card	"	1.501	500	140		640
8120	Vital information processor software, 4000 acct.	"	24.242	83,160	2,330		85,490
8140	8000 acct.	"	33.333	124,740	3,200		127,940
8160	11,000 acct.	"	50.000	166,210	4,800		171,010
8180	33,000 acct.	"	72.727	249,490	6,980		256,470
8200	Autodial	"	2.000	1,260	190		1,450
8220	Ready key readers, low profile reader	"	4.000	1,000	380		1,380
8240	Vandal resistant reader	"	5.000	1,120	480		1,600
8260	PIN reader	"	8.000	1,760	770		2,530
8280	Door controller, multi function, four door	"	24.242	7,100	2,330		9,430
8300	Two door	"	16.000	5,040	1,540		6,580
8320	Enclosure	"	1.000	130	96.00		230
8340	Networked systems, MS DOS PC software 16 doors	"	18.182	3,060	1,750		4,810
8360	Remote	"	18.182	3,060	1,750		4,810
8380	Windows PC software upgrade	"	12.121	1,010	1,160		2,170
8400	Central network, single site controller, MS DOS	"	40.000	14,300	3,840		18,140
8420	Multi site	"	61.538	28,190	5,910		34,100
8440	Single, windows	"	40.000	14,350	3,840		18,190
8460	Ancillary products, Wiegand interface	"	2.000	340	190		530
8480	Alarm module	"	1.751	1,320	170		1,490
8500	Reader mounting bracket	"	1.000	68.00	96.00		160

FIRE SAFETY

ID Code	Component Descriptions	Unit of Meas.	Manhr / Unit	Material Cost	Labor Cost	Equipment Cost	Total Cost
	Description	**Output**		**Unit Costs**			

28 - 31001   **FIRE ALARM SYSTEMS**   **28 - 31001**

ID Code	Component Descriptions	Unit of Meas.	Manhr / Unit	Material Cost	Labor Cost	Equipment Cost	Total Cost
1000	Master fire alarm box, pedestal mounted	EA	16.000	10,600	1,540		12,140
1020	Master fire alarm box	"	6.015	5,450	580		6,030
1040	Box light	"	0.500	190	48.00		240
1060	Ground assembly for box	"	0.667	150	64.00		210
1080	Bracket for pole type box	"	0.727	200	70.00		270
1090	Pull station						
1100	Waterproof	EA	0.500	97.00	48.00		140
1110	Manual	"	0.400	72.00	38.50		110
1120	Horn, waterproof	"	1.000	140	96.00		240
1140	Interior alarm	"	0.727	91.00	70.00		160
1160	Coded transmitter, automatic	"	2.000	1,390	190		1,580
1180	Control panel, 8 zone	"	8.000	3,200	770		3,970
1200	Battery charger and cabinet	"	2.000	1,070	190		1,260
1240	Batteries, nickel cadmium or lead calcium	"	5.000	800	480		1,280
2500	CO$_2$ pressure switch connection	"	0.727	150	70.00		220
3000	Annunciator panels						
3020	Fire detection annunciator, remote type, 8 zone	EA	1.818	540	170		710
3100	12 zone	"	2.000	690	190		880
3120	16 zone	"	2.500	870	240		1,110
4000	Fire alarm systems						
4010	Bell	EA	0.615	160	59.00		220
4020	Weatherproof bell	"	0.667	110	64.00		170
4030	Horn	"	0.727	97.00	70.00		170
4040	Siren	"	2.000	980	190		1,170
4050	Chime	"	0.615	120	59.00		180
4060	Audio/visual	"	0.727	180	70.00		250
4070	Strobe light	"	0.727	160	70.00		230
4080	Smoke detector	"	0.667	260	64.00		320
4090	Heat detector	"	0.500	45.50	48.00		94.00
4100	Thermal detector	"	0.500	42.25	48.00		90.00
4110	Ionization detector	"	0.533	210	51.00		260
4120	Duct detector	"	2.759	710	260		970
4130	Test switch	"	0.500	120	48.00		170
4140	Remote indicator	"	0.571	75.00	55.00		130
4150	Door holder	"	0.727	260	70.00		330
4160	Telephone jack	"	0.296	4.92	28.50		33.50
4170	Fireman phone	"	1.000	640	96.00		740
4180	Speaker	"	0.800	130	77.00		210

FIRE SAFETY

ID Code	Description — Component Descriptions	Output — Unit of Meas.	Output — Manhr / Unit	Unit Costs — Material Cost	Unit Costs — Labor Cost	Unit Costs — Equipment Cost	Unit Costs — Total Cost
28 - 31001	**FIRE ALARM SYSTEMS, Cont'd...**						**28 - 31001**
4185	Remote fire alarm annunciator panel						
4190	24 zone	EA	6.667	3,330	640		3,970
4200	48 zone	"	13.008	6,660	1,250		7,910
4205	Control panel						
4210	12 zone	EA	2.963	2,520	280		2,800
4220	16 zone	"	4.444	2,690	430		3,120
4230	24 zone	"	6.667	4,110	640		4,750
4240	48 zone	"	16.000	7,680	1,540		9,220
4250	Power supply	"	1.509	480	140		620
4260	Status command	"	5.000	13,000	480		13,480
4270	Printer	"	1.509	4,150	140		4,290
4280	Transponder	"	0.899	210	86.00		300
4290	Transformer	"	0.667	300	64.00		360
4300	Transceiver	"	0.727	420	70.00		490
4310	Relays	"	0.500	160	48.00		210
4320	Flow switch	"	2.000	530	190		720
4330	Tamper switch	"	2.963	330	280		610
4340	End of line resistor	"	0.348	23.25	33.50		57.00
4350	Printed circuit card	"	0.500	210	48.00		260
4360	Central processing unit	"	6.154	14,220	590		14,810
4370	UPS backup to CPU	"	8.999	26,040	860		26,900
8020	Smoke detector, fixed temp. & rate of rise comb.	"	1.600	420	150		570
28 - 31101	**FIRE ALARMS**						**28 - 31101**
1000	Fire alarm duct detector accessory, 24v DC power	EA	0.500	56.00	48.00		100
1100	24v Remote inducator LED	"	0.500	81.00	48.00		130
1120	Test probe adapter	"	0.500	60.00	48.00		110
1140	Sampling tube, 3 foot	"	0.500	22.25	48.00		70.00
1150	6 foot	"	0.500	44.75	48.00		93.00
1160	10 foot	"	0.750	100	72.00		170
1200	Form C 5 amp relay contacts (2)	"	1.000	110	96.00		210
2000	Panel accessory, electronic type, McCullom loop alarm	"	4.000	840	380		1,220
2020	Dome light assembly, single	"	0.350	33.50	33.50		67.00
2040	Double	"	0.450	43.50	43.25		87.00
2060	Alarm tone module, wail	"	0.500	110	48.00		160
2100	Whoop	"	0.500	120	48.00		170
2120	Chime	"	0.500	59.00	48.00		110
2140	Power transformer	"	0.750	250	72.00		320

FIRE SAFETY

ID Code	Component Descriptions	Unit of Meas.	Manhr / Unit	Material Cost	Labor Cost	Equipment Cost	Total Cost
	Description	**Output**		**Unit Costs**			
28 - 31101		**FIRE ALARMS, Cont'd...**					**28 - 31101**
2160	Replacement power supply	EA	1.000	830	96.00		930
2180	Trouble circuit module	"	0.750	400	72.00		470
2220	Trouble silence switch	"	0.350	24.50	33.50		58.00
2240	Bell silence switch	"	0.350	24.50	33.50		58.00
2260	Voltage regulator	"	0.400	29.25	38.50		68.00
2280	Replacement control panel PC board	"	2.000	960	190		1,150
2290	Power supply	"	1.000	400	96.00		500
2300	Pre-amplifier module	"	0.750	98.00	72.00		170
2320	Input switching module	"	0.750	120	72.00		190
2340	Audio zone selector card	"	0.750	230	72.00		300
2360	Backup switching module	"	0.350	94.00	33.50		130
2380	Interface PC board	"	0.400	85.00	38.50		120
2390	Amplifier supervisory module	"	0.500	150	48.00		200
2400	Fireman telephone main control board	"	1.000	400	96.00		500
2420	Dual telephone zone module	"	0.500	170	48.00		220
2440	Diodes	"	0.350	11.25	33.50		44.75
2460	Fuses and fuse holder	"	0.350	18.00	33.50		52.00
3000	Metal face plate, engraving per character	"					2.24
3020	Black plastic tags w/engraving per character	"					1.09
3040	Addressable single zone class A zone module	"	1.501	700	140		840
3060	Class B zone module, dual zone	"	1.501	380	140		520
3080	Quad zone	"	1.501	780	140		920
3100	Single class A bell module	"	1.000	610	96.00		710
3120	Dual class B	"	1.000	500	96.00		600
3140	Fire alarm march time coder module	"	0.750	290	72.00		360
3160	Flow switch module	"	1.000	310	96.00		410
3180	Tamper	"	1.000	310	96.00		410
3200	40 watt audio module	"	1.501	1,450	140		1,590
3220	Auxiliary power supply module	"	1.000	580	96.00		680
3240	Time delay module	"	0.750	250	72.00		320
3260	Digital voice message repeater	"	5.000	2,910	480		3,390
3280	Firefighter telephone system, 1 zone	"	3.008	1,840	290		2,130
3300	4 zone	"	5.000	2,460	480		2,940
3320	8 zone	"	6.015	2,800	580		3,380
3340	12 zone	"	8.000	3,130	770		3,900
3360	16 zone	"	10.000	3,490	960		4,450
3380	20 zone	"	12.121	4,360	1,160		5,520
3400	24 zone	"	13.333	4,700	1,280		5,980

FIRE SAFETY

ID Code	Component Descriptions	Unit of Meas.	Manhr / Unit	Material Cost	Labor Cost	Equipment Cost	Total Cost
	Description	**Output**		**Unit Costs**			
28 - 31101		**FIRE ALARMS, Cont'd...**					**28 - 31101**
3420	30 zone	EA	16.000	5,220	1,540		6,760
3440	40 zone	"	18.182	6,390	1,750		8,140
3460	48 zone	"	20.000	7,090	1,920		9,010
3480	Portable telephone handset w/plug and cord	"	0.500	330	48.00		380
3490	Sound powered handset w/plug and cord	"	0.500	360	48.00		410
3500	Remote break glass telephone jack	"	0.500	100	48.00		150
3520	Handset storage rack for (4) portable handsets	"	1.000	280	96.00		380
3540	Manual pull station w/hammer and chain	"	1.000	170	96.00		270
3560	Dust tight enclosure	"	2.000	390	190		580
3580	Light emitting diode	"	1.000	150	96.00		250
4000	Telephone jack	"	1.250	160	120		280
4020	Explosion proof enclosure	"	2.500	610	240		850
4040	Radio master box w/antenna and pedestal	"	16.000	21,250	1,540		22,790
4060	Halon control panel w/abort sw., emerg release						
4080	Ground fault and batteries (6 amp. hr.)	EA	8.000	2,500	770		3,270
4100	Halon releasing device module	"	1.000	290	96.00		390
4120	Time delay module	"	1.000	250	96.00		350
4140	Dual 4 PDT relay	"	0.667	270	64.00		330
4160	Quad switch panel	"	0.500	200	48.00		250
4180	Utility disconnect relay w/bypass switch	"	0.500	200	48.00		250
4200	18 amp. hr. batteries	"	1.000	740	96.00		840
6000	Emergency manual halon release station	"	0.750	110	72.00		180
6020	Abort station	"	0.750	100	72.00		170
6040	Addressable fire alarm control panels, 6 zone	"	5.000	11,410	480		11,890
6060	12 zone	"	10.000	14,580	960		15,540
6080	16 zone	"	14.035	16,660	1,350		18,010
6100	20 zone	"	18.182	18,120	1,750		19,870
6120	24 zone	"	21.053	19,900	2,020		21,920
6140	30 zone	"	25.000	21,250	2,400		23,650
6160	34 zone	"	30.769	25,720	2,950		28,670
6180	40 zone	"	38.095	28,520	3,660		32,180
6200	48 zone	"	44.444	31,760	4,270		36,030
6220	Remote annunciator panel, 6 zone	"	3.008	2,190	290		2,480
6240	12 zone	"	5.000	3,510	480		3,990
6260	16 zone	"	8.000	4,820	770		5,590
6280	20 zone	"	10.000	6,130	960		7,090
6290	24 zone	"	12.121	6,360	1,160		7,520
6300	30 zone	"	14.035	7,240	1,350		8,590

FIRE SAFETY

ID Code	Description — Component Descriptions	Output — Unit of Meas.	Output — Manhr / Unit	Unit Costs — Material Cost	Unit Costs — Labor Cost	Unit Costs — Equipment Cost	Unit Costs — Total Cost
28 - 31101	**FIRE ALARMS, Cont'd...**						**28 - 31101**
6320	34 zone	EA	16.000	7,890	1,540		9,430
6340	40 zone	"	18.182	8,330	1,750		10,080
6360	48 zone	"	20.000	9,650	1,920		11,570
6380	Smoke detectors	"	1.501	240	140		380
6400	Manual pull stations	"	1.501	530	140		670

DIVISION 31
EARTHWORK

EARTHWORK, EXCAVATION & FILL

ID Code	Component Descriptions	Unit of Meas.	Manhr / Unit	Material Cost	Labor Cost	Equipment Cost	Total Cost
	Description	**Output**		**Unit Costs**			
31 - 22130	**ROUGH GRADING**					**31 - 22130**	
1000	Site grading, cut & fill, sandy clay, 200' haul, 75 hp	CY	0.032		2.34	2.72	5.06
1100	Spread topsoil by equipment on site	"	0.036		2.60	3.02	5.62
1200	Site grading (cut and fill to 6") less than 1 acre						
1300	75 hp dozer	CY	0.053		3.90	4.53	8.44
1400	1.5 c.y. backhoe/loader	"	0.080		5.86	6.80	12.75
31 - 23163	**BULK EXCAVATION**					**31 - 23163**	
1700	Hydraulic excavator						
1720	1 c.y. capacity						
1740	Light material	CY	0.040		2.93	2.50	5.43
1760	Medium material	"	0.048		3.51	3.00	6.51
1780	Wet material	"	0.060		4.39	3.75	8.14
1790	Blasted rock	"	0.069		5.02	4.28	9.31
2000	Wheel mounted front-end loader						
2020	7/8 c.y. capacity						
2040	Light material	CY	0.020		1.95	2.50	4.45
2060	Medium material	"	0.023		2.22	2.85	5.08
2080	Wet material	"	0.027		2.60	3.33	5.93
2100	Blasted rock	"	0.032		3.12	4.00	7.12
2600	Track mounted front-end loader						
2620	1-1/2 c.y. capacity						
2640	Light material	CY	0.013		1.30	1.66	2.96
2660	Medium material	"	0.015		1.41	1.81	3.23
2680	Wet material	"	0.016		1.56	2.00	3.56
2700	Blasted rock	"	0.018		1.73	2.22	3.95
31 - 23164	**BUILDING EXCAVATION**					**31 - 23164**	
0090	Structural excavation, unclassified earth						
0100	3/8 c.y. backhoe	CY	0.107		10.50	13.25	23.75
0110	3/4 c.y. backhoe	"	0.080		7.80	10.00	17.75
0120	1 c.y. backhoe	"	0.067		6.50	8.33	14.75
0600	Foundation backfill and compaction by machine	"	0.160		15.50	20.00	35.50
31 - 23165	**HAND EXCAVATION**					**31 - 23165**	
0980	Excavation						
1000	To 2' deep						
1020	Normal soil	CY	0.889		66.00		66.00
1040	Sand and gravel	"	0.800		59.00		59.00
1060	Medium clay	"	1.000		74.00		74.00

EARTHWORK, EXCAVATION & FILL

ID Code	Description / Component Descriptions	Output Unit of Meas.	Output Manhr / Unit	Unit Costs Material Cost	Unit Costs Labor Cost	Unit Costs Equipment Cost	Unit Costs Total Cost
31 - 23165	**HAND EXCAVATION, Cont'd...**						**31 - 23165**
1080	Heavy clay	CY	1.143		84.00		84.00
1100	Loose rock	"	1.333		98.00		98.00
1200	To 6' deep						
1220	Normal soil	CY	1.143		84.00		84.00
1240	Sand and gravel	"	1.000		74.00		74.00
1260	Medium clay	"	1.333		98.00		98.00
1280	Heavy clay	"	1.600		120		120
1300	Loose rock	"	2.000		150		150
2200	Compaction of backfill around structures or in trench						
2220	By hand with air tamper	CY	0.571		42.25		42.25
2240	By hand with vibrating plate tamper	"	0.533		39.25		39.25
2250	1 ton roller	"	0.400		29.25	34.00	63.00
5400	Miscellaneous hand labor						
5440	Trim slopes, sides of excavation	SF	0.001		0.09		0.09
5450	Trim bottom of excavation	"	0.002		0.11		0.11
5460	Excavation around obstructions and services	CY	2.667		200		200
31 - 23167	**UTILITY EXCAVATION**						**31 - 23167**
2080	Trencher, sandy clay, 8" wide trench						
2100	18" deep	LF	0.018		1.30	1.51	2.81
2200	24" deep	"	0.020		1.46	1.70	3.16
2300	36" deep	"	0.023		1.67	1.94	3.61
6080	Trench backfill, 95% compaction						
7000	Tamp by hand	CY	0.500		37.00		37.00
7050	Vibratory compaction	"	0.400		29.50		29.50
7060	Trench backfilling, with borrow sand, place & compact	"	0.400	27.25	29.50		57.00
31 - 23169	**HAULING MATERIAL**						**31 - 23169**
0090	Haul material by 10 c.y. dump truck, round trip distance						
0100	1 mile	CY	0.044		3.25	3.77	7.03
0110	2 mile	"	0.053		3.90	4.53	8.44
0120	5 mile	"	0.073		5.33	6.18	11.50
0130	10 mile	"	0.080		5.86	6.80	12.75
0140	20 mile	"	0.089		6.51	7.55	14.00
0150	30 mile	"	0.107		7.81	9.06	17.00

EARTHWORK, EXCAVATION & FILL

ID Code	Component Descriptions	Unit of Meas.	Manhr / Unit	Material Cost	Labor Cost	Equipment Cost	Total Cost
	Description	**Output**		**Unit Costs**			
31 - 23336		**TRENCHING**					**31 - 23336**
0100	Trenching and continuous footing excavation						
0980	By gradall						
1000	1 c.y. capacity						
1020	Light soil	CY	0.023		2.22	2.85	5.08
1040	Medium soil	"	0.025		2.40	3.07	5.47
1060	Heavy/wet soil	"	0.027		2.60	3.33	5.93
1080	Loose rock	"	0.029		2.83	3.63	6.47
1090	Blasted rock	"	0.031		3.00	3.84	6.84
1095	By hydraulic excavator						
1100	1/2 c.y. capacity						
1120	Light soil	CY	0.027		2.60	3.33	5.93
1140	Medium soil	"	0.029		2.83	3.63	6.47
1160	Heavy/wet soil	"	0.032		3.12	4.00	7.12
1180	Loose rock	"	0.036		3.46	4.44	7.91
1190	Blasted rock	"	0.040		3.90	5.00	8.90
3000	Hand excavation						
3100	Bulk, wheeled 100'						
3120	Normal soil	CY	0.889		66.00		66.00
3140	Sand or gravel	"	0.800		59.00		59.00
3160	Medium clay	"	1.143		84.00		84.00
3180	Heavy clay	"	1.600		120		120
3200	Loose rock	"	2.000		150		150
3300	Trenches, up to 2' deep						
3320	Normal soil	CY	1.000		74.00		74.00
3340	Sand or gravel	"	0.889		66.00		66.00
3360	Medium clay	"	1.333		98.00		98.00
3380	Heavy clay	"	2.000		150		150
3390	Loose rock	"	2.667		200		200
3400	Trenches, to 6' deep						
3420	Normal soil	CY	1.143		84.00		84.00
3440	Sand or gravel	"	1.000		74.00		74.00
3460	Medium clay	"	1.600		120		120
3480	Heavy clay	"	2.667		200		200
3500	Loose rock	"	4.000		300		300
3590	Backfill trenches						
3600	With compaction						
3620	By hand	CY	0.667		49.25		49.25
3640	By 60 hp tracked dozer	"	0.020		1.46	1.70	3.16

EARTHWORK, EXCAVATION & FILL

ID Code	Component Descriptions	Unit of Meas.	Manhr / Unit	Material Cost	Labor Cost	Equipment Cost	Total Cost
	Description	**Output**		**Unit Costs**			
31 - 23336	**TRENCHING, Cont'd...**					**31 - 23336**	
3650	By 200 hp tracked dozer	CY	0.009		0.86	1.11	1.97
3660	By small front-end loader	"	0.023		1.67	1.94	3.61
3890	Backfill trenches, sand bedding, no compaction						
3900	By hand	CY	0.667	23.75	49.25		73.00
3940	By small front-end loader	"	0.023	27.75	2.22	2.85	32.75

SOIL STABILIZATION & TREATMENT

ID Code	Component Descriptions	Unit of Meas.	Manhr / Unit	Material Cost	Labor Cost	Equipment Cost	Total Cost
31 - 32005	**SOIL STABILIZATION**					**31 - 32005**	
0100	Straw bale secured with rebar	LF	0.027	7.92	1.96		9.88
0120	Filter barrier, 18" high filter fabric	"	0.080	1.91	5.90		7.81
0130	Sediment fence, 36" fabric with 6" mesh	"	0.100	4.53	7.38		12.00
31 - 41330	**TRENCH SHEETING**					**31 - 41330**	
0980	Closed timber, including pull and salvage, excavation						
1000	8' deep	SF	0.064	3.17	4.72	6.51	14.50
2000	20' deep	"	0.098	4.54	7.26	10.00	21.75

DIVISION 33
UTILITIES

SANITARY SEWER

ID Code	Component Descriptions	Unit of Meas.	Manhr / Unit	Material Cost	Labor Cost	Equipment Cost	Total Cost
	Description	**Output**		**Unit Costs**			

33 - 39133 MANHOLES **33 - 39133**

ID Code	Component Descriptions	Unit of Meas.	Manhr / Unit	Material Cost	Labor Cost	Equipment Cost	Total Cost
0100	Precast sections, 48" dia.						
0110	Base section	EA	2.000	450	150	130	720
0120	1'0" riser	"	1.600	130	120	100	350
0130	1'4" riser	"	1.714	160	130	110	390
0140	2'8" riser	"	1.846	230	140	120	480
0150	4'0" riser	"	2.000	430	150	130	700
0160	2'8" cone top	"	2.400	280	180	150	610
0170	Precast manholes, 48" dia.						
0180	4' deep	EA	4.800	920	350	300	1,570
0200	6' deep	"	6.000	1,410	440	380	2,220
0250	7' deep	"	6.857	1,610	500	430	2,540
0260	8' deep	"	8.000	1,810	590	500	2,900
0280	10' deep	"	9.600	2,030	700	600	3,330
1000	Cast-in-place, 48" dia., with frame and cover						
1100	5' deep	EA	12.000	780	880	750	2,410
1120	6' deep	"	13.714	1,030	1,010	860	2,890
1140	8' deep	"	16.000	1,510	1,170	1,000	3,680
1160	10' deep	"	19.200	1,760	1,410	1,200	4,370
1480	Brick manholes, 48" dia. with cover, 8" thick						
1500	4' deep	EA	8.000	830	720		1,550
1501	6' deep	"	8.889	1,050	800		1,850
1505	8' deep	"	10.000	1,340	900		2,240
1510	10' deep	"	11.429	1,670	1,030		2,700
1600	12' deep	"	13.333	2,100	1,200		3,300
1620	14' deep	"	16.000	2,550	1,440		3,990
4200	Frames and covers, 24" diameter						
4210	300 lb	EA	0.800	470	59.00		530
4220	400 lb	"	0.889	500	66.00		570
4230	500 lb	"	1.143	570	84.00		650
4240	Watertight, 350 lb	"	2.667	600	200		800
4250	For heavy equipment, 1200 lb	"	4.000	1,300	300		1,600
4980	Steps for manholes						
5000	7" x 9"	EA	0.160	20.00	11.75		31.75
5020	8" x 9"	"	0.178	25.25	13.00		38.25

POWER & COMMUNICATIONS

ID Code	Component Descriptions	Unit of Meas.	Manhr / Unit	Material Cost	Labor Cost	Equipment Cost	Total Cost
	Description	**Output**		**Unit Costs**			
33 - 71160	**UTILITY POLES & FITTINGS**						**33 - 71160**
0980	Wood pole, creosoted						
1000	25'	EA	2.353	760	230		990
1010	30'	"	2.963	910	280		1,190
1020	35'	"	3.478	1,210	330		1,540
1030	40'	"	3.791	1,460	360		1,820
1040	45'	"	6.957	1,680	670		2,350
1050	50'	"	7.207	1,980	690		2,670
1060	55'	"	7.547	2,280	720		3,000
1065	Treated, wood preservative, 6"x6"						
1070	8'	EA	0.500	140	48.00		190
1080	10'	"	0.800	200	77.00		280
1090	12'	"	0.889	210	85.00		300
1100	14'	"	1.333	270	130		400
1120	16'	"	1.600	320	150		470
1140	18'	"	2.000	360	190		550
1150	20'	"	2.000	460	190		650
1155	Aluminum, brushed, no base						
1160	8'	EA	2.000	850	190		1,040
1170	10'	"	2.667	980	260		1,240
1180	15'	"	2.759	1,100	260		1,360
1190	20'	"	3.200	1,330	310		1,640
1200	25'	"	3.810	1,770	370		2,140
1210	30'	"	4.396	2,650	420		3,070
1220	35'	"	5.000	3,110	480		3,590
1230	40'	"	6.250	3,990	600		4,590
1235	Steel, no base						
1240	10'	EA	2.500	1,030	240		1,270
1245	15'	"	2.963	1,130	280		1,410
1250	20'	"	3.810	1,510	370		1,880
1260	25'	"	4.520	1,700	430		2,130
1280	30'	"	5.096	2,180	490		2,670
1300	35'	"	6.250	2,540	600		3,140
2000	Concrete, no base						
2020	13'	EA	5.517	1,260	530		1,790
2040	16'	"	7.273	1,760	700		2,460
2060	18'	"	8.791	2,120	840		2,960
2080	25'	"	10.000	2,590	960		3,550
2100	30'	"	12.121	3,450	1,160		4,610

POWER & COMMUNICATIONS

ID Code	Component Descriptions	Unit of Meas.	Manhr / Unit	Material Cost	Labor Cost	Equipment Cost	Total Cost
	Description	**Output**		**Unit Costs**			
33 - 71160	**UTILITY POLES & FITTINGS, Cont'd...**					**33 - 71160**	
2120	35'	EA	14.035	4,450	1,350		5,800
2140	40'	"	16.000	5,190	1,540		6,730
2160	45'	"	17.021	6,170	1,630		7,800
2180	50'	"	18.182	7,650	1,750		9,400
2200	55'	"	19.048	8,520	1,830		10,350
2220	60'	"	20.000	9,750	1,920		11,670
5020	Pole line hardware						
5040	Wood crossarm						
5060	4'	EA	1.333	120	130		250
5062	8'	"	1.667	240	160		400
5064	10'	"	2.051	400	200		600
5120	Angle steel brace						
5140	1 piece	EA	0.250	14.50	24.00		38.50
5250	2 piece	"	0.348	31.25	33.50		65.00
5270	Eye nut, 5/8"	"	0.050	3.59	4.80		8.39
5280	Bolt (14-16"), 5/8"	"	0.200	26.75	19.25		46.00
5420	Transformer, ground connection	"	0.250	8.52	24.00		32.50
5440	Stirrup	"	0.308	19.00	29.50		48.50
5460	Secondary lead support	"	0.400	29.00	38.50		68.00
5600	Spool insulator	"	0.200	6.20	19.25		25.50
5620	Guy grip, preformed						
5640	7/16"	EA	0.145	3.68	14.00		17.75
5660	1/2"	"	0.145	6.37	14.00		20.25
5680	Hook	"	0.250	4.60	24.00		28.50
5690	Strain insulator	"	0.364	37.25	35.00		72.00
5695	Wire						
5700	5/16"	LF	0.005	2.05	0.48		2.53
5710	7/16"	"	0.006	2.04	0.57		2.61
5720	1/2"	"	0.008	4.91	0.76		5.67
5730	Soft drawn ground, copper, #8	"	0.008	0.63	0.76		1.39
5740	Ground clamp	EA	0.308	5.13	29.50		34.75
5810	Perforated strapping for conduit, 1-1/2"	LF	0.145	3.84	14.00		17.75
5820	Hot line clamp	EA	0.800	21.00	77.00		98.00
5880	Lightning arrester						
5890	3kv	EA	1.000	680	96.00		780
5900	10kv	"	1.600	1,040	150		1,190
5910	30kv	"	2.000	1,910	190		2,100
5920	36kv	"	2.500	3,870	240		4,110

POWER & COMMUNICATIONS

ID Code	Component Descriptions	Unit of Meas.	Manhr / Unit	Material Cost	Labor Cost	Equipment Cost	Total Cost
33 - 71160	**UTILITY POLES & FITTINGS, Cont'd...**						**33 - 71160**
6000	Fittings						
6010	Plastic molding	LF	0.145	4.51	14.00		18.50
6020	Molding staples	EA	0.050	1.07	4.80		5.87
6030	Ground wires staples	"	0.030	0.46	2.89		3.35
6040	Copper butt plate	"	0.296	1.29	28.50		29.75
6050	Anchor bond clamp	"	0.145	5.65	14.00		19.75
6055	Guy wire						
6060	1/4"	LF	0.030	0.64	2.89		3.53
6070	3/8"	"	0.050	0.99	4.80		5.79
6075	Guy grip						
6080	1/4"	EA	0.050	11.00	4.80		15.75
6090	3/8"	"	0.050	12.75	4.80		17.50
33 - 71190	**ELECTRIC MANHOLES**						**33 - 71190**
0980	Precast, handhole, 4' deep						
1000	2'x2'	EA	3.478	610	330		940
1020	3'x3'	"	5.556	810	530		1,340
1040	4'x4'	"	10.256	1,750	980		2,730
1060	Power manhole, complete, precast, 8' deep						
1080	4'x4'	EA	14.035	2,510	1,350		3,860
1100	6'x6'	"	20.000	3,350	1,920		5,270
1140	8'x8'	"	21.053	3,970	2,020		5,990
1180	6' deep, 9' x 12'	"	25.000	4,380	2,400		6,780
1980	Cast-in-place, power manhole, 8' deep						
2000	4'x4'	EA	14.035	2,970	1,350		4,320
2020	6'x6'	"	20.000	3,830	1,920		5,750
2040	8'x8'	"	21.053	4,260	2,020		6,280
33 - 71392	**HIGH-VOLTAGE CABLE**						**33 - 71392**
1000	High voltage XLP copper cable, shielded, 5000v						
1010	#6 awg	LF	0.013	2.38	1.24		3.62
1020	#4 awg	"	0.016	2.74	1.53		4.27
1030	#2 awg	"	0.019	3.57	1.82		5.39
1040	#1 awg	"	0.021	3.78	2.02		5.80
1050	#1/0 awg	"	0.024	4.47	2.29		6.76
1060	#2/0 awg	"	0.029	7.52	2.79		10.25
1070	#3/0 awg	"	0.034	7.65	3.26		11.00
1080	#4/0 awg	"	0.036	8.06	3.49		11.50
1090	#250 awg	"	0.043	9.33	4.15		13.50

POWER & COMMUNICATIONS

ID Code	Component Descriptions	Unit of Meas.	Manhr / Unit	Material Cost	Labor Cost	Equipment Cost	Total Cost
	Description	**Output**		**Unit Costs**			
33 - 71392	**HIGH-VOLTAGE CABLE, Cont'd...**						**33 - 71392**
1100	#300 awg	LF	0.048	10.75	4.65		15.50
1110	#350 awg	"	0.053	12.00	5.12		17.00
1210	#500 awg	"	0.073	17.25	6.98		24.25
1220	#750 awg	"	0.080	25.25	7.68		33.00
1230	Ungrounded, 15,000v						
1240	#1 awg	LF	0.031	5.70	2.95		8.65
1250	#1/0 awg	"	0.034	6.74	3.26		10.00
1260	#2/0 awg	"	0.036	7.73	3.49		11.25
1265	#3/0 awg	"	0.040	8.93	3.84		12.75
1280	#4/0 awg	"	0.046	9.88	4.38		14.25
1290	#250 awg	"	0.048	11.00	4.65		15.75
1300	#300 awg	"	0.053	12.50	5.12		17.50
1310	#350 awg	"	0.062	14.00	5.90		20.00
1320	#500 awg	"	0.080	18.25	7.68		26.00
1330	#750 awg	"	0.098	27.25	9.36		36.50
1340	#1000 awg	"	0.123	40.00	11.75		52.00
1345	Aluminum cable, shielded, 5000v						
1350	#6 awg	LF	0.011	1.97	1.05		3.02
1360	#4 awg	"	0.013	2.14	1.24		3.38
1370	#2 awg	"	0.015	2.39	1.43		3.82
1380	#1 awg	"	0.017	2.61	1.63		4.24
1382	#1/0 awg	"	0.019	2.88	1.82		4.70
1384	#2/0 awg	"	0.020	3.19	1.92		5.11
1386	#3/0 awg	"	0.021	3.63	2.02		5.65
1388	#4/0 awg	"	0.024	4.01	2.29		6.30
1390	#250 awg	"	0.026	4.28	2.47		6.75
1400	#300 awg	"	0.031	4.66	2.95		7.61
1402	#350 awg	"	0.034	5.06	3.26		8.32
1404	#500 awg	"	0.036	6.09	3.49		9.58
1406	#750 awg	"	0.044	7.89	4.26		12.25
1408	#1000 awg	"	0.050	8.92	4.80		13.75
1409	Ungrounded, 15,000v						
1410	#1 awg	LF	0.021	3.59	2.02		5.61
1412	#1/0 awg	"	0.025	3.79	2.40		6.19
1414	#2/0 awg	"	0.027	4.08	2.60		6.68
1416	#3/0 awg	"	0.028	4.47	2.69		7.16
1418	#4/0 awg	"	0.029	4.84	2.79		7.63
1420	#250 awg	"	0.031	5.19	2.95		8.14

POWER & COMMUNICATIONS

ID Code	Component Descriptions	Unit of Meas.	Manhr / Unit	Material Cost	Labor Cost	Equipment Cost	Total Cost
		Description	**Output**	**Unit Costs**			
33 - 71392	**HIGH-VOLTAGE CABLE, Cont'd...**						**33 - 71392**
1422	#300 awg	LF	0.032	5.62	3.07		8.69
1424	#350 awg	"	0.036	6.04	3.49		9.53
1426	#500 awg	"	0.043	7.07	4.15		11.25
1430	#750 awg	"	0.052	9.27	4.95		14.25
1432	#1000 awg	"	0.064	12.25	6.14		18.50
1434	Indoor terminations, 5000v						
1436	#6 - #4	EA	0.157	60.00	15.00		75.00
1438	#2 - #2/0	"	0.157	70.00	15.00		85.00
1440	#3/0 - #250	"	0.157	90.00	15.00		110
1442	#300 - #750	"	2.759	99.00	260		360
1444	#1000	"	3.810	120	370		490
1448	In-line splice, 5000v						
1450	#6 - #4/0	EA	3.810	140	370		510
1460	#250 - #500	"	10.000	160	960		1,120
1470	#750 - #1000	"	13.008	210	1,250		1,460
1475	T-splice, 5000v						
1480	#2 - #4/0	EA	11.994	140	1,150		1,290
1490	#250 - #500	"	20.000	160	1,920		2,080
1500	#750 - #1000	"	25.000	210	2,400		2,610
1505	Indoor terminations, 15,000v						
1510	#2 - #2/0	EA	3.478	80.00	330		410
1520	#3/0 - #500	"	5.333	99.00	510		610
1530	#750 - #1000	"	6.154	120	590		710
1535	In-line splice, 15,000v						
1540	#2 - #4/0	EA	8.999	140	860		1,000
1550	#250 - #500	"	11.994	190	1,150		1,340
1560	#750 - #1000	"	18.018	240	1,730		1,970
1565	T-splice, 15,000v						
1570	#4	EA	18.018	140	1,730		1,870
1580	#250 - #500	"	29.963	190	2,880		3,070
1590	#750 - #1000	"	44.944	240	4,310		4,550
1595	Compression lugs, 15,000v						
1600	#4	EA	0.400	6.83	38.50		45.25
1610	#2	"	0.533	7.88	51.00		59.00
1620	#1	"	0.533	8.59	51.00		60.00
1630	#1/0	"	0.667	13.25	64.00		77.00
1640	#2/0	"	0.667	13.75	64.00		78.00
1650	#3/0	"	0.851	15.50	82.00		98.00

POWER & COMMUNICATIONS

ID Code	Component Descriptions	Unit of Meas.	Manhr / Unit	Material Cost	Labor Cost	Equipment Cost	Total Cost
	Description	**Output**		**Unit Costs**			
33 - 71392	**HIGH-VOLTAGE CABLE, Cont'd...**						**33 - 71392**
1670	#4/0	EA	0.851	17.00	82.00		99.00
1680	#250	"	0.952	20.00	91.00		110
1690	#300	"	0.952	23.25	91.00		110
1700	#350	"	1.159	24.00	110		130
1710	#500	"	1.250	36.25	120		160
1720	#750	"	1.509	58.00	140		200
1730	#1000	"	1.905	84.00	180		260
1735	Compression splices, 15,000v						
1740	#4	EA	0.667	7.68	64.00		72.00
1750	#2	"	0.727	8.42	70.00		78.00
1810	#1	"	0.899	9.63	86.00		96.00
1820	#1/0	"	1.000	10.25	96.00		110
1830	#2/0	"	1.159	11.00	110		120
1840	#3/0	"	1.250	12.00	120		130
1850	#4/0	"	1.404	13.00	130		140
1860	#250	"	1.509	14.25	140		150
1870	#350	"	1.739	16.25	170		190
1880	#500	"	2.000	24.00	190		210
1890	#750	"	2.500	39.50	240		280
33 - 71394	**SUPPORTS & CONNECTORS**						**33 - 71394**
1000	Cable supports for conduit						
1010	1-1/2"	EA	0.348	4.02	33.50		37.50
1020	2"	"	0.348	5.58	33.50		39.00
1030	2-1/2"	"	0.400	6.24	38.50		44.75
1040	3"	"	0.400	8.07	38.50		46.50
1050	3-1/2"	"	0.500	10.50	48.00		59.00
1060	4"	"	0.500	12.75	48.00		61.00
1070	5"	"	0.667	23.00	64.00		87.00
1080	6"	"	0.727	48.50	70.00		120
1085	Split bolt connectors						
1090	#10	EA	0.200	3.28	19.25		22.50
1100	#8	"	0.200	3.86	19.25		23.00
1110	#6	"	0.200	4.26	19.25		23.50
1120	#4	"	0.400	4.96	38.50		43.50
1130	#3	"	0.400	7.10	38.50		45.50
1140	#2	"	0.400	8.05	38.50		46.50
1150	#1/0	"	0.667	11.00	64.00		75.00

POWER & COMMUNICATIONS

ID Code	Component Descriptions	Unit of Meas.	Manhr / Unit	Material Cost	Labor Cost	Equipment Cost	Total Cost
	Description	**Output**		**Unit Costs**			
33 - 71394	**SUPPORTS & CONNECTORS, Cont'd...**						**33 - 71394**
1160	#2/0	EA	0.667	16.50	64.00		81.00
1170	#3/0	"	0.667	25.25	64.00		89.00
1180	#4/0	"	0.667	28.50	64.00		93.00
1190	#250	"	1.000	29.50	96.00		130
1200	#350	"	1.000	52.00	96.00		150
1210	#500	"	1.000	69.00	96.00		160
1220	#750	"	1.509	120	140		260
1230	#1000	"	1.509	160	140		300
1235	Single barrel lugs						
1240	#6	EA	0.250	1.22	24.00		25.25
1260	#1/0	"	0.500	2.39	48.00		50.00
1270	#250	"	0.667	5.76	64.00		70.00
1280	#350	"	0.667	7.50	64.00		72.00
1290	#500	"	0.667	14.50	64.00		79.00
1300	#600	"	0.899	15.25	86.00		100
1310	#800	"	0.899	17.50	86.00		100
1320	#1000	"	0.899	21.00	86.00		110
1325	Double barrel lugs						
1330	#1/0	EA	0.899	4.71	86.00		91.00
1340	#250	"	1.290	13.75	120		130
1350	#350	"	1.290	19.50	120		140
1360	#600	"	1.905	29.75	180		210
1370	#800	"	1.905	34.00	180		210
1380	#1000	"	1.905	34.75	180		210
1385	Three barrel lugs						
1390	#2/0	EA	1.290	38.00	120		160
1400	#250	"	1.905	73.00	180		250
1410	#350	"	1.905	120	180		300
1420	#600	"	2.667	130	260		390
1430	#800	"	2.667	210	260		470
1440	#1000	"	2.667	300	260		560
1445	Four barrel lugs						
1450	#250	EA	2.759	81.00	260		340
1460	#350	"	2.759	130	260		390
1470	#600	"	3.478	150	330		480
1480	#800	"	3.478	230	330		560
1485	Compression conductor adapters						
1490	#6	EA	0.296	11.00	28.50		39.50

POWER & COMMUNICATIONS

ID Code	Description / Component Descriptions	Output Unit of Meas.	Output Manhr / Unit	Unit Costs Material Cost	Unit Costs Labor Cost	Unit Costs Equipment Cost	Unit Costs Total Cost
33 - 71394	**SUPPORTS & CONNECTORS, Cont'd...**						**33 - 71394**
1500	#4	EA	0.348	11.50	33.50		45.00
1510	#2	"	0.444	12.00	42.75		55.00
1520	#1	"	0.444	14.00	42.75		57.00
1530	#1/0	"	0.533	14.50	51.00		66.00
1540	#250	"	0.800	27.50	77.00		100
1550	#350	"	0.851	32.75	82.00		110
1560	#500	"	1.096	42.50	110		150
1570	#750	"	1.143	58.00	110		170
1575	Terminal blocks, 2 screw						
1580	3 circuit	EA	0.200	21.50	19.25		40.75
1590	6 circuit	"	0.200	29.75	19.25		49.00
1600	8 circuit	"	0.200	34.75	19.25		54.00
1610	10 circuit	"	0.296	40.25	28.50		69.00
1620	12 circuit	"	0.296	45.50	28.50		74.00
1630	18 circuit	"	0.296	61.00	28.50		90.00
1810	24 circuit	"	0.348	77.00	33.50		110
1820	36 circuit	"	0.348	110	33.50		140
1825	Compression splice						
1830	#8 awg	EA	0.381	3.60	36.50		40.00
1840	#6 awg	"	0.276	4.27	26.50		30.75
1850	#4 awg	"	0.276	4.53	26.50		31.00
1860	#2 awg	"	0.533	7.01	51.00		58.00
1870	#1 awg	"	0.533	9.88	51.00		61.00
1880	#1/0 awg	"	0.533	12.00	51.00		63.00
1890	#2/0 awg	"	0.851	12.75	82.00		95.00
1900	#3/0 awg	"	0.851	15.00	82.00		97.00
1910	#4/0 awg	"	0.851	16.00	82.00		98.00
1920	#250 awg	"	1.356	17.00	130		150
1930	#300 awg	"	1.356	18.50	130		150
1940	#350 awg	"	1.404	18.75	130		150
1950	#400 awg	"	1.404	25.50	130		160
1960	#500 awg	"	1.509	29.75	140		170
1970	#600 awg	"	1.509	45.00	140		190
1980	#750 awg	"	1.739	47.75	170		220
1990	#1000 awg	"	1.739	61.00	170		230

COMMUNICATIONS STRUCTURES

ID Code	Component Descriptions	Unit of Meas.	Manhr / Unit	Material Cost	Labor Cost	Equipment Cost	Total Cost
	Description	**Output**		**Unit Costs**			
33 - 81160	**ANTENNAS AND TOWERS**						**33 - 81160**
0980	Guy cable, alumaweld						
1000	1x3, 7/32"	LF	0.050	0.54	4.80		5.34
1020	1x3, 1/4"	"	0.050	0.64	4.80		5.44
1040	1x3, 25/64"	"	0.059	0.90	5.68		6.58
1060	1x19, 1/2"	"	0.070	2.25	6.67		8.92
1080	1x7, 35/64"	"	0.080	2.57	7.68		10.25
1100	1x19, 13/16"	"	0.100	2.78	9.60		12.50
1480	Preformed alumaweld end grip						
1500	1/4" cable	EA	0.100	4.11	9.60		13.75
1520	3/8" cable	"	0.100	5.43	9.60		15.00
1540	1/2" cable	"	0.145	6.81	14.00		20.75
1560	9/16" cable	"	0.200	8.47	19.25		27.75
1580	5/8" cable	"	0.250	10.50	24.00		34.50
2040	Fiberglass guy rod, white epoxy coated						
2060	1/4" dia.	LF	0.145	2.78	14.00		16.75
2080	3/8" dia	"	0.145	4.16	14.00		18.25
2100	1/2" dia	"	0.200	5.56	19.25		24.75
2120	5/8" dia	"	0.250	6.95	24.00		31.00
2130	Preformed glass grip end grip, guy rod						
2140	1/4" dia.	EA	0.145	15.25	14.00		29.25
2160	3/8" dia.	"	0.200	17.75	19.25		37.00
2180	1/2" dia.	"	0.250	21.75	24.00		45.75
2200	5/8" dia.	"	0.250	24.50	24.00		48.50
2480	Spelter socket end grip, 1/4" dia. guy rod						
2500	Standard strength	EA	0.500	45.50	48.00		94.00
2520	High performance	"	0.500	56.00	48.00		100
2530	3/8" dia. guy rod						
2540	Standard strength	EA	0.348	44.25	33.50		78.00
2560	High performance	"	0.500	56.00	48.00		100
3040	Timber pole, Douglas Fir						
3060	80-85 ft	EA	19.512	4,250	1,870		6,120
3080	90-95 ft	"	22.222	5,320	2,130		7,450
3140	Southern yellow pine						
3160	35-45 ft	EA	10.959	2,540	1,050		3,590
3180	50-55 ft	"	14.035	3,550	1,350		4,900

DIVISION 41
HANDLING EQUIPMENT

HOISTS AND CRANES

ID Code	Description — Component Descriptions	Unit of Meas.	Manhr / Unit	Material Cost	Labor Cost	Equipment Cost	Total Cost
41 - 22133	**INDUSTRIAL HOISTS**						**41 - 22133**
1000	Industrial hoists, electric, light to medium duty						
1010	500 lb	EA	4.000	6,050	380		6,430
1020	1000 lb	"	4.211	9,570	400		9,970
1030	2000 lb	"	4.444	9,900	430		10,330
1040	3000 lb	"	4.706	11,810	450		12,260
1050	4000 lb	"	5.000	12,080	480		12,560
1060	5000 lb	"	5.333	13,040	510		13,550
1070	6000 lb	"	5.517	16,590	530		17,120
1080	7500 lb	"	5.714	18,590	550		19,140
1090	10,000 lb	"	5.926	48,270	570		48,840
1100	15,000 lb	"	6.154	60,560	590		61,150
1110	20,000 lb	"	6.667	71,650	640		72,290
1120	25,000 lb	"	7.273	72,580	700		73,280
1130	30,000 lb	"	8.000	74,240	770		75,010
1200	Heavy duty						
1210	500 lb	EA	4.000	16,000	380		16,380
1220	1000 lb	"	4.211	22,390	400		22,790
1240	2000 lb	"	4.444	24,750	430		25,180
1250	3000 lb	"	4.706	25,660	450		26,110
1260	4000 lb	"	5.000	26,800	480		27,280
1270	5000 lb	"	5.333	27,330	510		27,840
1280	6000 lb	"	5.517	29,610	530		30,140
1290	7500 lb	"	5.714	34,010	550		34,560
1300	10,000 lb	"	5.926	36,290	570		36,860
1310	15,000 lb	"	6.154	44,030	590		44,620
1320	20,000 lb	"	6.667	53,440	640		54,080
1330	25,000 lb	"	7.273	59,970	700		60,670
1340	30,000 lb	"	8.000	66,650	770		67,420

DIVISION 48
ELECTRICAL POWER GENERATION

SOLAR

ID Code	Description — Component Descriptions	Output		Unit Costs			
		Unit of Meas.	Manhr / Unit	Material Cost	Labor Cost	Equipment Cost	Total Cost
48 - 14110	**SOLAR ELECTRICAL SYSTEMS**						**48 - 14110**
1000	Photovoltaic, Full Grid-Tie System						
1010	Panel array, inverter, mounts, racks, conduit, etc.,						
1020	Minimum	EA					24,220
1030	Average	"					32,290
1040	Maximum	"					40,360
1050	4,000 Watt						
1060	Minimum	EA					32,290
1070	Average	"					43,050
1080	Maximum	"					53,810
1090	5,000 Watt						
1100	Minimum	EA					40,360
1110	Average	"					53,810
1120	Maximum	"					67,260
2000	Photovoltaic Components						
3000	Polycristalline Rigid Panel, 200 watt						
3010	Minimum	EA	0.286	540	27.50		570
3020	Average	"	0.286	610	27.50		640
3030	Maximum	"	0.286	310	27.50		340
3040	215 Watt						
3050	Minimum	EA	0.286	580	27.50		610
3060	Average	"	0.286	650	27.50		680
3070	Maximum	"	0.286	700	27.50		730
3080	230 Watt						
3090	Minimum	EA	0.286	630	27.50		660
3100	Average	"	0.286	700	27.50		730
3110	Maximum	"	0.286	770	27.50		800
3120	245 Watt						
3130	Minimum	EA	0.286	670	27.50		700
3140	Average	"	0.286	740	27.50		770
3150	Maximum	"	0.286	810	27.50		840
4000	Anodized aluminum rail, 8'						
4010	Minimum	EA	0.178	45.75	17.00		63.00
4020	Average	"	0.178	48.50	17.00		66.00
4030	Maximum	"	0.178	51.00	17.00		68.00
4040	10'						
4050	Minimum	EA	0.200	57.00	19.25		76.00
4060	Average	"	0.200	61.00	19.25		80.00
4070	Maximum	"	0.200	65.00	19.25		84.00

SOLAR

ID Code	Description Component Descriptions	Output Unit of Meas.	Output Manhr / Unit	Unit Costs Material Cost	Unit Costs Labor Cost	Unit Costs Equipment Cost	Unit Costs Total Cost
48 - 14110	**SOLAR ELECTRICAL SYSTEMS, Cont'd...**						**48 - 14110**
4080	12'						
4090	Minimum	EA	0.229	70.00	22.00		92.00
4100	Average	"	0.229	73.00	22.00		95.00
4110	Maximum	"	0.229	75.00	22.00		97.00
4500	Panel mounts, aluminum, mount tile, flush						
4510	Minimum	EA	0.533	37.75	51.00		89.00
4520	Average	"	0.533	40.25	51.00		91.00
4530	Maximum	"	0.533	43.00	51.00		94.00
4540	Standard						
4550	Minimum	EA	0.533	29.50	51.00		81.00
4560	Average	"	0.533	32.25	51.00		83.00
4570	Maximum	"	0.533	35.00	51.00		86.00
4800	Rail clamp, mid-clamp						
4810	Minimum	EA					9.41
4820	Average	"					10.75
4830	Maximum	"					12.00
4840	End-clamp						
4850	Minimum	EA					12.00
4860	Average	"					13.75
4870	Maximum	"					15.25
4880	Anodized aluminum, rail splice kit						
4890	Minimum	EA					18.75
4900	Average	"					21.50
4910	Maximum	"					24.25
5000	Power Distribution Panel						
5010	Minimum	EA	2.000	160	190		350
5020	Average	"	2.000	190	190		380
5030	Maximum	"	2.000	230	190		420
6000	Inverters						
6010	Light capacity, micro inverter, 190 Watt						
6020	Minimum	EA	0.267	240	25.50		270
6030	Average	"	0.267	270	25.50		300
6040	Maximum	"	0.267	300	25.50		330
6500	Medium capacity, inverter, 1000 Watt						
6510	Minimum	EA	4.000	2,420	380		2,800
6520	Average	"	4.000	2,690	380		3,070
6530	Maximum	"	4.000	2,960	380		3,340
6540	2500 Watt						

SOLAR

ID Code	Description — Component Descriptions	Output — Unit of Meas.	Output — Manhr / Unit	Unit Costs — Material Cost	Unit Costs — Labor Cost	Unit Costs — Equipment Cost	Unit Costs — Total Cost
48 - 14110	**SOLAR ELECTRICAL SYSTEMS, Cont'd...**						**48 - 14110**
6550	Minimum	EA	4.000	4,040	380		4,420
6560	Average	"	4.000	4,300	380		4,680
6570	Maximum	"	4.000	4,570	380		4,950
6580	5000 Watt						
6590	Minimum	EA	4.000	6,460	380		6,840
6600	Average	"	4.000	6,730	380		7,110
6610	Maximum	"	4.000	7,000	380		7,380
7000	Circuits						
7010	Combiner Box, 12 circuit						
7020	Minimum	EA	2.667	740	260		1,000
7030	Average	"	2.667	880	260		1,140
7040	Maximum	"	2.667	980	260		1,240
7050	28 circuit						
7060	Minimum	EA	2.667	1,190	260		1,450
7070	Average	"	2.667	1,340	260		1,600
7080	Maximum	"	2.667	1,490	260		1,750
7200	Solar Circuit Breaker, 15A, 150 VDC						
7210	Minimum	EA					14.75
7220	Average	"					17.50
7230	Maximum	"					20.25
7240	Wires and Conductors						
8230	Cable, bare copper, 10 AWG, 500' coils						
8240	Minimum	EA					210
8250	Average	"					220
8260	Maximum	"					240
8270	Multi-Contact branch connector, MC4						
8280	Minimum	EA	0.267	37.75	25.50		63.00
8300	Average	"	0.267	40.25	25.50		66.00
8320	Maximum	"	0.267	43.00	25.50		69.00

Square Foot Tables

The following Square Foot Tables list hundreds of actual projects for ten building types, each with associated building size, total square foot building cost and percentage of project costs for total mechanical and electrical components. This data provides an overview of construction costs by building type. These costs are for actual projects. The variations within similar building types may be due, among other factors, to size, location, quality and specified components, materials and processes. Depending upon all such factors, specific building costs can vary significantly and may not necessarily fall within the range of costs as presented. The data has been updated to reflect current construction costs.

All prices are updated to January 1, 2023 and are national averages.
For a more in-depth information contact Design, Cost and Data at 800-533-5680, or go to www.DCD.com

335

PROJECT	DESCRIPTION	CITY	STATE	SIZE	$/SF	NOTES
	Commercial					
Bank	School Credit Union Administration Building	Katy	TX	30,700	$272.52	New
	FineMark National Bank & Trust	Fort Myers	FL	20,039	$514.12	New
	Florida Shores Bank	Pompano Beach	FL	11,697	$654.35	New
	Mobiloil Credit Union	Vidor	TX	9,252	$458.90	New
	Beaumont Community Credit Union	Beaumont	TX	3,267	$520.24	New
Office	Allendale Town Center	Allendale	NJ	80,226	$48.03	Addition/Renovation
	Roanoke Electric Cooperative	Ahoskie	NC	52,752	$273.95	New
	Regional Aviation & Training Center	Currituck	NC	39,930	$347.52	New
	Transportation/Warehouse Facility	Monroe	GA	32,400	$241.24	New
	Collection System Operations Facility	Walnut Creek	CA	27,179	$476.59	New
	ULTA - (Shell Only)	Pensacola	FL	10,850	$123.31	Renovation
	Daycare Center	Clawson	MI	4,270	$195.26	Adaptive Reuse
	Campgrounds Office & Retail	Cincinnati	OH	2,300	$427.76	New
Parking	Palm Avenue Parking Garage	Sarasota	FL	287,040	$73.08	New
	Reynolds Street Parking Deck	Augusta	GA	214,000	$86.93	New
	Awty Int. School Parking Structure	Houston	TX	174,582	$76.96	New
Retail	SNG Center - Mixed-Use	Fargo	ND	143,860	$129.34	New
	Roof & Lifeway/Steinmart Renovation	Pensacola	FL	88,299	$45.43	Renovation
	No Frills Supermarket	Omaha	NE	61,000	$138.38	New
	West Oaks Mall Redevelopment	Houston	TX	49,800	$246.26	Renovation
	World of Decor	Deerfield Beach	FL	47,500	$187.95	New
	Sarasota Yacht Club	Sarasota	FL	41,332	$552.22	New
	Karschs Village Market	Barnhart	MO	35,384	$107.86	New
	Fresh Thyme Farmers Market	Fishers	IN	28,784	$236.50	New
	Nashville Hangar Inc.	Nashville	TN	28,702	$247.98	New
	Party Time Plus	Billings	MT	26,000	$125.82	New
	Marshalls - (Shell Only)	Pensacola	FL	25,990	$60.20	Renovation
	Ed Hicks Mercedes-Benz USA	Corpus Christi	TX	25,273	$329.07	New
	Montana Honda & Marine	Billings	MT	22,963	$143.00	Addition
	Fresh Market - (Shell Only)	Pensacola	FL	21,000	$103.49	Renovation
	Theatre Exchange Interior Fit Up	Manitoba	CA	19,344	$166.92	Renovation
	DSW Shoes Renovation	Pensacola	FL	18,000	$115.86	Renovation
	Dormans Lighting & Design	Lutherville	MD	15,220	$157.07	Addition
	The Groves Exterior Renovation	Farmington	MI	15,137	$88.79	Renovation
	Don Gibson Theatre	Shelby	NC	13,386	$468.15	Renovation
	Tri Ford Showroom Expansion	Highland	IL	12,881	$153.11	Addition/Renovation
	Fiat of LeHigh Valley	Easton	PA	11,905	$196.08	New
	Sicardi Art Gallery	Houston	TX	6,175	$308.44	New
	Childrens Mercy Hospital Gift Shop	Kansas City	MO	5,010	$388.74	Tenant Build-out
Restaurant	LaMar Cebicheria Peruana Restaurant	San Francisco	CA	11,000	$397.45	Tenant Build-out
	Ulele Restaurant	Tampa	FL	8,905	$963.30	Adaptive Reuse
	Youells Oyster House	Allentown	PA	6,107	$316.22	New
	Mellow Mushroom Highlands Shell	Louisville	KY	5,802	$128.18	New
	Mellow Mushroom Highlands TBO	Louisville	KY	5,802	$212.59	Tenant Build-out
	Mellow Mushroom Pizza	Wilder	KY	5,500	$304.06	New
	Liberty Microbrewery	Plymouth	MI	3,425	$218.76	Addition
	700 South Deli	Linthicum	MD	3,200	$269.55	Tenant Build-out
	New York Pizza Department (NYPD)	Tempe	AZ	2,338	$412.39	Tenant Build-out
	Airport Restaurant Build Out	Eglin Air Force Base	FL	2,320	$355.02	Tenant Build-out

All prices are updated to January 1, 2023 and are national averages.
For a more in-depth information contact Design, Cost and Data at 800-533-5680, or go to www.DCD.com

337

PROJECT	DESCRIPTION	CITY	STATE	SIZE	$/SF	NOTES
	Civic/Government					
Civic Center	Lincoln Center	Fort Collins	CO	38,160	$271.90	Addition/Renovation
	Rockport City Services Building	Rockport	TX	20,062	$281.28	New
	Sinclair Park Community Centre	Manitoba	CA	17,007	$398.26	Addition/Renovation
	Mt. Olive City Hall Complex	Mount Olive	IL	14,360	$152.72	New
	Teaneck Municipal Complex	Teaneck	NJ	12,870	$292.66	Addition/Renovation
	Cobb Community Center Additions	Pensacola	FL	5,200	$407.32	Addition/Renovation
	Newtown Municipal Center	Newtown	OH	5,077	$211.94	Adaptive Reuse
	Nederland City Hall	Nederland	TX	4,983	$471.54	New
Correctional	County Sheriffs Office	Morgantown	WV	31,645	$373.71	New
	Nederland Public Safety Complex	Nederland	TX	21,189	$273.17	Adaptive Reuse
	Detention Center & Sheriffs Office	Spencer	IA	16,983	$529.76	New
	Ogle City Sheriff & Coroner Admin	Oregon	IL	15,377	$356.96	New
	Chautauqua City Jail & Sheriff	Sedan	KS	12,257	$376.47	New
Courthouse	Courthouse HVAC System Replacement	Gainesville	FL	101,000	$54.30	Renovation
	Courthouse Renovation & Restoration	Springfield	IL	47,720	$232.10	Renovation
Fire Department	College Station Fire Station #6	College Station	TX	25,133	$440.19	New
	Mt. Orab Fire Station	Village of Mt. Orab	OH	18,170	$227.13	New
	Richardson Fire Station No. 4	Richardson	TX	14,090	$516.83	New
	Willowfork Fire Station No. 2	Katy	TX	13,358	$379.89	New
	Joint Fire & Rescue Station	Newtown	OH	13,125	$244.87	Addition/Renovation
	Little Miami Fire & Rescue	Fairfax	OH	12,316	$290.77	New
	Fire Station No. 11	Fort Smith	AR	12,155	$440.45	New
	Pearisburg Fire Station	Pearisburg	VA	11,818	$283.48	New
	Ponderosa Fire Station No. 62	Spring	TX	11,163	$380.33	New
	El Dorado Hills Fire Station 84	El Dorado Hills	CA	10,869	$572.68	New
	Wayne Fire Department	Goshen	OH	10,000	$124.65	New
	Fire Station No. 40	Jacksonville	FL	9,703	$502.53	New
	Rosenberg Fire Station No. 3	Rosenberg	TX	8,479	$476.38	New
	Little Rock Fire Station No. 23	Little Rock	AR	8,291	$612.88	New
	Cleveland Volunteer Fire Station	Cleveland	MS	6,910	$428.43	New
Government	Council Center For Scouting	Fargo	ND	20,466	$256.95	New
	Camp Crook Ranger Station	Camp Crook	SD	4,880	$634.64	New
	Beaumont Municipal Tennis Center	Beaumont	TX	4,460	$464.37	Addition
	Florence Transit Hub	Florence	KY	3,115	$663.31	New
	Knox Area Rescue Ministries	Knoxville	TN	1,762	$873.57	New
	Entrance Station Lake Mead	Clark County	NV	480	$3,544.21	New
	Vehicle Charging Stations	Denton	TX	6 spaces	$9,093.41	New
Library	Dover Public Library	Dover	DE	46,424	$557.15	New
	Clinton-Macomb Public Library	Clinton Township	MI	24,723	$162.79	Adaptive Reuse
	Crozet Western Albemarle Library	Crozet	VA	23,199	$452.31	New
	Upper Tampa Bay Regional Library	Tampa	FL	13,630	$313.05	Addition/Renovation
	Palmetto Branch Library	Palmetto	GA	11,200	$624.75	New
	Regional Library Expansion	Valrico	FL	10,970	$383.76	Addition/Renovation
Miscellaneous	City of Pampa Animal Welfare	Pampa	TX	13,578	$363.55	New
	Royal Winnipeg Ballet Renovations	Manitoba	CA	13,237	$98.08	Renovation
	Senior Services - Kitchen Facility	Batavia	OH	6,000	$155.25	New
	Historical Site Locomotive Shelter	Bismarck	ND	2,100	$207.49	New
Office	Federal Building & Courthouse Modernization	Denver	CO	41,600	$449.92	Renovation
	Brazos County Tax Office	Bryan	TX	13,143	$387.10	New
	Illinois Water District Office	Lincoln	IL	8,974	$201.47	New
	Arkansas River Resource Center	Little Rock	AR	4,926	$739.69	New

All prices are updated to January 1, 2023 and are national averages.
For a more in-depth information contact Design, Cost and Data at 800-533-5680, or go to www.DCD.com

PROJECT	DESCRIPTION	CITY	STATE	SIZE	$/SF	NOTES
	Educational					
Athletic Facility	Physical Activity/Sports Science	Morgantown	WV	117,344	$307.16	New
	Indoor Football Practice Facility	Clemson	SC	81,992	$237.01	New
	Jesuit College Locker Room Addition	Dallas	TX	41,673	$283.38	Addition
	Multi-Purpose Gymnasium	Jacksonville	FL	22,844	$376.04	New
College Classroom	New Mexico Tech Geology Building	Socorro	NM	86,813	$409.65	New
	CSU Concourse & Training Room	Fort Collins	CO	53,050	$191.35	Addition/Renovation
	Jack Williamson Liberal Arts Center	Portales	NM	52,480	$307.47	Renovation
	UNLV Literature & Law Building	Las Vegas	NV	44,830	$255.64	Renovation
	Southern State Community College	Mount Orab	OH	43,833	$245.95	New
	SERT Building Iowa Lakes College	Estherville	IA	42,940	$168.57	Adaptive Reuse
Elementary	Cibolo Valley Elementary School	Cibolo	TX	153,130	$332.96	New
	Hill Farm Elementary School	Bryant	AR	88,800	$398.29	New
	CREC International Magnet School	South Windsor	CT	63,923	$547.43	New
	Pineville Elementary School	Pineville	WV	51,650	$291.51	New
	Crownpoint Elementary School	Crownpoint	NM	48,592	$543.17	New
	Janney Elementary School Addition	Washington	DC	10,000	$802.81	Addition
High School	Hmong College Prep Academy	Saint Paul	MN	154,434	$113.99	Addition/Renovation
	HFC High School North Building	Flossmoor	IL	136,555	$279.03	Addition/Renovation
	Somerset High School	Von Ormy	TX	125,800	$242.34	New
	Takoma Education Campus	Washington	DC	119,000	$299.73	Renovation
	High School Addition & Renovation	Decatur	GA	83,816	$348.01	Addition/Renovation
	Palmer Catholic Academy	Ponte Vedra Beach	FL	34,209	$188.57	New
	Elmwood High School Addition	Elmwood Park	IL	31,630	$466.92	Addition/Renovation
	High School Fine Arts Building	Heber Springs	AR	30,505	$545.33	New
	Alamo Heights High School Fine Arts	San Antonio	TX	25,536	$438.23	Addition/Renovation
	St. Patrick Catholic School	Jacksonville	FL	23,227	$402.23	New
	Springdale School Alteration	Corbett	OR	13,680	$171.61	Renovation
	Goddard School Addition Renovation	Anderson Township	OH	12,489	$114.35	Addition/Renovation
	ISD Outdoor Education Center	Sabine Pass	TX	10,193	$434.56	New
	Indian Mountain School Student Center	Lakeville	CT	9,335	$417.19	Addition
	First Impressions Academy	Fayetteville	NC	7,752	$240.04	New
	High School South Campus Field	Cincinnati	OH	5,580	$321.93	New
Middle School	Timberline Middle School	Waukee	IA	187,375	$216.07	New
	Red Bank Middle School	Chattanooga	TN	158,637	$357.76	New
	Jaime Escalante Middle School	Pharr	TX	156,538	$267.88	New
	Conservatory Green ECE-8 School	Denver	CO	113,616	$238.90	New
	Midland School Addition	Floral	AR	30,150	$321.24	Addition
Laboratory/Research	Science & Technology Building	Fayetteville	NC	65,048	$677.74	New
	NSU/US Geological Survey	Davie	FL	24,000	$140.83	Tenant Build-out
	Northeast Technology Center	Pryor	OK	11,909	$422.46	New
	Environmental Education Center	Bushkill Township	PA	9,275	$732.45	New
	Research & Education Center	Homestead	FL	5,760	$924.31	New
Multi-Purpose	BGSU Student Rec Center	Bowling Green	OH	179,549	$88.78	Renovation
	Kennedy Center Theatre/Studio Arts	Clinton	NY	96,100	$448.35	New
	Classroom & Administration Building	Houston	TX	65,234	$289.82	New
	NM State U Pete Domenici Building	Las Cruces	NM	53,341	$365.32	Addition/Renovation
	Widener University Freedom Hall	Chester	PA	36,700	$432.43	New
	Alumni Hall, Lincoln Park	Midland	PA	29,027	$352.63	New
	GSU Piedmont North Dining Hall	Atlanta	GA	12,300	$495.10	Addition
	NAU Dining Hall Expansion Phase II	Flagstaff	AZ	10,096	$631.40	New
	Neighborhood Resource Center	Richmond	TX	6,935	$272.57	New
	Heber Springs Cafeteria Remodel	Heber Springs	AR	5,585	$307.16	Renovation

All prices are updated to January 1, 2023 and are national averages.
For a more in-depth information contact Design, Cost and Data at 800-533-5680, or go to www.DCD.com

339

PROJECT	DESCRIPTION	CITY	STATE	SIZE	$/SF	NOTES

Hotels

Hotels	Omni Dallas Hotel	Dallas	TX	1,161,450	$557.92	New
	John Ascuagas Nugget Hotel/Casino	Sparks	NV	449,820	$190.34	Addition
	Le Centre On Fourth Embassy Suites	Louisville	KY	408,229	$150.63	Adaptive Reuse
	Minneapolis Marriott West	Minneapolis	MN	237,362	$217.42	New
	Sheraton Centre Park Hotel	Dallas	TX	231,031	$311.15	New
	AmeriSuites	Chicago	IL	191,600	$196.74	Addition/Renovation
	Hampton Inn & Suites Hotel	Chicago	IL	162,000	$214.85	New
	Sheraton Harbor Island Hotel Tower	San Diego	CA	144,126	$261.95	Addition
	Compri Hotel	Los Angeles	CA	110,150	$231.99	New
	Best Western Columbia Hotel	San Diego	CA	108,040	$179.47	New
	Spooky Nook Warehouse Hotel	Manheim	PA	92,726	$170.91	Adaptive Reuse
	The Atrium Motel	Norfolk	VA	75,889	$183.53	New
	Staybridge Hotel At Preston Ridge	Alpharetta	GA	74,607	$235.33	New
	Hampton Inn & Suites	Allentown	PA	71,686	$196.26	New
	Hampton Inn Hotel	Carol Stream	IL	71,000	$241.02	New
	The Inn On Lake Superior	Duluth	MN	65,345	$193.88	New
	The Lancaster Hotel	Houston	TX	64,310	$405.71	Renovation
	Fairfield Inn	Helena	MT	31,009	$189.46	New
	The Edison Hotel	Miami	FL	28,875	$157.07	Renovation
	Western Executive Inn	Billings	MT	21,984	$140.23	New
	Country Hearth Inn	Preston	MN	21,028	$164.65	New
	Lawrence Welk Resort Hotel	Escondido	CA	19,874	$187.80	Addition
	Hanalei Hotel Conference Center	San Diego	CA	8,587	$298.22	Addition
	Summit At Vail, Multi-Purpose Lodge	Vail	CO	6,000	$356.06	New

PROJECT	DESCRIPTION	CITY	STATE	SIZE	$/SF	NOTES
		Industrial				
Manufacturing	Brentwood Industries Manufacturing	Reading	PA	205,000	$55.29	New
	Lee Steel Corporate Plant	Romulus	MI	200,625	$120.25	New
	Siemens Westinghouse Fuel Cell Facility	Munhall	PA	191,090	$130.94	New
	Manufacturing Plant & Headquarters	Lansing	MI	188,975	$126.96	Addition
	Concepts Direct	Longmont	CO	117,900	$154.85	New
	Nypro Inc.	Clinton	MA	102,475	$158.85	Addition
	SWF Industrial	Wrightsville	PA	76,218	$97.87	New
	Headquarters & Manufacturing Facility	Lower Nazareth Township	PA	62,980	$46.87	Renovation
	Aerzen USA (Office/Manufacturing)	Coatesville	PA	40,000	$239.19	New
	Lee Steel Corporate Expansion	Wyoming	MI	34,821	$101.62	New
	Prescott Aerospace	Prescott	AZ	31,400	$146.40	New
	Battery Innovation Center	Newberry	IN	30,080	$571.17	New
	American Steel	Billings	MT	25,957	$106.18	New
	Phillip S. Luttazi Town Garage	Dover	MA	21,913	$182.70	New
	ITT Flygt - Industrial Facility	Milford	OH	16,991	$173.24	New
	Broadmoor Golf Maintenance	Colorado Springs	CO	16,064	$314.06	New
	Cooper B-Line Expansion	Highland	IL	15,290	$221.89	Addition/Renovation
	Robberson Ford Collision Center	Bend	OR	15,089	$182.14	New
	Brown Industrial Building	Truckee	CA	13,345	$179.93	New
	CTC Vehicle Maintenance Shops	Killeen	TX	11,250	$170.14	New
	Storage & Shop Facility	Billings	MT	8,763	$117.41	New
	Central Plant with Equipment Bay	Mesa	AZ	8,500	$1,210.59	New
Office	Woodlands Business Center	Richmond	VA	48,000	$105.17	New
	Wiregrass Research Center	Headland	AL	9,740	$324.80	New
Office/Warehouse	American Superconductor	Devens	MA	354,000	$223.50	New
	Castcon Stone Inc.	Saxonburg	PA	47,000	$149.70	New
	Minnesota DNR Headquarters	Tower	MN	37,802	$221.70	New
	Office & Warehouse	Miami	FL	14,815	$181.63	New
	DOT Office & Maintenance Building	Hillsboro	OH	10,876	$320.40	New
Warehouse	Distribution Center	Windsor	CT	303,750	$45.57	New
	Zany Brainy Distribution Center	Bridgeport	NJ	250,000	$54.42	New
	Galderma - Warehouse	Fort Worth	TX	70,000	$117.24	New
	Manzana Products Warehouse	Sebastopol	CA	41,395	$87.32	New
	Tactical Equip Maintenance Facility	Fort Campbell	KY	35,290	$315.05	New
	Sonoma Wine Company Canopy	Graton	CA	26,000	$74.12	New
	Administration/Chemical Storage Building	Killeen	TX	23,837	$448.16	New
	DOT Truck Storage Building	Hillsboro	OH	18,400	$122.27	New
	F.I. Storage Facility	Kentwood	MI	13,125	$89.91	New
	50 Columbia Drive Warehouse	Pooler	GA	10,000	$112.84	New
	DOT Salt Storage Building	Hillsboro	OH	9,100	$123.27	New
	Organizational Storage Facility	Fort Campbell	KY	8,040	$187.04	New
	Dwan Maintenance Building	Bloomington	MN	7,240	$252.61	Addition/Renovation
	Maintenance/Storage Building	Batavia	OH	7,200	$108.65	New
	DOT Cold Storage Building	Hillsboro	OH	5,040	$131.44	New
	Job Corp Warehouse	Hartford	CT	3,800	$623.17	New
	DOT Materials Storage Building	Hillsboro	OH	1,920	$146.96	New
	Aerial Vehicle Storage	Fort Campbell	KY	1,800	$280.85	New
	Petro, Oil, Lubricant Storage	Fort Campbell	KY	640	$392.61	New
	Hazardous Waste Storage Building	Fort Campbell	KY	640	$419.58	New

All prices are updated to January 1, 2023 and are national averages.
For a more in-depth information contact Design, Cost and Data at 800-533-5680, or go to www.DCD.com

341

PROJECT	DESCRIPTION	CITY	STATE	SIZE	$/SF	NOTES

Medical

PROJECT	DESCRIPTION	CITY	STATE	SIZE	$/SF	NOTES
Clinic	HealthCare Emergency/Trauma Center	Topeka	KS	115,000	$529.53	Addition
	Sadler Clinic (Shell Only)	Conroe	TX	61,599	$164.85	New
	Pinellas County Health Department	Largo	FL	54,965	$343.51	Retrofit
	Sanford Moorhead Clinic	Moorhead	MN	49,250	$336.64	New
	County Health Department	Port Charlotte	FL	47,564	$381.70	New
	Sadler Clinic	Conroe	TX	41,066	$156.17	Tenant Build-out
	PineMed Medical Plaza	The Woodlands	TX	30,398	$174.16	New
	Outpatient Specialty Clinic	Vancouver	WA	20,139	$433.33	New
	Ambulatory Surgery Center	Stroudsburg	PA	19,929	$465.08	Addition/Renovation
	Thundermist Health Center	West Warwick	RI	18,217	$268.15	Adaptive Reuse
	E Texas Community Health Services	Nacogdoches	TX	12,500	$126.99	Retrofit
	North Mobile Health Center	Mt. Vernon	AL	6,765	$378.84	New
Dental Office	Construct Dental Clinic Roseburg	Roseburg	OR	7,750	$659.22	New
	Kitchens Pediatric Dental Clinic	Little Rock	AR	6,068	$396.83	New
	Dental Office Shell & Parking	Olympia	WA	5,302	$225.16	New
	Evans Family Dental	Austin	TX	2,354	$275.95	Tenant Build-out
Hospital	Union Hospital Addition	Terre Haute	IN	492,348	$414.74	Addition
	Regional Medical Center	Lafayette	LA	410,273	$694.13	New
	Houston Medical Pavilion	Warner Robins	GA	180,000	$81.77	Adaptive Reuse
	Langley AFB Hospital Renovation	Langley Air Force Base	VA	160,000	$693.00	Renovation
	Cass Regional Medical Center	Harrisonville	MO	137,524	$487.29	New
	Texas Spine & Joint Hospital	Tyler	TX	115,789	$322.74	Addition/Renovation
	Oktibbeha County Hospital Expansion	Starkville	MS	87,116	$422.90	New
	Childrens Mercy Hospital	Independence	MO	54,682	$407.46	New
	Oktibbeha County Hospital Renovation	Starkville	MS	30,263	$211.08	Renovation
	El Rio Community Health Center	Tucson	AZ	26,998	$311.21	New
	Pondella Public Health Center	Fort Myers	FL	26,400	$387.64	Renovation
	Topeka Ear Nose & Throat	Topeka	KS	24,073	$401.91	New
	Rapha Primary Care	Fayetteville	NC	19,907	$172.54	Renovation
	UNM Hospitals North Valley Center	Albuquerque	NM	16,500	$406.36	New
	El Rio Community Health Center	Tucson	AZ	14,000	$353.32	New
	VA Medical Center Area G Renovation	Houston	TX	12,000	$429.92	Renovation
	Surgical Suite Expansion	Dobbs Ferry	NY	9,000	$490.71	Renovation
	Legacy Emergency Room	Allen	TX	8,432	$813.87	New
	Oral & Maxillofacial Surgery Center	Fayetteville	NC	2,214	$692.50	Renovation
Nursing Home/Rehab	Senior Living Community	Hoschton	GA	56,251	$180.47	New
	Retirement Community	Carlisle	PA	47,075	$203.22	Addition/Renovation
	Assisted Living & Memory Center	Dacula	GA	38,221	$209.92	New
	Homestead Village Nursing Care	Lancaster	PA	28,149	$137.45	Renovation
	Jewish Services For The Aging	Tucson	AZ	24,993	$255.87	New
	Short-Term Rehabilitation	Olathe	KS	13,800	$344.21	Addition
	St. Katharine Retirement Center	El Reno	OK	12,000	$406.81	New/Addition
Office	Orthopedic Hospital/Medical Office	Allentown	PA	79,807	$257.84	Renovation
	Tomball Medical Office Building	Tomball	TX	54,380	$175.32	New
	Olathe Health Education Center	Olathe	KS	50,258	$392.05	New
	Home & Hospice Care	Providence	RI	47,734	$247.30	Renovation
	NE Georgia Medical Plaza 400	Dawsonville	GA	26,997	$345.04	Adaptive Reuse
	VA Medical Center/Pharmacy	Waco	TX	19,171	$200.89	Renovation
	Cancer Specialists of North Florida	Jacksonville	FL	18,654	$366.46	New
	MJHS Hospice Residence	N.Y.C.	NY	12,500	$189.63	Renovation
	Medical Office Building	Pelham	NH	8,399	$341.26	New
	Podiatry Group	Marietta	GA	6,768	$69.40	Renovation
	Marietta Podiatry Group	Marietta	GA	4,400	$316.93	New

All prices are updated to January 1, 2023 and are national averages.
For a more in-depth information contact Design, Cost and Data at 800-533-5680, or go to www.DCD.com

PROJECT	DESCRIPTION	CITY	STATE	SIZE	$/SF	NOTES
	Office					
Office	Restaurant Support Center	Lenexa	KS	186,465	$335.55	New
	5000 NASA Boulevard	Fairmont	WY	132,000	$357.79	New
	Rockford Construction Office	Grand Rapids	MI	71,144	$142.96	Adaptive Reuse
	Woodlawn Office Bldg. (Shell Only)	Louisville	KY	60,000	$180.85	New
	Rosecrance Ware Center	Rockford	IL	44,800	$178.49	Adaptive Reuse
	Professional Center (Shell)	White Marsh	MD	43,025	$227.52	New
	Swan Skyline Office Plaza (Shell)	Tucson	AZ	37,200	$153.46	New
	Infinite Energy Phase IV	Gainesville	FL	36,500	$402.24	New
	Freedom Plaza Building	Cookeville	TN	28,488	$314.49	New
	Landmark Professional Building	Clayton	NC	27,231	$298.21	New
	FC Gulf Freeway Building (Shell)	Houston	TX	24,084	$339.48	New
	White Street Building (Shell Only)	Marietta	GA	23,809	$277.04	New
	Columbia Shores Office Condo	Vancouver	WA	22,574	$229.92	New
	Office & Design Studio	Chicago	IL	20,244	$161.42	Tenant Build-out
	Pinnacle III Office Tenant Finish	Leawood	KS	18,409	$76.85	Tenant Build-out
	Commerce Park (Shell)	Suwanee	GA	17,097	$157.61	New
	PCWA Business Center Interior	Auburn	CA	12,085	$79.94	Renovation
	Tenth Avenue Holdings Offices	N.Y.C.	NY	11,000	$112.21	Tenant Build-out
	Office Park - Building A (Shell)	Fort Collins	CO	10,000	$344.59	New
	Longshoremens Welfare Fund Building	Savannah	GA	8,160	$505.93	New
	Garry Street Office Building	Manitoba	CA	7,506	$201.87	Renovation
	Reserve Advisors	Milwaukee	WI	5,300	$64.72	Tenant Build-out
	510 Armory Street Office	Boston	MA	5,100	$123.92	Renovation
	Offices of Bonsall Shafferman	Bethlehem	PA	4,950	$91.65	Tenant Build-out
	FCT Capital Partners	Houston	TX	4,100	$107.22	Tenant Build-out
	Martin Rogers Associates Office	Wilkes-Barre	PA	4,000	$171.57	Addition/Renovation
	Cowan & Kohne Financial	Suwanee	GA	3,713	$161.56	Tenant Build-out
	212 Archer Street Office	Bel Air	MD	3,600	$228.54	New
	Utilities Analyses Inc.	Suwanee	GA	3,513	$149.87	Tenant Build-out
	Visual Lizard Interior Fit-Up	Manitoba	CA	2,430	$120.24	Tenant Build-out
	Richardson State Farm	Houston	TX	2,200	$145.38	Tenant Build-out
Mixed-Use	Office/Retail/Parking Mixed-Use	Jackson	MS	228,407	$329.88	New
	Korte & Luitjohan Office & Shop	Highland	IL	26,000	$158.51	New
Medical Office	Evanston Medical Office Building	Evanston	WY	9,157	$363.33	New
	Advanced Medical Group	Suwanee	GA	4,433	$154.28	Tenant Build-out
	Bothell Dental Office Build Out	Bothell	WA	2,203	$256.18	Tenant Build-out
Headquarters	CONSUL Energy Corporation Headquarters	Southpointe, Canonsburg	PA	317,500	$305.98	New
	Fairmont Supply Corporate Headquarters	Southpointe, Canonsburg	PA	75,255	$202.21	New
	Practice Velocity Corporate Headquarters	Machesney Park	IL	64,318	$127.06	Adaptive Reuse
	Enterprise Integration Headquarters	Jacksonville	FL	57,723	$79.05	Renovation
	Linear Technology	Cary	NC	20,000	$410.64	New
	PIPS Technology Inc.	Knoxville	TN	19,884	$326.41	New
	Lee Steel Corporate Offices	Novi	MI	15,781	$216.95	Renovation
	Weaver Cooke Headquarters	Goldsboro	NC	15,464	$390.37	New
	In Capital Holdings	Boca Raton	FL	13,000	$219.39	Tenant Build-out
	ACCION Regional Headquarters	Albuquerque	NM	7,580	$419.99	New
Civic Office	Miss Department of Environmental Quality	Jackson	MS	121,170	$122.35	Renovation
	County Central Office Complex	Pensacola	FL	74,630	$370.80	New
	State of WV Office Building	Fairmont	WV	70,442	$330.38	New
	JAX Chamber of Commerce Renovation	Jacksonville	FL	20,110	$272.78	Renovation

All prices are updated to January 1, 2023 and are national averages.
For a more in-depth information contact Design, Cost and Data at 800-533-5680, or go to www.DCD.com

343

PROJECT	DESCRIPTION	CITY	STATE	SIZE	$/SF	NOTES
		Recreational				
Educational	The Pavilion at Ole Miss	Oxford	MS	235,301	$580.74	New
	University Laker Turf Building	Allendale	MI	137,662	$192.55	New
	CSU Recreation Center	Chico	CA	110,245	$599.18	New
	Intramural Recreation Penn State	State College	PA	59,303	$531.51	Addition/Renovation
	Center For Women's Athletics	Fayetteville	AR	39,183	$473.29	New
	Pickens Recreation Center	Pickens	SC	20,400	$236.27	New
	High School Concessions & Press Box	Loganville	GA	1,506	$607.03	New
Health Club	Brooklyn Yard Fitness Club (Shell)	Portland	OR	63,987	$148.75	New
	Title Boxing Club	Cedar Hill	TX	4,330	$92.21	Tenant Build-out
Recreational	Community Recreation Center	Williston	ND	223,787	$578.10	New
	Spirit Lake Casino & Resort	St. Michael	ND	112,277	$78.41	Renovation
	Phipps Tropical Forest	Pittsburgh	PA	80,000	$421.93	New
	Family Recreation Center	Colonie	NY	70,256	$308.37	New
	Youth Activity Center	Joplin	MO	62,056	$130.59	New
	New Holland Recreational Center	New Holland	PA	51,256	$170.28	Renovation
	Community College Recreation Center	Cedar Rapids	IA	43,500	$214.87	New
	Anderson Recreation Center	Anderson	SC	34,282	$481.79	New
	East Park Community Center	Nashville	TN	33,000	$386.47	New
	Church Family Life Center	Clemson	SC	31,509	$417.05	New/Renovation
	C.K. Ray Recreation Center	Conroe	TX	30,380	$196.83	Addition/Renovation
	St. Raphael Athletic & Wellness Center	Pawtucket	RI	30,268	$340.77	New
	The Forge For Families	Houston	TX	29,860	$321.29	New
	Christian Life Center	Birmingham	MI	26,966	$439.75	Addition
	Children's Sports Center	Woodbury	MN	26,219	$157.27	New
	Trinity River Audubon Center	Dallas	TX	20,791	$1,212.63	New
	Baptist Church Activity Center	Indianapolis	IN	16,636	$194.43	New
	Boys & Girls Club Syracuse	Syracuse	NY	12,107	$237.64	Addition
	Job Corp Recreational Building	Hartford	CT	11,300	$266.65	New
	Presbyterian Family Life Center	Strawberry Plains	TN	11,236	$216.67	New
	McDaniel Yacht Basin	North East	MD	7,620	$244.02	New
	Bicentennial Park	Cincinnati	OH	4,050	$1,143.11	New
	Bahosky Softball Complex	Bronx	NY	3,800	$618.19	New
Swimming Center	Resort & Indoor Waterpark	Cortland	NY	175,060	$345.36	New
	The Aquatic Center	Tunica	MS	45,008	$411.50	New
	Spirit Lake Phase 4	St. Michael	ND	26,630	$470.48	Addition/Renovation
	Family Aquatic Center	Beachwood	OH	7,500	$1,528.96	New
	Community Aquatic Park & Center	Billings	MT	6,730	$961.60	New
Theater	Cinema & IMAX Theatre	Lansing	MI	13,750	$290.67	New
	Academy Theater	N.Y.C.	NY	4,593	$295.31	Renovation
YMCA	David D. Hunting YMCA	Grand Rapids	MI	162,966	$264.89	New
	YMCA Recreational Center	Ann Arbor	MI	83,377	$360.54	New
	Floyd Co. YMCA & Aquatic Center	New Albany	IN	82,324	$438.99	New
	Wade Walker Park Family YMCA	Stone Mountain	GA	59,134	$458.20	New
	Alexandria YMCA	Alexandria	MN	55,150	$221.70	New
	Greater Nashua YMCA	Nashua	NH	49,980	$272.86	New
	Lancaster YMCA Harrisburg Ave.	Lancaster	PA	42,502	$464.04	New
	Greater Kingsport Family YMCA	Kingsport	TN	40,007	$331.46	New
	Eastside YMCA	Knoxville	TN	39,984	$329.33	New
	Highland County Family YMCA	Hillsboro	OH	33,228	$196.23	New
	Houston Texans YMCA	Houston	TX	31,628	$465.63	New
	Cypress Creek YMCA	Houston	TX	25,699	$196.39	Addition/Renovation

     All prices are updated to January 1, 2023 and are national averages.
For a more in-depth information contact Design, Cost and Data at 800-533-5680, or go to www.DCD.com

PROJECT	DESCRIPTION	CITY	STATE	SIZE	$/SF	NOTES

Religious

PROJECT	DESCRIPTION	CITY	STATE	SIZE	$/SF	NOTES
Church	First United Methodist Church	Orlando	FL	121,536	$366.54	New
	Beautiful Savior Lutheran Church	Plymouth	MN	69,700	$149.91	New
	Solid Rock Baptist Church	Berlin	NJ	58,359	$135.55	New
	North Side Baptist Church	Greenville	SC	48,087	$346.15	New
	St. Martha Catholic Church	Porter	TX	46,748	$683.93	New
	Gracepoint Gospel Fellowship Church	Ramapo	NY	46,595	$254.39	New
	Good Shepherd Methodist Church	Odessa	TX	41,003	$405.17	New
	Davisville Church Addition	Southampton	PA	36,090	$239.54	Addition
	Immaculate Catholic Church	Columbia	IL	34,000	$334.96	New
	Keystone Community Church	Ada	MI	29,775	$197.23	New
	Grace Church	Des Moines	IA	29,296	$267.59	New
	Good Shepherd Church	Naperville	IL	27,869	$265.07	Addition/Renovation
	River Hills Baptist Church	Corpus Christi	TX	27,404	$401.30	New
	Good Shepherd Catholic Church	Smithville	MO	24,810	$275.68	New
	Prince of Peace Catholic Church	Chesapeake	VA	24,740	$345.42	Addition/Renovation
	Sanctuary Addition Christian Church	Oklahoma City	OK	23,820	$207.05	Addition
	St. Sylvester Catholic Church	Gulf Breeze	FL	22,000	$542.90	New
	St. Peters Catholic Sanctuary	Fallbrook	CA	20,764	$598.21	New
	Notre Dame Catholic Church	Houston	TX	20,280	$551.86	New
	St Eugene Catholic Church	Oklahoma City	OK	20,000	$604.87	New
	Chapin Presbyterian Church	Chapin	SC	19,900	$441.03	New
	St. Michaels Catholic Church	Glen Allen	VA	19,770	$473.15	New
	Shrine of Holy Spirit	Branson	MO	19,200	$450.27	New
	Wildwood United Methodist Church	Magnolia	TX	19,000	$321.73	New
	Episcopal Church of the Nativity	Scottsdale	AZ	18,288	$160.91	Adaptive Reuse
	Hardin Church of Christ	Knoxville	TN	17,149	$161.76	New
	First United Methodist Church	Crossville	TN	15,816	$633.13	New
	Good Shepherd Episcopal Church	Silver Spring	MD	15,200	$377.78	Addition/Renovation
	Ascension Catholic Church	LaPlace	LA	15,057	$498.69	New
	St. Patrick Catholic Church	Jacksonville	FL	14,139	$356.45	New
	Our Lady of Guadalupe Catholic Church	Rosenberg	TX	12,910	$549.80	New
	St. Timothy's Episcopal Church	Creve Coeur	MO	12,682	$274.09	New/Renovation
	St. Paul Lutheran Church	Pomaria	SC	12,072	$448.77	New
	Covenant Baptist Church	Florida City	FL	10,725	$334.85	New
	First United Methodist Church	Katy	TX	10,503	$406.69	Addition/Renovation
	Lake Ann United Methodist Church	Lake Ann	MI	9,975	$281.02	New
	United Methodist Church	Odenton	MD	8,783	$439.10	New
	Kent R. Hance Chapel at Texas Tech	Lubbock	TX	6,530	$797.05	New
	Haven for Hope Chapel	San Antonio	TX	2,232	$598.68	New
Multi-Purpose	Baptist Church Multi-Purpose Bldg	Maryville	TN	41,656	$149.38	New
	Good Shepherd Parish Center	San Diego	CA	28,752	$206.03	New
	Christian Life Center	Kansas City	MO	26,320	$400.15	New
	Baptist Church Outreach Center	Fort Smith	AR	25,000	$351.35	New
	St Rafael Administration Building	San Diego	CA	24,276	$221.90	New
	Catholic Church Social Hall	Chula Vista	CA	23,596	$369.21	New
	United Methodist Church	West Chester	PA	11,935	$322.86	Addition/Renovation
	Student Ministry Center	Knoxville	TN	11,700	$399.99	New
	Catholic Pastoral Ministries Center	Spring	TX	10,135	$513.53	New
	Christian Renewal Center	Dickinson	TX	8,500	$282.33	New
	Holy Family St Lawrence Parish Center	Essex Junction	VT	7,900	$296.08	New
	Presbyterian Church Addition	Gap	PA	7,414	$339.88	Addition/Renovation
	New Hope Church Addition/Alteration	Saint Louis	MO	5,564	$260.24	Renovation

All prices are updated to January 1, 2023 and are national averages.
For a more in-depth information contact Design, Cost and Data at 800-533-5680, or go to www.DCD.com

345

PROJECT	DESCRIPTION	CITY	STATE	SIZE	$/SF	NOTES
	Residential					
Apartment	Solace Apartments	Virginia Beach	VA	331,681	$131.73	New
	1221 Broadway Lofts	San Antonio	TX	205,137	$186.05	Adaptive Reuse
	Sustainable Fellwood Phase I	Savannah	GA	124,037	$174.05	New
	Kelly Cullen Community	San Francisco	CA	98,385	$721.16	Adaptive Reuse
	Bachelors Enlisted Quarters	Camp Williams	UT	76,253	$327.91	New
	Mockingbird Terrace Homes	Louisville	KY	71,110	$208.40	New
	Homeless Men's Residential	San Antonio	TX	67,908	$345.45	New
	Homeless Women's/Family Residence	San Antonio	TX	60,182	$332.88	New
	Magnolia Place	Lancaster	PA	39,714	$209.93	New
	Elkins First Ward Apartments	Elkins	WV	27,000	$150.70	Adaptive Reuse
	Young Burlington Apartments	Los Angeles	CA	24,399	$270.08	New
	The Lofts at 300 Bowman	Dickson City	PA	23,900	$112.58	Adaptive Reuse
	Peaceful Paths Emergency Svc Campus	Gainesville	FL	22,535	$193.26	New
	Wylie House - Ronald McDonald House	Kansas City	MO	21,885	$245.13	New
	Dogwood Manor Apartments	Oak Ridge	TN	19,975	$222.32	New
	Anderson Village Multi-Family	Austin	TX	12,500	$371.15	New
	Salvation Army Sally's House	Houston	TX	7,812	$323.89	Addition
	Stones River Apartment Complex	Murfreesboro	TN	7,548	$299.52	Addition
	Sunshine Park Apartments Renovation	Gainesville	FL	2,252	$158.76	Renovation
Assisted Living	Kenmore Apartments Senior Housing	Chicago	IL	90,528	$277.71	Renovation
	Country Meadows Retirement	Allentown	PA	53,237	$212.49	New
	Creekside Village Assisted Living	Harrisburg	PA	16,150	$184.67	New
	Landis Homes Retirement Community	Lititz	PA	14,255	$87.28	Renovation
Dormitory	NSU Graduate Student Housing	Davie	FL	203,500	$285.32	Renovation
	Rider University Student Housing	Lawrenceville	NJ	50,500	$306.89	New
	JWU Biscayne Commons Dormitory	Miami	FL	40,048	$336.63	New
	College Residence Dorm	Bloomfield	NJ	25,980	$414.50	Renovation
Single-Family Home	Island Residence	Grosse Ile	MI	19,237	$912.19	New
	MG Residence Restoration	Williamston	MI	9,768	$97.93	Renovation
	Concepcion House	Coral Gables	FL	6,067	$478.89	New
	Leal House	Miami	FL	5,935	$336.26	New/Renovation
	Monserrate Street Residence	Coral Gables	FL	5,885	$594.01	New
	Private Residence	Newburgh	IN	5,566	$477.10	New
	Private Residence	Austin	MN	5,489	$206.83	New
	Private Residence	Lake Wallenpaupack	PA	4,845	$432.20	New
	Island in the Grove	Boca Raton	FL	4,701	$491.57	New
	Private Residence	Benson	AZ	3,660	$248.72	New
	Fairhope Green Home	Fairhope	AL	3,610	$280.76	New
	PATH Concept House	Omaha	NE	3,490	$118.32	New
	Private Residence	La Jolla	CA	3,420	$482.85	New
	Solar House - Private Residence	Fly Creek	NY	3,304	$250.36	New
	Elliott Residence	Fort Collins	CO	3,300	$368.67	New
	Renfrew House	Manitoba	CA	3,206	$268.81	New
	Rosado I Hansen Residence	Tucson	AZ	3,175	$192.79	New
	306 W. Waldburg Residence	Savannah	GA	2,588	$237.33	New
	Nutter Green Home	Milford	OH	2,289	$227.12	New
	Guest House Residence	Ahwatuckee	AZ	1,913	$529.24	New
	Kiwi House	Baton Rouge	LA	1,515	$221.85	New
	Private Residence Renovation	Shavertown	PA	810	$229.90	Renovation

   All prices are updated to January 1, 2023 and are national averages.
For a more in-depth information contact Design, Cost and Data at 800-533-5680, or go to www.DCD.com

Metro Area Multipliers

The costs as presented in this book attempt to represent national averages. Costs, however, vary among regions, states and even between adjacent localities.

In order to more closely approximate the probable costs for specific locations throughout the U.S., this table of Metro Area Multipliers is provided. These adjustment factors can be used to modify costs obtained from this book to help account for regional variations of construction costs and to provide a more accurate estimate for specific areas. The factors are formulated by comparing costs in a specific area to the costs presented in this Costbook. An example of how to use these factors is shown below. Whenever local current costs are known, whether material prices or labor rates, they should be used when more accuracy is required.

Cost from Costbook Pages   X   Metro Area Multiplier   =   Adjusted Cost

For example, a project estimated to cost $1,000,000 using the Costbook pages can be adjusted to more closely approximate the cost in Los Angeles:

$1,000,000   X   1.27   = $1,270,000

State	Metropolitan Area	Multiplier
AK	ANCHORAGE	1.22
	FAIRBANKS	1.22
	JUNEAU	1.22
	KETCHIKAN	1.22
	KODIAK	1.22
	NOME	1.22
	SITKA	1.22
AL	ANNISTON	0.76
	AUBURN	0.75
	BIRMINGHAM	0.77
	DECATUR	0.78
	DOTHAN	0.75
	FLORENCE	0.79
	GADSDEN	0.76
	HUNTSVILLE	0.76
	MOBILE	0.76
	MONTGOMERY	0.76
	PRATTVILLE	0.76
	SELMA	0.76
	OPELIKA	0.75
	TUSCALOOSA	0.78
AR	BATESVILLE	0.74
	CONWAY	0.74
	EL DORADO	0.74
	FAYETTEVILLE	0.73
	FORT SMITH	0.73
	HOT SPRINGS	0.74
	JONESBORO	0.73
	LITTLE ROCK	0.76
	PINE BLUFF	0.74
	ROGERS	0.73
	SPRINGDALE	0.73
	TEXARKANA	0.74
AZ	CASA GRANDE	0.79
	CLIFTON	0.79
	FLAGSTAFF	0.80
	LAKE HAVASU CITY	0.79
	MESA	0.82
	PHOENIX	0.82
	SIERRA VISTA	0.79
	TUCSON	0.76
	YUMA	0.77
CA	ANAHEIM	1.25
	BAKERSFIELD	1.31
	CHICO	1.30
	EUREKA	1.30
	FRESNO	1.33
	LOS ANGELES	1.27
	MEXICALI	1.30
	MODESTO	1.33
	OAKLAND	1.33
	REDDING	1.30

State	Metropolitan Area	Multiplier
CA	RIVERSIDE	1.27
	SACRAMENTO	1.30
	SALINAS	1.33
	SAN BERNARDINO	1.28
	SAN DIEGO	1.25
	SAN FRANCISCO	1.33
	SAN JOSE	1.33
	SAN LUIS OBISPO	1.27
	SANTA BARBARA	1.27
	SANTA CRUZ	1.33
	SANTA ROSA	1.30
	STOCKTON	1.33
	TULARE	0.94
	VALLEJO-FAIRFIELD-NAPA	1.30
	VENTURA	1.26
	VISALIA	0.94
	WATSONVILLE	1.33
	YOLO	1.30
	YUBA CITY	1.30
CO	BOULDER	0.88
	COLORADO SPRINGS	0.87
	DENVER	0.87
	DURANGO	0.86
	FORT COLLINS	0.85
	GRAND JUNCTION	0.85
	GREELEY	0.86
	LONGMONT	0.88
	LOVELAND	0.85
	PUEBLO	0.85
	STERLING	0.86
CT	BRIDGEPORT	1.27
	DANBURY	1.27
	HARTFORD	1.26
	MANCHESTER	1.26
	MERIDEN	1.25
	MIDDLETOWN	1.26
	NEW HAVEN	1.25
	NEW LONDON	1.25
	NORWALK	1.27
	NORWICH	1.25
	STAMFORD	1.27
	TORRINGTON	1.26
	WATERBURY	1.25
DE	DOVER	1.17
	NEWARK	1.17
	WILMINGTON	1.17
DC	WASHINGTON DC	0.99
FL	BOCA RATON	0.83
	BRADENTON	0.81
	CAPE CORAL	0.80
	CLEARWATER	0.81
	DAYTONA BEACH	0.81

State	Metropolitan Area	Multiplier
FL	FORT LAUDERDALE	0.84
	FORT MYERS	0.80
	FORT PIERCE	0.80
	FORT WALTON BEACH	0.78
	GAINESVILLE	0.81
	JACKSONVILLE	0.81
	LAKELAND	0.78
	MELBOURNE	0.91
	MIAMI	0.83
	NAPLES	0.81
	OCALA	0.80
	ORLANDO	0.80
	PALM BAY	0.91
	PANAMA CITY	0.78
	PENSACOLA	0.78
	PORT ST. LUCIE	0.80
	PUNTA GORDA	0.79
	SARASOTA	0.80
	ST PETERSBURG	0.81
	TALLAHASSEE	0.79
	TAMPA	0.82
	TITUSVILLE	0.91
	WEST PALM BEACH	0.83
	WINTER HAVEN	0.78
GA	ALBANY	0.80
	ATHENS	0.78
	ATLANTA	0.77
	AUGUSTA	0.79
	COLUMBUS	0.75
	MACON	0.77
	MARIETTA	0.78
	ROME	0.78
	SAVANNAH	0.79
	VALDOSTA	0.78
HI	HILO	1.30
	HONOLULU	1.30
	MAUI	1.30
IA	BURLINGTON	0.91
	CEDAR FALLS	0.94
	CEDAR RAPIDS	0.97
	COUNCIL BLUFFS	0.98
	DAVENPORT	1.05
	DES MOINES	1.01
	DUBUQUE	1.01
	IOWA CITY	0.99
	FT. DODGE	0.98
	MARSHALLTOWN	0.98
	MASON CITY	0.98
	OTTAMWA	0.98
	SIOUX CITY	0.96
	WATERLOO	0.94

State	Metropolitan Area	Multiplier
ID	BOISE	0.83
	LEWISTON	0.85
	POCATELLO	0.87
	SALMON	0.85
	ST. ANTHONY	0.85
	ST. MARIES	0.85
	TWIN FALLS	0.85
IL	ALTON	1.08
	BLOOMINGTON	1.18
	CHAMPAIGN	1.14
	CHICAGO	1.36
	DE KALB	1.18
	DECATUR	1.12
	EAST ST. LOUIS	1.18
	FREEPORT	1.18
	HARRISBURG	1.18
	KANKAKEE	1.32
	MOLINE	1.18
	NORMAL	1.18
	PEKIN	1.15
	PEORIA	1.15
	ROCKFORD	1.18
	SPRINGFIELD	1.10
	URBANA	1.14
IN	BLOOMINGTON	1.01
	COLUMBUS	1.04
	ELKHART	1.13
	EVANSVILLE	0.99
	FORT WAYNE	1.01
	GARY	1.14
	GOSHEN	1.04
	INDIANAPOLIS	1.01
	KOKOMO	1.01
	LAFAYETTE	1.01
	MUNCIE	1.01
	NEW ALBANY	1.04
	SOUTH BEND	1.14
	TERRE HAUTE	0.99
KS	DODGE CITY	1.02
	HUTCHINSON	1.02
	KANSAS CITY	1.06
	LAWRENCE	1.04
	MANHATTAN	1.02
	OLATHE	1.02
	SALINA	1.02
	TOPEKA	1.03
	WICHITA	0.93
KY	BOWLING GREEN	0.95
	HOPKINSVILLE	0.95
	LEXINGTON	0.96
	LOUISVILLE	0.94
	MAYFIELD	0.95

State	Metropolitan Area	Multiplier
KY	MAYSVILLE	0.95
	OWENSBORO	0.95
	PADUCAH	0.95
LA	ALEXANDRIA	0.79
	BATON ROUGE	0.80
	BOSSIER CITY	0.80
	HOUMA	0.80
	LAFAYETTE	0.80
	LAKE CHARLES	0.79
	MONROE	0.79
	NEW IBERIA	0.80
	NEW ORLEANS	0.80
	SHREVEPORT	0.81
MA	ATTLEBORO	1.29
	BARNSTABLE	1.30
	BOSTON	1.30
	BROCKTON	1.33
	FALL RIVER	1.29
	FARMINGHAM	1.29
	FITCHBURG	1.30
	FRAMINGHAM	1.29
	HAVERHILL	1.29
	LAWRENCE	1.30
	LEOMINSTER	1.30
	LOWELL	1.30
	NEW BEDFORD	1.30
	NORTHAMPTON	1.29
	PITTSFIELD	1.29
	SPRINGFIELD	1.18
	WORCESTER	1.30
	YARMOUTH	1.30
MD	ABERDEEN	0.88
	ANNAPOLIS	0.88
	BALTIMORE	0.87
	CUMBERLAND	0.89
	FREDERICK	0.88
	HAGERSTOWN	0.89
	LEXINGTON PARK	0.88
	ROCKVILLE	0.88
	SALISBURY	0.88
	AUGUSTA	0.82
ME	AUBURN	0.83
	BANGOR	0.81
	BIDDEFORD	0.82
	FARMINGTON	0.82
	LEWISTON	0.83
	PORTLAND	0.83
	SANFORD	0.82
MI	ANN ARBOR	1.11
	BATTLE CREEK	1.00
	BAY CITY	1.01
	BENTON HARBOR	1.00

State	Metropolitan Area	Multiplier
MI	CHEBOYGAN	1.01
	DETROIT	1.12
	EAST LANSING	1.04
	FLINT	1.07
	GRAND RAPIDS	0.85
	HOLLAND	0.91
	JACKSON	1.06
	KALAMAZOO	1.00
	LANSING	1.04
	MARQUETTE	1.01
	MIDLAND	0.98
	MUSKEGON	0.90
	SAGINAW	1.00
MN	AUSTIN	1.12
	DULUTH	1.11
	GRAND FORKS	1.12
	GRAND RAPIDS	1.12
	MANKATO	1.12
	MINNEAPOLIS	1.16
	MOORHEAD	1.12
	ROCHESTER	1.12
	ST CLOUD	1.12
	ST PAUL	1.16
MO	COLUMBIA	1.08
	FARMINGTON	1.05
	JEFFERSON CITY	1.05
	JOPLIN	0.95
	KANSAS CITY	1.12
	POPLAR BLUFF	1.05
	SAINT JOSEPH	1.10
	SAINT LOUIS	1.09
	SPRINGFIELD	0.95
MS	BILOXI	0.76
	COLUMBUS	0.78
	GREENVILLE	0.78
	GULFPORT	0.76
	HATTIESBURG	0.78
	HELENA	0.78
	JACKSON	0.78
	MERIDIAN	0.78
	PASCAGOULA	0.80
	VICKSBURG	0.78
MT	BILLINGS	0.95
	BUTTE	0.94
	GLASGOW	0.94
	GREAT FALLS	0.93
	HAVRE	0.94
	HELENA	0.94
	KALISPELL	0.94
	MILES CITY	0.94
	MISSOULA	0.95

State	Metropolitan Area	Multiplier
NC	ASHEVILLE	0.72
	CHAPEL HILL	0.71
	CHARLOTTE	0.72
	DURHAM	0.71
	FAYETTEVILLE	0.72
	GASTONIA	0.71
	GOLDSBORO	0.71
	GREENSBORO	0.71
	GREENVILLE	0.71
	HICKORY	0.72
	HIGH POINT	0.71
	JACKSONVILLE	0.71
	LENOIR	0.73
	MORGANTON	0.73
	KANNAPOLIS	0.71
	RALEIGH	0.71
	ROCKY MOUNT	0.71
	WILMINGTON	0.71
	WINSTON-SALEM	0.71
ND	BISMARCK	0.89
	DICKINSON	0.90
	FARGO	0.91
	GRAND FORKS	0.89
	JAMESTOWN	0.90
	MINOT	0.90
	WILLISTON	0.90
NE	BEATRICE	0.85
	CHADRON	0.86
	COLUMBUS	0.86
	GRAND ISLAND	0.86
	LINCOLN	0.82
	NORFOLK	0.86
	NORTH PLATTE	0.86
	OMAHA	0.90
	SCOTTSBLUFF	0.86
	VALENTINE	0.86
NH	BERLIN	0.90
	CLAREMONT	0.90
	CONCORD	0.90
	LACONIA	0.90
	MANCHESTER	0.90
	NASHUA	0.90
	PORTSMOUTH	0.91
	ROCHESTER	0.90
	KEENE	0.90
NJ	ATLANTIC	1.26
	BERGEN	1.32
	BRIDGETON	1.25
	CAMDEN	1.27
	CAPE MAY	1.25
	FLANDERS	1.27
	HAMMONTON	1.27

State	Metropolitan Area	Multiplier
NJ	HUNTERDON	1.28
	JERSEY CITY	1.32
	LAKEWOOD	1.27
	MIDDLESEX	1.29
	MILLVILLE	1.25
	MONMOUTH	1.29
	NEW BRUNSWICK	1.27
	NEWARK	1.31
	OCEAN	1.25
	PASSAIC	1.32
	PATERSON	1.27
	SOMERSET	1.28
	TRENTON	1.20
	VINELAND	1.25
	WASHINGTON	1.27
	WILLINGBORO	1.27
NM	ALAMOGORDO	0.86
	ALBUQUERQUE	0.88
	CARLSBAD	0.86
	FARMINGTON	0.86
	GALLUP	0.86
	LAS CRUCES	0.85
	LAS VEGAS	0.86
	SANTA FE	0.85
NV	ELY	1.20
	HAWTHORNE	1.20
	HENDERSON	1.20
	LAS VEGAS	1.22
	RENO	1.19
	SPARKS	1.20
	WINNEMUCCA	1.20
NY	ALBANY	1.09
	AMSTERDAM	1.18
	BINGHAMTON	1.06
	BUFFALO	1.04
	DUTCHESS COUNTY	1.40
	ELMIRA	1.07
	GLENS FALLS	1.08
	ITHACA	1.18
	JAMESTOWN	1.05
	KINGSTON	1.18
	LOCKPORT	1.18
	LONG ISLAND	1.18
	MALONE	1.18
	NASSAU	1.44
	NEW YORK CITY	1.50
	NEWBURGH	1.40
	NIAGARA FALLS	1.16
	ROCHESTER	1.07
	ROME	1.07
	SCHENECTADY	1.09
	SUFFOLK	1.44

State	Metropolitan Area	Multiplier
NY	SYRACUSE	1.06
	TROY	1.09
	UTICA	1.07
	WATERTOWN	1.18
	WHITE PLAINS	1.18
OH	AKRON	1.02
	CANTON	0.99
	CINCINNATI	0.99
	CLEVELAND	1.05
	COLUMBUS	0.99
	DAYTON	0.98
	ELYRIA	1.05
	FINDLAY	1.00
	HAMILTON	0.98
	LIMA	0.97
	LORAIN	1.05
	MANSFIELD	0.99
	MARION	1.00
	MASSILLON	0.99
	MIDDLETOWN	0.98
	PORTSMOUTH	1.00
	SPRINGFIELD	0.98
	STEUBENVILLE	1.00
	TOLEDO	1.02
	WARREN	1.01
	YOUNGSTOWN	1.02
OK	BARTLESVILLE	0.81
	ENID	0.79
	LAWTON	0.82
	MUSKOGEE	0.81
	NORMAN	0.81
	OKLAHOMA CITY	0.80
	PONCA CITY	0.81
	TULSA	0.81
OR	ASHLAND	1.01
	BEND	1.05
	CORVALLIS	1.04
	EUGENE	1.05
	MEDFORD	1.01
	PENDLETON	1.05
	PORTLAND	1.10
	SALEM	1.07
	SPRINGFIELD	1.05
	THE DALLES	1.05
PA	ALLENTOWN	1.11
	ALTOONA	1.07
	BETHLEHEM	1.13
	CARLISLE	1.04
	EASTON	1.13
	ERIE	1.05
	HARRISBURG	1.05
	HAZLETON	1.08

State	Metropolitan Area	Multiplier
PA	JOHNSTOWN	1.06
	LANCASTER	1.02
	LEBANON	1.02
	NEW CASTLE	1.08
	PHILADELPHIA	1.26
	PITTSBURGH	1.09
	READING	1.12
	SCRANTON	1.08
	SHARON	1.11
	STATE COLLEGE	1.07
	WILKES-BARRE	1.08
	WILLIAMSPORT	1.08
	YORK	1.04
PR	MAYAGUEZ	0.58
	PONCE	0.58
	SAN JUAN	0.58
RI	NEWPORT	1.20
	PAWTUCKET	1.20
	PROVIDENCE	1.20
	WARWICK	1.20
	WESTERLY	1.20
	WOONSOCKET	1.20
SC	AIKEN	0.94
	ANDERSON	0.70
	AUGUSTA	0.73
	CHARLESTON	0.71
	COLUMBIA	0.71
	FLORENCE	0.72
	GREENVILLE	0.71
	MYRTLE BEACH	0.70
	NORTH CHARLESTON	0.71
	ROCK HILL	0.73
	SPARTANBURG	0.72
	SUMTER	0.71
SD	ABERDEEN	0.80
	BROOKINGS	0.80
	MITCHELL	0.80
	PIERRE	0.80
	RAPID CITY	0.77
	SIOUX FALLS	0.83
	WATERTOWN	0.80
TN	CHATTANOOGA	0.78
	CLARKSVILLE	0.73
	JACKSON	0.76
	JOHNSON CITY	0.77
	KNOXVILLE	0.75
	MEMPHIS	0.78
	NASHVILLE	0.76
TX	ABILENE	0.74
	AMARILLO	0.74
	ARLINGTON	0.73
	AUSTIN	0.77

State	Metropolitan Area	Multiplier
TX	BEAUMONT	0.75
	BRAZORIA	0.75
	BROWNSVILLE	0.71
	BRYAN	0.75
	COLLEGE STATION	0.75
	CORPUS CHRISTI	0.74
	DALLAS	0.74
	DENISON	0.74
	EDINBURG	0.72
	EL PASO	0.73
	FORT WORTH	0.73
	GALVESTON	0.75
	HARLINGEN	0.71
	HOUSTON	0.71
	KILLEEN	0.73
	LAREDO	0.72
	LONGVIEW	0.73
	LUBBOCK	0.75
	MARSHALL	0.69
	MCALLEN	0.72
	MIDLAND	0.74
	MISSION	0.72
	ODESSA	0.74
	PORT ARTHUR	0.75
	SAN ANGELO	0.73
	SAN ANTONIO	0.76
	SAN BENITO	0.71
	SAN MARCOS	0.75
	SHERMAN	0.74
	TEMPLE	0.73
	TEXARKANA	0.73
	TEXAS CITY	0.75
	TYLER	0.72
	VICTORIA	0.74
	WACO	0.73
	WICHITA FALLS	0.72
UT	LOGAN	0.93
	OGDEN	0.77
	OREM	0.76
	PROVO	0.76
	SALT LAKE CITY	0.77
VA	ALEXANDRIA	
	ARLINGTON	0.80
	CHARLOTTESVILLE	0.79
	LYNCHBURG	0.80
	NEWPORT NEWS	0.81
	NORFOLK	0.81
	PETERSBURG	0.79
	RICHMOND	0.79
	ROANOKE	0.81
	VIRGINIA BEACH	0.81

State	Metropolitan Area	Multiplier
VT	BARRE	0.79
	BRATTLEBORO	0.79
	BURLINGTON	0.79
	NEWPORT	0.79
	RUTLAND	0.79
	SPRINGFIELD	0.79
	ST. ALBANS	0.79
	ST. JOHNSBURY	0.79
WA	BELLEVUE	1.12
	BELLINGHAM	1.06
	BREMERTON	1.08
	EVERETT	1.11
	KENNEWICK	0.96
	OLYMPIA	1.10
	PASCO	0.95
	RICHLAND	0.96
	SEATTLE	1.12
	SPOKANE	0.91
	TACOMA	1.12
	VANCOUVER	1.04
	YAKIMA	0.99
WI	APPLETON	1.07
	BELOIT	1.09
	EAU CLAIRE	1.07
	FOND DU LAC	1.09
	GREEN BAY	1.06
	JANESVILLE	1.09
	KENOSHA	1.11
	LA CROSSE	1.07
	MADISON	1.09
	MILWAUKEE	1.13
	NEENAH	1.07
	OSHKOSH	1.07
	RACINE	1.11
	SHEBOYGAN	1.07
	SUPERIOR	1.09
	WAUKESHA	1.13
	WAUSAU	1.07
WV	BECKLEY	1.13
	CHARLESTON	1.09
	CLARKSBURG	1.09
	HUNTINGTON	1.11
	MORGANTOWN	1.09
	PARKERSBURG	1.07
	WHEELING	1.05
WY	CASPER	0.83
	CHEYENNE	0.84
	LARAMIE	0.83
	ROCK SPRINGS	0.83
	SHERIDAN	0.83